The New Prometheans

The New Prometheans

Faith, Science, and the Supernatural Mind in the Victorian Fin de Siècle

COURTENAY RAIA

The University of Chicago Press

CHICAGO AND LONDON

The University of Chicago Press, Chicago 60637
The University of Chicago Press, Ltd., London
© 2019 by The University of Chicago
Published 2019
Printed in the United States of America

28 27 26 25 24 23 22 21 20 19   1 2 3 4 5

ISBN-13: 978-0-226-63521-7 (cloth)
ISBN-13: 978-0-226-63535-4 (paper)
ISBN-13: 978-0-226-63549-1 (e-book)
DOI: https://doi.org/10.7208/chicago/9780226635491.001.0001

Library of Congress Cataloging-in-Publication Data

Names: Raia, Courtenay, author.
Title: The new Prometheans : faith, science, and the supernatural
mind in the Victorian fin de siècle / Courtenay Raia.
Description: Chicago ; London : The University of Chicago Press, 2019. |
Includes bibliographical references and index.
Identifiers: LCCN 2019012200 | ISBN 9780226635217 (cloth : alk. paper) |
ISBN 9780226635354 (pbk. : alk. paper) | ISBN 9780226635491 (e-book)
Subjects: LCSH: Parapsychology—England—History. | Supernatural—History. |
Crookes, William, 1832–1919. | Myers, F. W. H. (Frederic William Henry), 1843–1901. |
Lodge, Oliver, Sir, 1851–1940. | Lang, Andrew, 1844–1912.
Classification: LCC BF1028.5.G7 R35 2019 | DDC 130.941/09034—dc23
LC record available at https://lccn.loc.gov/2019012200

Thank you to my husband, Peter Grean, who helped this book tunnel into the light. And to Francesca Grean, who makes everything sunny and bright. And to all the dogs that warmed my feet after everyone said goodnight.

Contents

Preface

This book is about the first generation of psychical researchers, intellectuals working at the leading edge of experimental psychology in the late nineteenth century to test the limits of human consciousness. Their research agenda, which included telepathy, spirit communication, and other nonordinary experiences, resembled what we might think of today as "the paranormal," but if so, it was Stephen Hawking pulling up in that ghost van. The Society for Psychical Research (SPR), founded in 1882, attracted elite academics working at the heart of physics, psychology, anthropology, physiology, and statistics, among other fields. By enforcing the rules of science, but not its limiting ideas, the SPR was a fostering space for what Thomas Kuhn called "extraordinary science." Its focus remained on practical questions to be physically investigated: Could two minds communicate outside ordinary means? Was there an intentional force that could influence matter? But, as part of psychology, it was also entangled in deeper questions touching on human identity, merging with psychiatric curiosity about hypnotic states, split personality, and so-called religious delusions. There was, at heart, one horizon toward which all psychical inquiries converged: Was the mind merely a physical process, or was consciousness something fundamentally more? Yes or no? Are we body? Or soul? Even as the twentieth century was fast approaching, that question had been neither resolved nor set aside.

The idea of some potent, irreducible essence was both ancient and persistent: the continuing leaven of modern belief and the practical rule of ancestral magic. By that right, Tibetan lamas flew on prayer mats, South African Khoi spoke wordlessly across the Transvaal, and, now, in the middle of the self-consciously modern nineteenth century, séance spirits were erupting into drawing rooms across Europe, Britain, and America. Was this stalk of superstition simply too deep to uproot? Too comforting to let go? Too foundational for societies to decommission? Or was there something here that science could hold to be true? That question has long since been foreclosed upon within modern academia. The "supernatural mind" of this title belongs to our contemporary politics of knowledge, with its fixed boundary between nature and supernature. Psychical researchers deemed their area of interest to be "supernormal," crashing that semantic divide. The term was coined in 1886 by Frederic Myers, when telepathy was riding high on a wave of international corroboration. It remains a term of art for parapsychology but has failed to establish itself in the rest of science or even popular culture. As far as orthodox psychology is concerned, ESP is no more "normal" than a bigfoot sighting or abduction by unicorn. There is no point in making distinctions between various supernatural claims when the category itself negates all basis for critical judgment. Thus, "the supernatural" becomes, by the twentieth century, a region without standards left to its own *ideaphoria* like any branch of literature. To even concede to the possibility of an additional sense is for science to risk being overrun by every kind of magical, all-knowing flimflam.

This mental training makes it hard for academics to see psychical research as a genuine scientific effort. The reflexive exclusion of a "real" research object discredits anyone out to prove otherwise. But without a certain agnosticism, the seriousness of the pursuit unwinds into a kind of half-hearted playacting, organized around rituals rather than experiments, beliefs rather than questions. That does not square with who these researchers were, as a measure of either their personal authenticity or the scope of their intellectual achievement. This agnosticism isn't meant as a merely polite or patronizing gesture of tolerance (the "suspension of disbelief" one brings to a play, "making believe" it's not "make-believe"). The point is to clear the space of all projection. That is true from either side. Too much investment in the reality of the phenomena, and the frame of inquiry gets narrowed in other ways. Victorian psychical research can *only* be an investigative science. Any other agenda or context is seen to corrupt rather than reveal its historical identity. For this book, it remains important to let the psychical hypothesis hold the center of the maze. This was genuine research working toward a definite goal: to confirm or exclude the existence of a psychical faculty. But that goal itself was not a simple terminus; confirmation materialized new points of entry into a deeper labyrinth of concerns. We see that as soon as this psychical object begins to interact with other disciplines: physics, psychology, and anthropology are transformed. Science casts its net around the edges of a reimagined world. Psychical research promised to do more than just add a few new facts to the content of scientific knowledge; such data could explode the very *capacity* of science to know, crossing the powers of empiricism and epiphany.

To understand the genuinely civilizational stakes for these early psychical researchers, we have to see the big picture: the historical contingencies in which they moved. They stood at the edge of an expanding sea of consciousness, full of both possibilities and prohibitions as an emerging field of research. There had already been a century of strange testimony regarding these altered states, starting with Mesmer's healing animal magnetism and chased further down the rabbit hole by the Marquis of Puységur. This sunken inner landscape was roamed by drugged-out Romantic poets looking for Coleridge's under-consciousness, followed by magnetizers, mentalists, and mesmerists taking in their additional views. By the time academics got around to recognizing the unconscious, there were plenty of middle-class Victorians who already had a preliminary tour as séance participants, sitting in the twilight of their own consciousness while trance mediums channeled their dead. It would seem that the delirium once appropriate only to a Dionysian cultist had become the everyday chaw of polite conversation and the Sunday papers. British psychology in the 1870s made a point of steering clear of these alienated, mystical intelligences, but late in the decade, French psychiatric clinicians began to take that forbidden dive. It was not the utilitarian brain that they were studying here, but the broken mind, specifically the way physical brain damage ravaged mental function. This was still a biological paradigm, but its view of human intelligence was far more panoramic, taking in all the scenery below as medical hypnotists explored their patients' pathological unconscious.

The SPR established itself here, in the penumbra of abnormal psychology, working its way toward the center by maintaining strict disciplinary standards. But its research agenda was pointedly speculative: telepathy was first and foremost, followed by the more exotic species of hypnosis, as well as case studies of apparitions, spontaneous knowledge, and, eventually, séance mediumship, among other things. These research interests migrated beyond the SPR into more orthodox centers of research. Far from being a pseudoscience mortifying to the rest of experimental psychology, psychical research was its forward detachment and involved some of its most distinguished practitioners. This broader medical initiative, however, tended to ascribe psychical phenomena to some impersonal force, arising within matter. Its grim, evolutionary portrait of humanity remained unrelieved as it moved into the headlines. This pile-up of deviance was the car crash of consciousness from which people could not look away. As it turned out, the question "Why do we do what we do?" was as compulsive a listen as any religious origins story, a thread that continues with the popularity of personality quizzes and the corporate voodoo of the Myers-Briggs test. And yet for all the self-fascination that made psychology such an irresistible draw for Victorians, it was the field consensus that there probably wasn't such a self at all. Instead there was only the brain's automatic operating systems, sorted into "humiliating impulses" and "homicidal urges." This wasn't just a "naturalism" that purported to explain the physical world around us. This was something new, something that claimed to understand the secrets of our inner world, touching upon what Charles Taylor called the deepest sources of the self.

Psychical research kept the candle lit for this wounded subject. Once telepathy seemed predicated, they turned to séance mediumship to get at more existential questions. While some members had taken part in spiritual investigations at Cambridge in the mid-1870s, spiritualism and psychical research should be seen as vastly different social and intellectual projects. But still there were affinities. The séance was an investigative act with a commitment to evidence. Its centerpiece was the trance medium, providing a venue to explore extended models of consciousness, which could well include mind reading or even a surviving intelligence. These were questions of abiding philosophical interest. The spiritual repertoire involved levitation, tongues, and returning from the dead. This was reminiscent of the wonders of Christ, and the myths of ancient peoples witnessing a world brimming with *mana*. But this part of universal history was being written off for lack of evidence in the final, heady days of objectivity. The sensory textures of the séance room, its aromas, sounds, squeezes, and ghosts you could feel, were a way to make up any factual deficit. But spiritualists could not interest physicists in their testimony. To paraphrase Faraday, these physical spirits were no more welcome in science than a circular square.

The public offense taken by the institutional stonewalling of the people's evidence was never properly gaged by academic elites, then or now. Spiritualists offered facts gathered directly with their eyes and ears, or even held in their hands. To be told such things could not exist wasn't just to be barred from a scientific debate, it was being expelled from reality itself. As the spiritualist W. H. Harrison complained, the Royal Society was "high on the stilts" when it came to so vital a public interest. Tweak that a bit and we are looking up from a "flyover state." The sense of grievance that crept into the popular love affair with science hardened in place. By taking a seat at the séance table and holding hands, the Cambridge group made a limited but important concession to their fellow investigators: their experience mattered. With all due respect to scientific achievement, psychical researchers, like the spiritualists, would not consent to live within its theoretical prediction of the possible. As the frontiers of physics tightened around the "known unknowns" of mechanical theory, it lost touch with all other possible horizons—the greatest of which was consciousness itself.

By the time the SPR took up these interests once again in 1886, spiritualism was already sinking beneath the weight of its psychiatric diagnosis. Stigma would become part of the modern playbook for managing eccentric beliefs as religious freedom migrated into the commercial sphere, where it became breathtakingly unconventional. In this nineteenth-century civil order, there was no Torquemada to root out heresy, but labeling spiritualism a form of uterine insanity was effective in its own way. What spiritualists had managed to make familiar, had to be reframed. Walking arm and arm with a ghost? Channeling a Tibetan mahatma? Well, that's weird. This streak of supernatural innovation could not be stopped, but it could be shamed into hiding. The real leverage, however, was held over professionals who could be easily sanctioned for being "bizarre." An already divergent tendency gets exaggerated. Disciplines exerted more discipline; eccentric beliefs became more so when forced underground. And now, more than a hundred years on, modern magic has become so radically inclusive, there is no fringe, just an inbound stream of Babylonian astrology, quantum mechanics, Western *hermetica* and Eastern medicine, as well as the religious heritage from around the world. Modern seekers wanting to get free of modern life have drawn to them every technology that could fast-track transcendence or give them some edge over physical laws. And the cultural aesthetic seems to be bending ever more inexorably toward zombies, werewolves, witches, and vampires, as the chthonic spirit of the Romantics invades our televisions from the underworld. Two cultures, already at odds, resolved their conflict by drifting apart. From one end of the telescope, academics are seen knee-deep in their tiny footnotes, weighing every word, scrupulous bores; overhead, magical thinkers drift into outer space in a wobbling, shimmering bubble, getting curiouser and curiouser. No one seems to know what the other is talking about, and no one really cares.

The supernormal began to fold up fairly quickly in professional psychology, ending any potential for common ground between rule-bound science and magical possibility. It seems, especially after nearly a hundred and fifty years of trying, this is a cut and dry case of insufficient evidence. But the numbers are there. Or they are not, depending on one's background assumptions of conscious processes. Meta-analytical studies, including one published in 2018 by Etzel Cardenas in *American Psychologist*, seem to support a persistent nonzero probability value for psi. Is this a real effect being denied its due? Or a confirmation bias taken for evidence? The blurred data sharpens only upon deciding what can or cannot be true in science. But there were other factors that went into hardening this null hypothesis, seen more clearly in that earlier context. This was not just about getting rid of spooky forces in psychology; scientists in general wanted to get clear of reality as well. Instead of studying the physical world, the new science aimed only to decipher the rules of a mirage, shaped in the act of perception. Anything extrasensory broke the cognitive seal on skepticism. This was a thin slice of nature, with few moral implications and no physical facts. Without objectivity, the bothersome gusto of the Victorian reformer lost its moral footing in reality. The advantages of uncertainty are clear: when in doubt, leave science to the scientists and religion to the theologians.

By letting go of the reality of nature, science also set the supernatural free. The historic hinge of "heaven and earth" was finally uncoupled, making room for a growing diversity of religious beliefs. The supernatural could exist beyond or above the physical world, or only as a fiction of language. Truly separate spheres were transposable across many mental geographies and relations of power. But this secular model rests on a social experiment in subjectivism never tried before. Doubt made for good institutional boundaries, keeping any one team from knowing it all. But it did not make much common sense to Victorians looking for genuine answers in an age of discovery. Consciousness did not have to be called "a soul" to be implicated in the deepest part of human identity. No matter how fairly psychology may have adjudicated a psychical force or psi, the case has never been allowed to close. It is not just that parapsychologists continue to push the point but that ghosts themselves have been surprisingly persistent in the twenty-first century. They might even be having a resurgence, popping up in Japanese taxis, in the wake of a funeral, or making cameo appearances on the other side as near-death experiences. This kind of testimony triggers the contradiction at the heart of "separate spheres." While science holds all theological beliefs to be equally valid, there are no religious experiences that it can allow to be true. Phenomenology let go of objectivity. But empirical experience was kept on a tight leash. Professional psychologists rarely are drawn into any direct disputes. What cannot be deconflicted is best ignored. And yet, anecdote by anecdote, these stories weave together a world out of alignment with official accounts.

The reason to take stock of this psychical moment in the late nineteenth century does not hinge on the reality of its phenomena. This remains, regardless, an important part of the origin story of institutional science, told from the point of view of those who saw trouble ahead. As the twenty-first century seems to slide into its own epistemological crisis amid elite bewilderment, it may be useful to reexamine conditions at the source. The dominant impulse of nineteenth-century intellectual reformers was to bring moral and practical knowledge together, whether that agenda was pushed as social Darwinism or spiritual science. Even moderate, institutional voices did not conceive of bartering truth for friendship. The new, thoroughly modern cognition installed at the top of the twentieth century was, in the thinking of some psychical critics, a deliberate act of epistemological sabotage. The success of the "separate spheres" model has been most durable within its own architecture, the structured cultures of corporations and universities. But that wanes the farther out one goes, in terms of private life or geography. The cultural turbulence scientists thought to avoid with the retreat from objectivity has created other problems further down the stream. Subjectivism failed to maintain a binding sense of science's own moral significance, weakening its hold on the common ground. It is now left to be judged by its own epistemological standard: knowledge that is useful rather than objectively true. "Everything is subjective" has become the refuge of scoundrels against the tyranny of experts. Truth is *only* power. Science is back in the culture wars with no clear way to win them.

The psychical researchers profiled here may not have any workable solutions for divided knowledge, but at least they grasped the difficulty. They were willing to reckon deeply with what their colleagues wished to deny: science and social values cannot magically be untied. The supernormal fundamentally altered the conditions of knowledge so that science could potentially embrace that moral charge. Psychical research situated ordinary human experience at the intersection of two distinct realities, a mental and physical principle. But occasionally, the mind slipped out of its normal setting, going further into matter or becoming too free of it. In such moments, the empirical subject was confronted with a more fundamental reality that held a deeper truth of human existence. It was here that science could be put directly in touch with an implicate order. Instead of cutting the supernatural loose, the psychical project was a way to reel it in, slowly denying it any separate existence. This universe contained every wonder there was to be known and science was the way this knowledge could be shared.

Acknowledgments

This book is indebted to many people and institutions. I would like to thank UCLA for helping to fund my archival research, which ended up taking multiple trips and several years to complete. Victorians wrote a lot of letters. I workshopped this manuscript chapter by chapter through presentations at the AHA, HSS, PCCBS, and INCS and various colloquia. I am grateful for all the feedback from fellow presenters, as well as for their work and their insights. This book always got better in the wake of a colloquium or conference. I would also like to thank the archivists at the Cambridge University Library, London University College, and British Library, who seemed to always help me find more than I was looking for, as well as the Society for Psychical Research, for the use of their library and their adventurous intellectual spirit. Thank you also to the colleagues and critics who helped with this manuscript, especially Ted Porter for setting such high standards for the production of historical knowledge. He is the voice of caution that keeps me from mixing metaphors when I write. I am also grateful to Dean Richard Beene for his support of my scholarship and the opportunity to teach across all my areas of interest. That let the gestalt of a larger psychical story arise. Also, thank you to Claudia Verhoeven and Minsoo Kang for being such supportive colleagues, as well as my friends Christel and David, who got me to leave my desk. And thank you to my family, especially my sisters, Susan, Francine, and Rachel, who came to my rescue at various times. And finally, my big thanks to Karen Darling, who believed in this book and kept it on track even as I rolled over various deadlines and up a few steep hills.

Portions of this book draw on material from Courtenay Grean Raia, "From Ether Theory to Ether Theology: Oliver Lodge and the Physics of

Immortality," *Journal of the History of the Behavioral Sciences* 43 (1): 19–43, © 2007 Wiley Periodicals, Inc.; and Courtenay Raia-Grean, "Picturing the Supernatural: Spirit Photography, Radiant Matter, and the Spectacular Science of Sir William Crookes," in Minsoo Kang and Amy Woodson-Boulton, eds., *Visions of the Industrial Age, 1830–1914: Modernity and the Anxiety of Representation in Europe* (London: Routledge, 2008), 55–79, © Minsoo Kang and Amy Woodson-Boulton.

Introduction

The Victorian crisis of faith has been lost amid a scene of thriving worship, but there is more here than the fog of a war that never was. These psychical biographies help us dig into a deeper layer of discontent: not a polemic for one side or the other, but fears about the divided scheme of knowledge. No matter how "reconciled" religious and academic institutions were to sharing power, that peace existed at the level of the problem. It left the conditions of crisis unaltered: faith was a fundamentally discredited form of knowing, with no power of public enforcement, while science was a tragically limited one, with no claim on ultimate truth. Far from being complementary, each in their way ensured the inadequacy of the other. Worse for those looking ahead was the growing disparity in power. Already Darwin had rolled up on traditionally sacred ground: from human origins to social relations and, now, the contents of the human mind. Evolution was proving to be a strange and powerful kind of mythmaker, more degrading than inspiring. For the intellectual insurgents profiled here, the institutional harmony meant to protect the spiritual roots of civilization was ensuring their demise. The solution was not more religion but bigger science. The Society for Psychical Research (SPR) was founded in 1882 to unchain the scientific imagination, fielding interests as diverse as telepathy, automatic writing, apparitions, séances, multiple personalities, witch doctors, and curses, all on the fleeting trail of some indwelling higher power. Despite this mixing of physics and metaphysics, the SPR did not represent some nostalgic religious yearning but a bold new take on the future of secularization, an alternative route to the twentieth century yet to be reckoned by science. This mental or "psychical" faculty potentially possessed both an objective (causal) and subjective (perceptual) dimension that exceeded not only the physical explanations of science but also the limited sensory cognition on which those explanations rested. In other words, such proofs had implications not just for the body of modern knowledge, but for its brain as well, transforming what could be known and how. Extrasensory perception gave back some of the dimension lost in the fashioning of "Locke's punctual self." Here, the philosopher Charles Taylor described a modern subject suddenly condensed into an excessively metered point in time and space, the sovereign of a tiny kingdom.[3] Psychical research kept this empirical individual at the center of modern experience, but the purpose of its study was to chart pathways out of neural captivity, (with Myers even suggesting some future hypnotic navigation). Even as Victorian brain science whittled Locke's punctual self down to the cellular level (not just a captive, but a construct of matter), psychical research moved in the opposite, more expansive direction. This was a transpersonal psychology that dissolved traditional boundaries, putting minds in touch with other minds, and perhaps even bringing God within reach of scientific discovery. And in the first two decades of the Promethean enterprise, it was not entirely clear that the psychical researchers would not succeed.

The officers of the SPR held leadership positions in both the Royal Society and the British Association for the Advancement of Science, while the membership included academic elites and also artistic, social, and political luminaries. This roster of celebrities is usually what make the SPR historically notable, but the preoccupation with "who" has been allowed to eclipse the more interesting question of "what." This book returns to what was truly compelling about this endeavor: the experimental tour de force of the first few decades of its activity, marked by an astonishing theoretical depth and integrity of effort. Whatever hopes were held privately by its membership, the society itself aimed first and foremost to be an experimental research program, studiously replicating the norms of disciplinary culture in its publications, practices, and institutional structure. Its core epistemological modality was one of skepticism, not faith, while its mandate came straight from academic psychology as it stumbled forward into the abyss of the unconscious mind. There, psychiatric pathologies crossed paths with seemingly supernatural brain states, briefly potentiating an authentic science of the supernormal mind.[4] This is not to say that nineteenth-century psychical research was a scientific discipline by contemporary standards (though that could be said of any branch of academia in formation at the time). Rather, we must judge it by the academic standards that its leadership had been acculturated to, with the additional understanding that these leaders were the same establishment figures tasked to define and enforce the terms of orthodoxy elsewhere. They did more than just bring their "credentials" to psychical research. They brought their training, their influence, their methods, and their mindset as well, in the determination to address, once and for all, the proliferating chaos of mental forces and phenomena that the scientific worldview had been unable to either exclude or confirm.

I approach psychical research through the rigorous disciplinary gateways of chemistry (Sir William Crookes), psychology (Frederic Myers), physics (Sir Oliver Lodge), and anthropology (Andrew Lang). While all of my subjects knew each other personally, Lodge was on intimate terms with each, and serves as the central point of exchange drawing all four into a single historical system. All took a turn serving as society president, and all maintained an exceptionally high profile within their respective scientific communities, highlighting the helix that bound official and psychical scientific culture. But perhaps as important as their historical value is the way in which all four occupied the spotlight. The strong proselytizing dimension of their work brought them out of academic circles and into the public sphere, not only as psychical advocates, but also as people who wanted to stir things up and force a critical examination of science and society. Their ideological objectives were highly verbalized and thus wonderfully accessible to historical analysis. Simply put, there was no shortage of self-expression. First, their well-documented scientific work came, variously, in the form of scholarly papers, textbooks, laboratory notes, demonstrations, public addresses, recorded experiments, and opinion journalism. But there are also diaries, letters, poems, novels, mementos, and searching autobiographies. (In this sense, they loved to turn the spotlight inward as well.) This in no way detracts from their sincerity, but rather lets us authenticate their motives and come to a deeper understanding of the moral commitments that shaped their lives as well as their science.

What is perhaps most compelling about these researchers is their courage to confront head-on the new psychological demands of modernity and to find, within its framework of skepticism, continuing hope for some obtainable, yet transcendent, truth. Like many intellectuals of their generation, they found their metaphysical yearnings challenged by a new, stricter criterion for belief established by a host of scientific discoveries and an increasingly authoritative empiricism. This makes the midcentury "crisis of faith" an important ideological crucible for psychical researchers and a major historiographic current of this book.[5] While any such "war between science and religion" has been found to be more rhetoric than reality, what remains of interest is why belief in such a war was so pervasive, and, for some, so *preferred*.[6] The future founders of the SPR came of age reading Darwin and Spencer in a fever of agony and delight, unraveling the strands of naturalism and theism with their own intellectual labor while decrying the divide. They were all, to some extent, intellectual *bricoleurs* who made strategic use of "religious crisis" in complex ways, which helped them to originate powerful personal and professional identities and even to advance larger intellectual and generational agendas. Crisis does indeed equal opportunity, although such conversions are best seen over time. To fully tell this story, this book moves chronologically and thematically across the latter half of the nineteenth century, following the evolution of psychical research from its origins in the midcentury milieu of the "crisis of faith" to its theological flowering in the fin-de-siècle "crisis of meaning" (when skepticism about the nature of the self and its certainties took a more virulent turn).[7]

Chapter one sets the stage, examining how the rising prestige of scientific naturalism called into question the adequacy of faith as a basis of knowledge. This growing disconnect between sacred values and scientific theories was not new to the nineteenth century, but it had come home to roost with increasing intensity after the 1840s, as science began to fully enter civic culture and haunt the human self-imagination. This gave the robustly evidential narrative of Victorian séance spiritualism an irresistible appeal when it arrived on British shores in 1852. Spiritualism did more than offer spiritual "proof"; it offered personal empowerment, allowing individuals to invade the narratives of science and religion for their own purposes, while operating outside of any direct institutional control.

Chapter two homes in on the climactic encounter between science and spiritualism, when the celebrity chemist, Sir William Crookes, Fellow of the Royal Society and discoverer of the element thallium, undertook a highly publicized investigation of séance mediumship between 1871 and 1874. Crookes made this strange career calculus at a time when the (im-)possibilities of "scientific spiritualism" had yet to be fully clarified among science's various commercial, popular, vocational, and institutional factions. There were significant limits to such a project, and strolling arm in arm with a ghost was definitely one. Crookes himself visibly mapped these boundaries with his own highly publicized professional disgrace, rendering a much more pronounced line between respectable science and the quicksands of séance curiosity.[8] Yet, by the end of the decade, Crookes had managed to exceed his former glory with the invention of the Crookes tube, riding that same wave of pent-up metaphysical curiosity to national fame (only this time it involved his fellow physicists).

In chapter 3, I look at the early character formation of Frederic Myers, born at the crossroads of evangelism and English literary Romanticism, and seemingly driven from birth by an existential dissonance that blew him across every creed from Platonism to positivism, wanting something from each, but finding each deeply wanting. By exploring Myers's intensely religious upbringing and his subsequent painful renunciation of faith, we attend the birth drama of a uniquely Victorian secular consciousness: one that could not live with religion and could not live without it. Eventually Myers put his faith in scientific spiritualism, lured by the promise of William Crookes's "psychic force." But when this metaphysical turnstile became jammed by futility and scandal, Myers was finally left with no path forward. He would have to clear his own, which forms the subject of chapter 4.

Myers emerged from crisis to resolution with the establishment of the Society for Psychical Research in 1882, staked on ground cleared by Jean-Martin Charcot and French neuropsychiatry. The medical links between hysteria, hypnosis, and the subconscious mind that Charcot established in the late 1870s eventually expanded to include telepathy and even biologcal teleplasty by the mid-1880s.[9] But while Charcot seemed eager to investigate these bizarreries, he was reticent to ever fully recognize them. It would fall to Myers to be the first to provide the missing theoretical architecture for the aberrational data accumulating around the notion of the subconscious at Charcot's neurological clinic at the Pitié-Salpêtrière hospital. Myers's concept of the "subliminal self" preserved academic psychology's basic evolutionary understanding of the brain while transforming its assumptions about the mind-body relationship, extending the phenomena of consciousness to potentially include an interactive, psychophysical force.

Chapter 5 explores the partnership between physics and psychology through the psychical research program of Sir Oliver Lodge, the celebrated pioneer of wireless telegraphy and close collaborator of Frederic Myers. Like Myers, Lodge relied upon the authority of science to redeem his broken metaphysics, and his increasing anxiety concerning that authority forced his investigation of the ether into the deepest nature of sensory, and even extrasensory, perception.[10] It was science's inability to find some connection between matter and this underlying, universal plenum that led Lodge to join Myers at the Île Ribaud in 1894 to investigate the strange, corporeal extrusions of the séance medium Eusapia Palladino—searching for some link between "ether" and "ectoplasm." But perhaps more important than verifying the ether's objective existence was the need, for Lodge, to verify the subjective claims upon which such an observation might rest. What was really at stake for Lodge's science in the 1890s was the entire platform of scientific epistemology, weakened alongside the metaphysics of the self and the collective failure of science to manifest the deeper structural realities of physics. In this way, Lodge injected himself (and psychical research) into phenomenological debates that were, for most of his scientific contemporaries in the 1880s and 1890s, still a remote philosophical concern. Together with Myers, he built a model of consciousness with its own, distinct ontology, merging evolutionary psychology with the most recondite realm of force physics and metaphysics to instantiate a romantic subject at the heart of empirical science.

These same concerns about the character of scientific knowledge traveled across the anthropological writings and literary punditry of the famed Victorian folklorist Andrew Lang, the subject of chapter 6. His emphasis, however, was more explicitly with the capacity of science to supply cultural meaning rather than information. His interest in archetypal patterns of consciousness drew him to Myers's dynamic psychology and Lodge's entelechal evolution because of the ways in which they endowed mind (and memory) with power over matter. As a gifted folklorist and student of comparative religion, Lang grasped mythic function in ways most anthropologists influenced by Edward Burnett Tylor could not. While his utilitarian cohort were stuck viewing religion as a species of bad science, Lang intuited something else: it was contemporary science that would make for bad religion, an anemic mythos for the modern world. Lang made this case as a literary gadfly ridiculing the grim sociology of the realist novel and more soberly as a dissenting anthropologist who steadily refused to see all myth and magic in terms of their survival utility. Only in the 1890s, after Lang began to incorporate the data of psychical research, was he able to set human development free from the tyranny of Darwinian biology. He drew upon Lodge's vitalistic ether to give credence to the primal intuition that the world was infused with *mana*, arguing that religion itself arose from the mind's points of contact with this hidden power, documented around the world in the lore of levitation, apparitions, clairvoyance, and mystical events.

For all these investigators psychical research, far from being a passing fad or career stumble, was a lifetime commitment. The chapters on Myers, Lang, and Lodge each conclude with a statement of psychical theology, as it arose from sources deep within their character and evolved through several phases:

factual investigation giving rise to psychical theory, theory broadening into natural philosophy, and philosophy clearing the ground for a fully emancipated religious speculation. This final, mature phase of their thought was asserted on the basis of their independent intellectual authority (as opposed to being a formal position of the society). It was the culmination of decades of research and reflection, developing into sweeping narratives about humanity's spiritual identity and its relationship with a divine, bringing humanity up off its knees just in time for the twentieth century. This new religious subject did not await revelation but instead built up knowledge of God through his own instigation, accessing a suprasensory gateway that allowed the human mind to open up on the deepest levels of intellectual and even divine epiphany.

Chapter 7 opens back onto the larger cultural landscape to better contextualize psychical research, but this time with a focus on the late rather than the mid-Victorian period. Now the earlier skepticism about the knowledge of God has metastasized into skepticism about knowledge in general, giving rise to a crisis of subjectivity so profound that theological doubt was only one of its many forms. The phenomenalistic approach to psychology had begun pushing the modern self-examination toward the "dispiriting" conclusion that it did not after all exist. But psychical research accelerated in the opposite direction, rolling out an ongoing expansion of human consciousness with the most productive and daring decade of its history. Myers's subliminal self, Lang's psycho-folklore, and Lodge's ether theology all took shape in this period. While such late-Victorian metaphysics was eventually overtaken by phenomenology, a rash of recent projects seem to return to this ambitious spirit of the early SPR, which I discuss in my conclusion. For instance, the 2006 "Harvard Prayer Experiment," conducted at Massachusetts General Hospital and six other medical centers, aimed to test the power of third-party prayer, while Dr. Sam Parnia's AWARE project recently concluded an investigation of "near-death experiences" across a network of twenty-five hospitals. Along with this mind/body revaluation in the context of medicine, the intersection between mind and matter was also revisited in physics. The Princeton Engineering Anomalies Research (PEAR) program and the University of Arizona Laboratory for Advances in Consciousness and Health sponsored extensive research into the interaction of consciousness and physical systems as part of a sanctioned university effort. There are major funding organizations, such as the Templeton and Nour foundations, that provide grants to expressly encourage scientific curiosity regarding religious or metaphysical themes, which means that some such research initiatives will likely continue. And outside the university system, paranormal research societies (including the SPR) continue to attract lone academics to their programs. This search for something more has remained remarkably resilient in the face of mainstream science's seemingly intractable null hypothesis. But despite this strong scientific consensus that injudicable questions of transcendence should not be asked, there are still those who equally believe they should not be allowed to go unanswered.

THE SCIENCE OF PSYCHICAL RESEARCH

Understanding psychical research as an "authentic science" goes beyond merely appreciating its identification with academic psychology or its embrace of empirical methodology. Those are important formal elements, to be sure, but they say nothing of the inner life of the discipline, the actual science so contained. To reckon with this research requires reckoning with the researchers themselves, their ideas, and their practices. Certainly, at the level of ideology, empiricism was at the heart of their scientific self-concept. Frederic Myers could have thrown himself in with any number of fringe medical, mesmeric, or spiritualistic associations to use his metaphysical mind, and yet he persisted in courting the approbation of physiologically grounded neuropsychiatry; Oliver Lodge labored strenuously to keep his metaphysical theories tethered to the mechanical ether; and Lang's anthropology suffered British utilitarianism patiently for twenty years until he could emancipate his own idealist arguments on the facts of psychical research. But this experimental method at the core of psychical disciplinarity also constituted the central difficulty. It required the direct observation of the phenomena under consideration, meaning that the ultimate claim of psychical authenticity rested on the most problematic aspect of that claim. Can investigators who gave positive testimony regarding telepathy, telekinesis, spirit communication, or other such phenomena really be taken seriously? The background assumption (at least in academic discourse, which has firmly excluded parapsychology from its disciplinary structure) is no. Such phenomena should not be accorded any objectively real status. Therefore, it follows that those conducting such research must be either deceiving themselves or deceiving others. We dismiss out of hand what they are saying and wonder, instead, why they are saying it: are they duped, delusional, or dishonest? Psychical activity is thus immediately deflected away from scientific considerations and captured by interests of a more biographical, cultural, or even occult nature.

Such a judgment not only marginalizes the relevance of psychical research as an intellectual project, but, more comprehensively, taints the researchers themselves as intellectually and/or morally unreliable. This does not fit. Their professional contemporaries placed a great deal of confidence in these individuals, and many of their contributions are still highly valued today. They have littered the archives with their private correspondence, and not once does any chink appear in their investigative probity. Admittedly, though, it remains difficult to reconcile some of this testimony with these particular testifiers. I have, however, absolute conviction in their personal conviction, and without that, they would not have been nearly so interesting to me. It's not just that they can be relied on to report what they witnessed, giving us a seat at the (séance) table. We can see through their insight as well and learn from the depth of their questions. Their concerns were of an unusual magnitude, questions ordinarily encountered from a historical distance or a spiritual remove. Yet, they pursued them with boots on the ground, across disciplines and at the heart of their cultural moment. They readily assumed methodological constraints that forced them to dig deeper and work harder. Unlike most people just looking for answers, for them, the question took precedence.[11] They could not offset the demands of their own curiosity with weak arguments or unsubstantiated proofs. Yet even though psychical research (as pursued and defined by its core practitioners) submitted itself to a far stricter regime of scientific discipline than programs like spiritualism, Edward Cox's Psychological Society, mesmerism, or theosophy, it has been largely subsumed into that same historiography.[12] While such movements may have entwined with scientific themes and actors in fascinating ways, only psychical research constructively aspired to be a fully academic discipline. If we shift our focus from the character of the phenomena to the character of the individuals and the praxis of their research, the difference between midcentury "scientific spiritualism" and the SPR's codifying and consensus-building "Committee of Apparitions, Haunted Houses, etc." is readily discerned. Perhaps more important than this is to delve fully into the intellectual arguments through which psychical possibility was theoretically sustained.

My book aims to restore these neglected intellectual frameworks that legitimated psychical research for its practitioners, and thereby to recover the full scope of the Victorian scientific imagination as it ranged across this psychical terrain. By understanding this research as an extension of an authentically scientific curiosity, we recover more of the breadth of what Victorian scientists were actually curious about. To a surprising extent, psychical researchers slipped the occult back into mainstream disciplinary discourse through a vast network of publications and private letters that reached far beyond the SPR's immediate circle. So much serious interest put psychical possibility in play for the common culture, even if remotely, allowing it toleration from all but the most confirmed skeptics. Sister organizations spread out across America, Australia, Germany, Holland, Sweden, France, Italy, Norway, Scotland, Russia, and more, uniting corresponding members from around the world into a single research community. Nowhere, however, did these offshoots penetrate official science to the same degree as did the British flagship SPR, which fielded a very deep team at high social altitudes.[13] It counted among its early presidents a future prime minister, along with an archbishop, poet laureate, and many knights of the realm among its early members.[14] Related efforts on the continent and across the anglophone world could not defend the same kind of elite cultural space without soon becoming overheated by spiritual enthusiasm from below or frozen out by scientific asperity from above.

It was never the expectation that psychical research would be granted regular disciplinary status across all academic networks. The full psychical docket remained too problematic for most of academic science in that it included mediumistic trance, ectoplasmic excreta, ghostly apparitions, scrying, clairvoyance, and even some telekinetic phenomena associated with spiritualism, such as levitation and materialization. Yet these unrestricted horizons were essential to the honor of psychical free inquiry and worth the heat. However, as far as the society's principal inquiry of telepathy was concerned, there was a general toleration of "wait and see" for all but the most inveterate skeptics. This psychical object did not permanently foreclose future ties to the Royal Society and the British Association for the Advancement of Science, though that awaited some greater consensus. "Not now" was not "not ever," and the society conducted itself in such a way as to protect that future possibility. (Lodge observed that official resistance was actually softening in the 1890s, along with a growing toleration of vitalistic speculation in biology and physics.)[15] Given this zeitgeist, one might just as well say that the SPR was on the cutting edge of scientific curiosity rather than at its intellectual fringe. This edgy institutional geography made psychical research potentially more productive, not less, because it aligned with innovation versus consolidation. What some historians have labeled as a "pseudoscience" was striving to be the "extraordinary science" of its own time. The SPR was an active collaborator with Charcot's Société de Psychologie Physiologique de Paris in the mid-1880s, the most daring research effort into consciousness of its time, led by Charcot's particular protégés. (See chapter 4.)

These experimental acrobatics were given some leeway as an element of psychology so long as they remained tightly tethered to empirical standards of investigation. But for scientific watchdogs, the problem was not so easily overcome. It was not the methods that were objectionable so much as the nature of the "psychical" subject matter itself. The iron laws of thermodynamics excluded absolutely the action of mind on matter and forbade the prospect of unknown energies popping willy-nilly in and out of existence. This was a pronouncement whose authority had weighed on thousands of meticulous experiments quantifying mass and energy undertaken since Lavoisier. When it came to the clinical evidence offered by psychologists and investigators, the eye could be deceived, but the numbers could not lie.[16] The notion of a psychical force threatened to unbalance the energy equations upon which modern conservation relied. Take the troublesome séance phenomenon of a levitating table, for instance. Where had this addition of energy come from? Or, in the materialization of objects, how was such matter supplied so that the total mass of the system remained constant?[17] The problem ran even deeper than this wanton expenditure of energy and matter. The guidance asserted through psychic force stood to reverse the hard-and-fast tendency of entropy, restoring order and recapturing lost energy, and rendering meaningless the notion of deterministic laws. These were the facts that kept the universe in good working order, holding everything together from evolution to atomic theory, an orthodoxy ringed with support from all major academic and professional institutions. And it could all be brought down by a single datum of psychical proof, so great was the incompatibility between the laws of physics and psychic force.

Given these obstacles, the search for a psychic force might appear so quixotic and out of touch in hindsight that it is of only distant concern to the "real" history of science. Yet, protecting some spiritual or causal aspect of the human mind was a pervasive cultural and intellectual concern. This psychical campaign was waged in surprisingly close quarters with other disciplines, and in many ways its theoretical architecture arose to bridge genuinely problematic gaps in scientific explanation. After all, Victorian scientific naturalism offered a theory of evolution without DNA, electromagnetic radiation without electrons, in addition to glowing gases, fields of force, and a ubiquitous ether no one could catch or quite understand. And then there was the grim eschatology of cosmic heat death, radiating energy into the void until the universe wound down to a halt. For those who felt caught in this midcentury machinery, psychical research offered a legitimate way out.[18] Conservation did not deny the existence of a supernatural order, but it did banish all hope of accessing it, a cosmic seal potentially punctured by a range of psychical phenomena. For those religionists seeking a more amenable naturalism, this was a universe shot through with mysterious, replenishing powers, a veritable perpetual motion machine, allowing the living and the dead to coincide in a secular eternity. And it offered to effect a religious reconciliation with science in the boldest manner possible, by empowering science to fully assume theology's metaphysical burden.

As such, the SPR was part of a larger continuum of movements seeking to reattune naturalism and theism in a manner more befitting the modern era, yet its restricted cultural location sets it apart. It did not seek a comfortable berth in the permissive latitude of the private sphere. Victorian psychical research was the last great effort to integrate physics and metaphysics undertaken within the rising fortress of academia itself, making it the most systematic expression of what Peter Bowler has described as "the forgotten rapprochement" between science and religion that characterized British science in the late nineteenth and early twentieth centuries. It is also, of course, part of a longer, relatively submerged history in western philosophy: the struggle for synthesis. This runs alongside the more pronounced duality that travels across the last four hundred years (iterated and reiterated variously as mind versus body, science versus religion, spirit versus matter, romance versus reason), driving the recurring counterimpulse to overcome opposition and advance a new, more comprehensive totality.[19] Psychical research in some sense culminates this history, but also exceeds it, advancing some nexus between mind and matter against far greater odds. This was not the romantic brain science of the early nineteenth century, which allowed physiologists dissecting brain stems to make common cause with poets taking psychoactive drugs.[20] Nor was this the "romantic mechanism" of Paris where industrial energy and vitalistic force blended to bring machines "to life." Such conceits were worked out in the open-ended framework of *Naturphilosophie*, and receded alongside its declining fortunes as the Berlin Physical Society began to put the squeeze on the creative will in nature in the late 1830s.[21] Over the next two decades, thermodynamics worked out the fundamentally quantitative, mechanical nature of physical energy, which could be structured and converted across all known forces (whether gravitational, electrical, magnetic, or chemical). This quashed any such "romantic mood" uniting subject and object, matter and mind, within its theoretical domain with little hope of revival (or so one would assume given the hostile, mechanistic framework). But the SPR prevailed against these headwinds, pursuing a highly controversial idea, under ever-increasing institutional oversight, at an unprecedented scale of productivity.

But by the early twentieth century, the SPR came under increasing pressure to trim its sails. There were new winds blowing in academic psychology.

The transpersonal under-consciousness first raised by Coleridge in the early 1800s was being slowly drained of its mystery by psychophysics, leaving only a tangled skein of axons and neurons cathected by instinctual energies.[22] Psychophysics gave an essentially thermodynamic account of the biological brain, which itself gave rise to the epiphenomena of physical consciousness. It was within this framework correlating sensory stimulus to mental perception that Freud worked out his evolutionary id in the 1890s, and the behaviorists reduced free will to a stimulus response soon thereafter.[23] Such a psychology could not so easily brook the spiritual infringements posed by the psychical agenda. Those psychical researchers wishing to guarantee some form of academic tenancy for their work broke off into a purely statistical tributary (what would become today's parapsychology).[24] Those who remained did so amid a contracting sense of possibility, not just for psychical research, but for all of science, as the reach of human knowledge found itself fatally restrained. For, in addition to insinuating that consciousness was a product of material reality, psychophysics also implied that that "material reality" was likewise a product of consciousness (that is, all apparent externalities are in fact only synthetic presentations of the mind). This was the early insight of German phenomenology in the 1880s that paved the way for the eventual rise of subjectivism in science in the twentieth century. It was thus that materialism and antirealism simultaneously came to rule the modern discourses of the mind, and that science and religion decisively went their separate ways.

This weakening of truth claims gave phenomenology its allure for institutional elites ready to chart an unrestricted course for the twentieth century, one that would swing wide of any possible knowledge disputes. Two symbolic accounts of an unknowable reality mooted the need for competition or coordination of the rule of nature and the rule of God. This idea of two "non-overlapping magisteria" (NOMA) was a remarkable success, allowing hundreds of denominational universities to flourish in the modern West and God to politely reside within its politics.[25] But it turned out to be a more limited success than at first assumed Such a science withheld explicit support for a belief in God even as it demanded that people take on "faith" those facts issued under its own authority: facts people were unlikely to understand and to which they were not obliged to give their epistemic consent. Thus this alliance worked out between faith and reason at the elite level created a lasting cultural rift of another kind: that along the elite/popular divide. Given that most Victorians, including scientists, preferred to retain some friendly theistic slant on nature, the dénouement of dividing (and thus diminishing) religious and scientific "truths" into separate spheres fell short of the mark. This was accommodation and not reconciliation, an expedient pact of mutual uncertainty rather than two truths satisfactorily re-allied.

Psychical research aimed to preempt exactly this epistemic alienation: first in the rescue of metaphysics from the excesses of empirical authority, and then in the rescue of empirical authority through that same metaphysics (allowing human comprehension to transcend the limits of sensory cognition). Within this expansive perceptual field, the psychical subject would be able to hold the spiritual search for meaning and the philosophical problem of knowledge together in a single framework: the essential theme of reconciliation, recurring again and again in the writings of Myers, Lodge, and Lang. They ultimately understood science not as an instrument but as ideology, a worldview in which the moral, spiritual, and intellectual dimensions of "modern progress" must strive to coincide. But the latitude for psychical empiricism existed in the prospect that some intersection between ultimate reality and the world of appearances was judicable by science, a proposal swept aside by the twentieth-century regime of knowledge.[26] With phenomenology, the ancient quest for understanding begun by Anaximander and picking up speed through the Renaissance, Scientific Revolution and the Enlightenment appeared to have hit a skull wall beyond which the mind could not aspire. Though my subjects did not manage to effect their grand Victorian ambition of a gnostic science, there was honor in their attempt to exhaust that possibility, sinking their spades with unrelenting curiosity into the inky depths of the newly discovered subconscious mind, hoping to find a way out of the modern predicament.

CHAPTER ONE

The Culture of Proof
and the Crisis of Faith

In the second half of the nineteenth century, the Western world began its acceleration toward modernity. In the space of that fifty years, scientific knowledge entered fully into its physical frame. Bessemer steel poured into its scaffolds, its bridges and railways; coal fires became combustion engines, windmills became dynamos, and carriages turned into motor cars. This gave change its definite, visible architecture. But a parallel transformation was under way that would likewise make over the traditional mind. The theo-centric view that had cradled nature and the nation within the Kingdom of God was coming apart piece by piece as a manmade world was pushing its way into the midst of this God-given one. This had profoundly disturbed the resting place of private religious conscience, which had sheltered within Anglican natural theology. The harmonized understanding of heaven and earth was being swept away, and a new, replacement framework had yet to arise. Whether this framework was to be rebuilt through a reformed theol-ogy or secular metaphysics, or indeed abandoned altogether as an unwork-able divide, it was, for the first time, up to the people to decide. The forces of democratization had put ordinary individuals, their ideas, and their sense of possibility into play.

While the "Victorian crisis of faith" narrative has hung clouds of spiri-tual despair over this changing landscape, the rhetorical gloom belonged largely to the intellectual stratosphere. Scholars peering below have seen that this remained a profoundly religious age. What had changed was the nature of religious debate. The cultural ruckus over "how to get to heaven," under way since the Reformation, was now caught up in the gear-works of

this new scientific philosophy of knowledge. Religious belief now intersected with deeper questions about what *could* be known and the best way to
build such knowledge, complicating the terrain in which religious faith was
held. While scientists themselves did not particularly seek this competition
with theology, they did enforce a boundary between knowledge based on
faith and that based on reason, setting up their own distinct system for ar-
bitrating truth. And lurking within this new scientific understanding of the
physical world was the potential basis for developing new social guidelines
for modern living, calling into question the faith-based customs of the past.
The culture wars had begun.

Among the greatest weapons in this fight to assert new certainties, even
in matters of religious faith, was the persuasive power of science itself, and
especially the appeal to physical proof. Indeed, the appeal to proof was not
limited to any one side but was sought and accepted everywhere. It was a
common language that could be spoken by everyone and understood by all.
Evidence was in. Nothing captured the *demos* like demonstration. From the
sensational road shows of mentalists to the prestigious public lectures at the
Royal Institution, audiences from the late eighteenth century onward had
been encountering evidentiary spectacles: celebrity scientists made chemical
elements flame, smoke, phosphoresce, and flash with light; they discharged
electric currents from Leyden jars; they set discs gyrating within mysterious fields of force. Seeing was the new believing as Victorians began applying this passion for proof even to the afterlife, discovering that ghosts could
be subjected to the new evidence-based standards of the time. They clasped
hands eagerly to demarcate the revolutionary investigative space of the séance circle, hoping to witness at last "evidence of things unseen" and even to
touch "the substance of things hoped for."[1]

The implicit materialism of such a concrete, evidential "culture of proof"
can seem strikingly at odds with spiritualism's otherworldly agenda. However, it makes more sense if we understand this materialism as condition-
ing a set of empirical practices, and thus as mainly methodological in nature. It did not necessarily imply any deference to scientific theories or use
facts to enforce some metaphysical exclusion. Its particular brand of mid-
Victorian facticity is best interpreted within the context of a more naive
positivism. This outlook cheerfully identified with Auguste Comte's notion
that empirical science marked the high point in the evolution of knowledge, without taking on formal positivism's godless cultural agenda, that
is, Comte's "Religion of Humanity." Intellectual ascetics like George Eliot
and George Henry Lewes, high positivism's literary power couple, may have

felt duty-bound to expel religion from modern life as a necessary step in humanity's moral and social evolution, but most people were not so keen to abandon the divine order.[2] Likewise, the strict, antimetaphysical discipline of elite theoretical science had limited mass appeal because of its chastely physical view of natural philosophy.[3] Because naive positivism embraced experimental evidence, not theoretical explanation, as the most compelling aspect of scientific argument, it did not require the renunciation of God in the realm of either culture or philosophy. Evidence was also, not coincidentally, its most democratic aspect, privileging common sense and common senses. Popular empiricism set observational data free of the interpretive demands of recondite academic theories and other exclusionary qualifications on understanding, enlarging the public's personal authority in relation to professional scientific opinion. This allowed nonspecialists to bypass intellectual pronouncements inconsistent with their own values. In the case of spiritualistic phenomena, people could see for themselves the phenomenon in question and judge it accordingly, trumping any intellectual interdictions imposed from on high. Thus, even after institutional science put an end to overt clerical interference, it still had to contend with the tenacity of those unwilling to surrender their own, more personal claim to its talisman.[4] In the late 1860s and early 1870s, when spiritualism was approaching its zenith, the boundaries of scientific knowledge and scientific authority still lacked a clear, cultural consensus. Ideologues, entertainers, amateurs, eccentrics, and lay educators all vied with academics and specialists for their share of the narrative power of science. The openness of this terrain is important to understanding the ambitions of scientific spiritualists to assert their own rival scientific worldview, and the extent to which psychical researchers were actually able to do so. The choice for those seeking some yet-to-be-defined alternative to this still-to-be-consolidated orthodoxy was, in the words of Oliver Lodge, a choice between "two distinct conceptions of the universe: the one, that of a self-contained and self-sufficient universe, with no outlook into or links with anything beyond, and the other conception of a universe open to all manner of spiritual influences and permeated through and through with a Divine spirit."[5]

The philosophical materialism that imposed this strict seal upon the cosmos may have been a largely elite phenomenon exaggerated by polemicists, but the concerns it raised were far more diffuse. Such rhetoric drew a clear line between the ethos of scientific skepticism and the outcome of religious doubt for all the faithful to see, even those who would not normally be able to infer such a line for themselves. Even if one did not have doubts about one's own faith, one could never be sure about the faith of the person in the neighboring pew. The institutional power of the church might not visibly flag and cultural tradition might continue to compel a certain outward deference, but as long as the hearts of men were hidden, there would always be room for doubt. Here is where the eloquent apostasy of elites, often seen as driving a somewhat artificial or out-of-touch narrative, finds its broader significance. Such public "crisis confessions" made an emotional spectacle of an otherwise interior struggle, realizing for the public a somewhat mythical atheistic threat.[6] These elite misgivings put doubt on display, traveling socially and intellectually downstream to be taken up as part of a wider debate. Doubt, whether privately held or publicly multiplied through the media explosion of magazines, newspapers, pamphlets, essays, novels, and lay-interest journals, captured the Victorian imagination and entered prominently into its discourses. This anxiety was powered not by unbelief, but by the deep social and personal investment Victorians had in the beliefs of others, as well as the voice it gave to their own unconfessed tests of faith. The secret, subjective states of mind, where God had been inscribed since the Reformation as a matter of private conscience, had now come under new kinds of pressure as theological narratives encountered secular discourses. This gave the idea of a war between science and religion its popular resonance, even if the public's continued allegiance to both remained outwardly undisturbed.

Understood this way, the crisis of faith becomes an interior, psychological crisis concerned as much with the subjective act of faith as with its object: the religious claims themselves. This reflexive approach offers a productive shift away from the reality of a religious crisis to the more historically relevant notion of the perception of one. The substance of religious crisis need not derive from the actual headcount of confirmed atheists, but may arise from the potential, indwelling doubts of the faithful themselves. This inward turn toward private conviction also heightens the epistemological dimensions of the crisis, wherein doubt recognized a fundamental insufficiency in one's knowledge. In the context of the culture of proof, the insufficiency of faith is implicit in the distinction between believing something to be true and proving it to be true. This elevation of evidence-based knowledge could not help but stand in silent detraction against all assertions that were not themselves evidential: for instance, religious assertions based on faith. For the many who took pride in this positive age, no matter how tightly they held to their spirituality, somewhere (in the forefront or back of one's mind) this evidential ethos had opened up a ledger keeping track of one's running debt of proof. Séance spiritualism was a way of crediting this account.

Such metaphysical facts could justify the strength of one's private faith and also lend public support to others of potentially weak faith. Spiritualism's avid identification with proof was a play for, and defense against, its power, explaining in part why spiritualism's vicissitudes track so closely with the heyday of scientific naturalism.

THE SANITY OF "THE SPIRIT CRAZE"

What eventually became known as "the spirit craze" began in Hydesvillle, New York, in 1848, when two teenage girls and a ghost exchanged raps on a barn wall that divided this world from the next.[7] In this thoroughly improvised communication, the Fox sisters had found a way to make the dead speak in a staccato percussion signifying yes and no. Soon tables were tipping all across Europe, England, and America as curiosity seekers hailed ghosts of their own who tapped out their communiqués on the kitchen floor. The table-turning phenomenon arrived in England in 1852, where there was already a well-developed discourse of mesmerism ready to receive it, yet, it was the spiritualist narrative that soon took the initiative. Mesmerism's comparatively staid notion of an exotic mental force simply could not compete with this new, exhilarating movement promising a cameo of heaven on earth.[8] It dawned with a sense of quivering expectation, "pervading all classes, all sects, that the world stands upon some great spiritual revelation."[9] These tables were not just turning through some kind of mental kinesis, they were tapping out words and letters transmitted by a ghostly intelligence: a breakthrough in modern communication technology that eclipsed even "the new railway of thought," the electrical telegraph.[10] By 1853, the *planchette*, a gliding board with a pen attached for automatic writing, improved upon this rather cumbersome table-tipping device and gave further impetus to the fast-developing interest in spiritualism.

Spiritualism steadily assimilated the investigative armature of mesmerism in pursuit of its own, otherwise controversial agenda: the chain of clasped hands circulating magnetic fluid became the séance circle; the mesmerist's powers of clairvoyance, levitation, crystal vision, and thought reading were all attributed to the spiritual medium; and the hypnotist's ability to possess and even operate the nervous system of his subject was inverted to become the trance state in which the medium herself became possessed and operated upon by the agency of spirits. Even as spiritualism appropriated the set pieces of mesmerism, mesmerism likewise annexed the phenomena of spiritualism into its own area of concern. Phenomena such as hauntings, apparitions, trance mediumship, and levitations were taken up by the chief organ of scientific mesmerism, the *Zoist*, and framed as purely psychophysical phenomena.[11] The continuum between mental and spiritual agencies that connected these inquiries conveniently blurred the line between philosophy and metaphysics, uniting the mysteries of mind to the mysteries of death, while maintaining two points of entry. Mesmerists and spiritualists shared the belief that human consciousness could potentially be the point of contact with some divine order, keeping alive the hope that engagement, perhaps even scientific engagement, with something more ultimate was still possible.

The Victorians, as consumers of Romantic horror fiction, had already demonstrated a certain taste for occult spirits in the earlier part of the century, but investigative séance spiritualism constituted a distinct endeavor. The ghosts drawn by gothic writers like Horace Walpole and Anne Radcliffe were continental lotharios; they went about sexually terrorizing British maidens in the morally suspect Mediterranean. These amorous aristocrats were barely suitable for purposes of literary entertainment, let alone for the high-minded aspirations of this new undertaking. By way of contrast, spiritualists emphasized the facticity and moral forthrightness of their ghosts, setting them apart from these nefarious, fictional varieties emanating from the gothic imagination. The *bricolage* characteristic of spiritualism drew primarily from mesmeric, scientific, and nonconforming religious discourses rather than from literary ones, readily assuming their moral mantle of progress along with aspects of their utopianism. In keeping with this enlightened spirit, the familiar diabolical interpretation of spirit activity that could have been mined from religious and folk traditions was forcibly excluded from Victorian spiritualism's identity formation. Such accusations of diabolism, however, were frequently directed against the movement from outside its ranks. Rev. Thomas Lake Harris warned, that the Devil was no longer "bound within the confines of the invisible world" but had "ruptured the odylic spheres of the human race let loose for a season."[12] This kind of demonization, constructed by Harris as the furtherance of an already objectionable mesmerism, was characteristic of the early reception of spiritualism. Its novelty was such that moral conservatives grappled confusedly with its oncoming tide not quite knowing how to name it. While they called upon traditional satanic tropes to identify the threat, they also sensed something strange and new, giving rise to paranoid rants covering all bases of the spiritualist agenda. Rev. W. H. Ferris accused the movement simultaneously of old-fashioned satanism, newfangled Christian universalism, deism, atheism, pantheism, Epicureanism, and free love.[13] The authors of titles such as *Thoughts on Satanic Influence or Modern Spiritualism Reconsidered* (1854) and *Spiritualism: A Satanic Delusion and a Sign of the Times* (1856) understood that there was something modern to spiritualism's hubris, even while it had its roots in an age-old sin: temptation by the Devil.[14]

These reactionary narratives were for the most part successfully overwritten by the furious wave of publishing activity defending and evangelizing spiritualism, dashed off by various writers in the 1850s and written in a tone of heroic, unblemished virtue. Far from being out to make some secret, self-serving Faustian bargain, spiritualists sought, in the words of Catherine Crowe, "progress in spiritual knowledge, together with the earnest and wide spread desire, especially in this country [England], to improve the moral condition and alleviate the sufferings of humanity."[15] Crowe's assurances aside, this did not mean spiritualism posed no threat to traditional Christianity. A theme running across much of its literature was the quest for a new religious revelation, which implied a certain eagerness to cull unwanted elements from theological orthodoxy. What spiritualists desired, according to their leading apologist of the 1850s, was to get "closer to the truth of God," which meant challenging the "deluded faith of bigoted minds" whose conflict with spiritualism stemmed from their cowardly "desire not to be undeceived."[16] While such attitudes were rarely intended as a refutation of Christianity, they did boldly signal the dawn of a new era in terms of mode of worship and structures of power. Spiritualism's religious attitudes tended toward the antiauthoritarian, going so far as to deny the existence of the Devil and to suggest that Christ's miraculous nature might pertain to us all should we choose to discover it. (Catherine Crowe's anticlericalism boldly attributed humanity's hitherto stunted spiritual growth to clerical mercenaries intent on making "the salvation of men's souls a means of living.")[17]

Given the diversity of its subgroups, historians have tended to divide spiritualism into a variety of "spiritualisms." However, its doctrines were the product of a highly discursive culture in which core values and beliefs were debated and propagated through an open press. This allowed for a loose ideological consensus to evolve among the varieties of spiritualism: Anglican, theosophist, psychological, Unitarian, socialistic, plebeian, and scientific. From its inception, these narratives tended to culturally position spiritualism as a highly specialized form of intellectual activity, drawing upon the attributes of both science and religion. Whether it fully belonged to one or the other, both, or neither, the revolutionary "science of nature and of God" (as it was described by one leading spiritual pastor in 1851) was difficult to categorize.[18]

But it was abundantly clear to its followers that spiritualism charted the course of modern social, spiritual, and intellectual progress, holding out its promise of advancement to all.

This last point distinguishes the original Victorian project from successive iterations of modern spiritualism, which have since reabsorbed some of the necromantic themes excluded by this first wave. Because modern magic is so polymorphous and anti-institutional in nature, it can be somewhat difficult for us to grasp the earnest, programmatic, and uniquely indigenous character of Victorian spiritualism.[19] Already in such late nineteenth-century hermetic revivals as theosophy and the Hermetic Order of the Golden Dawn, the arcane, countercultural aspect of the magical tradition began to reassert itself. Such programs tended to be secretive and ceremonial, reasserting the hermetic ethos of personal development, ritual enactment, and the pursuit of esoteric wisdom. Madame Blavatsky claimed her "secret doctrine" came from the letters of mysterious Indian mahatmas known only to herself, while the Golden Dawn's theurgical rites to raise the gods were for privileged initiates only.[20]

Such cultural aesthetics may have been attractive to avant-garde artists and intellectuals, such as Aleister Crowley and William Butler Yeats, becoming a source of artistic potency and alchemical self-transformation, but they were not "progressive" in the way that term is commonly valued by liberalism's Enlightenment agenda. This kind of occult personal empowerment has been somewhat democratized by the New Age movement with its popularization of divination practices like modern Tarot, the I Ching, numerology, and astrology but such magic still operates outside the scientific worldview, having lost Victorian spiritualists' hope of claiming science for themselves. The homegrown and forward-looking nature of Victorian spiritualism also has little overlap with fringe movements like Wicca, Anthroposophy, Kabbalah studies, and the New Kadampa Tradition, which draw their doctrines from another time and place in order to critique Western "progress" and materialism (rather than to reaffirm and redirect it, as did spiritualism). Mysticism's countercultural entanglement in the 1960s with psychedelic drugs, libertine sexuality, rock 'n' roll, and the Western submission to Eastern wisdom has only further distanced it from the dominant Christian culture as a morally suspect enterprise.[21] Victorian spiritualism was able to run largely in harness with domestic denominations of Christianity and keep terms with its codes of social respectability. Whatever taint might apply to its beliefs and practices was contained largely within the persons of the mediums themselves, who, relative to their station, tended to individually benefit from spiritualism's economic and social opportunities.

Still, a potentially strong undertow ran beneath "progressive" spiritualism (the darkened rooms, the working-class women, the humbug, the mercenaries, and, of course, the undead). Underneath its trappings of respectability and plausibility, the ties to earlier, more forbidden traditions were never entirely severed. The very nature of the knowledge spiritualists pursued (extrasensory, mystical, postmortal, and prognosticating) had been dismissed by skeptics from Locke onward as a mark of modern philosophical maturity. The modern mind accepted the limits of knowledge; spiritualists did not. Like the earlier intellectual programs of Christian gnosticism, Neoplatonism, and renaissance alchemy, spiritualism worked to bring forth an absolute cosmic revelation uniting divine and natural philosophy. Spiritual mediums transmuted substances, levitated objects, and materialized ghosts in continuation of the most ancient shamanic rites. Those critics who decried them on intellectual grounds saw spiritual beliefs and practices as a mentally regressive force, made all the more dangerous when asserted as fact and entwined with science. But spiritualism managed to counter all of this occult, social, and intellectual pollution, at least early on, with an overpowering narrative of moral entitlement. Given the degree of opposition raised against it at both ends of the ideological spectrum (from traditional theologians to strident naturalists), this astounding fact should not go unrecognized or unexplained. This ability to bring together the trifecta of religion, science, and social progress was crucial to spiritualism's rapid assimilation into the Victorian cultural landscape, where it insinuated itself into the heart of the British home and managed to defend its position there for decades. For most Victorians (and many people today) the possibility that spirit communication *might* be true was just too attractive for them to concede that it was irrefutably false. Thus, spiritualism found its widest accommodation not as a settled "reality" but rather as a tantalizing possibility to be discreetly cheered from the sidelines and to be awarded the same protection as any other intensely interesting prospect.

As a practice, spiritualism acquired its greatest immunity from the protected cultural sphere in which it unfolded—a site where consumer agency met private religious conscience and where all efforts at institutional policing had little power. As long as spiritualism could sustain its brand of respectability, it benefited from the middle-class prerogative of self-responsible citizenship set free in liberalism's marketplace of ideas. Spiritualists published their own journals to claim their share of this marketplace, furthering the pedagogical, forthright personality of their movement through an open press. Individuals who maintained a strenuous public objection to spiritualism were relatively few. They tended to be either religious traditionalists or scientific watchdogs mobilized by the threat that spiritualism posed to their own domains of knowledge. Cultural critics writing for *Punch*, the *Pall Mall Gazette*, and the *Times* added ridicule to scorn to amuse their sophisticated readership, bringing a certain zest to this educated condescension. Otherwise, most people, at least through the early 1870s, adopted a kind of wait-and-see toleration for spiritualism. (Even Charles Darwin and George Eliot indulged in a séance or two.) Its popularity with social luminaries like the Balfours, the Countess of Caithness, and Lady Mount-Temple further eased the way for middle-class participation, with a steady supply of trickle-down respectability. If the aristocrats themselves were in doubt, they could look to the example of European and Russian royalty, who were known to have opened their own palace doors. Queen Victoria herself reportedly attempted contact with her beloved Albert.[22] Thus, despite the withering words of one put-upon *Times* reporter who deemed spiritualism to be "nothing more than a farrago of impotent conclusions, garnished by a mass of the most monstrous rubbish," its cultural cachet was able to cover its social expenses. The "grossest folly" had come to England to spread its "moral infection" and found a place to stay in some of the finest homes in England.[23]

As early as 1853, the popularity of table turning had become so widespread that the celebrated scientist Michael Faraday felt the need to stage a cultural intervention, launching his own investigation to discredit the phenomenon once and for all. In the preface to his published study, Faraday complained that skeptics were becoming so rare "that their number is almost as nothing to the great body who have believed and borne testimony."[24] He devised a mechanism that recorded contact pressure between the sitters' hands and the table surface, clearly demonstrating that tables turned through the application of manual force. Yet his expert findings, complete with illustrations, did little to persuade people out of their convictions. As de Gasparin wrote mockingly, "The tables turn in spite of M Faraday. Their death has been predicted, it has been demonstrated, but they continue to turn."[25] The problem facing Faraday and other censorious rationalists was that spiritual beliefs had stubbornly taken root in the direct observational experience of the table-turners. To a certain extent, scientists themselves had established the terms of this popular resistance to their intellectual authority through their earlier campaigns of show and tell, giving the visual a certain ascendancy over the conceptual with lay audiences from the start. As recent scholarship has shown, the power dynamics of the lecture-demonstrations that popularized science beginning in the late eighteenth century were more complex than the mere top-down transmission of settled ideas. Popular astonishment could be part of knowledge-making itself. An impressive demonstration could elevate the renown and reputation of a scientist, and thus amplify the persuasiveness of his theoretical proposal. One such case in point: when Alessandro Volta could gain no academic traction against Galvani's "animal electricity" through theoretical arguments alone, he added instructions for assembling a giant voltaic pile, encouraging readers to experience for themselves the "shocking truth" of bimetallic electricity.[26] James Joule's kinetic theory of heat was likewise snubbed by academic science until the dogged brewer built a machine demonstrating this mechanical equivalent to be unequivocally real.

Thus, the kind of humble evidential challenge posed by table turning to elite theoretical complacence had a precedent, even in the making of scientific orthodoxy. This gave a frisson of possibility to spiritual efforts that might not be readily apparent to historical hindsight. These earlier, nonacademic tributaries of knowledge conditioned the lingering expectation that science might include a supporting (or even active) role for the laity. In addition to this flourishing amateur tradition of the first half of the century, primary education began to actively proselytize science in the 1830s, presenting it, not as a standoffish, academic discourse, but as an inviting tale of wonder. Middle-class maternal educators and governesses mixed their lessons with a nature walk or some other hands-on activity to inculcate scientific learning and practice together. The economic boom of the 1850s further personalized science by turning it into a form of consumer entertainment. Hobbyists and enthusiasts used their new wealth and leisure to pursue interests as diverse as trick photography, insectivorous plants, dinosaur bones, exotic lepidoptera, phrenology, and now spiritualism. This curiosity and initiative can be seen still in the "willing game," a favorite party entertainment of the 1870s in which Victorians tested their personal mediumistic capabilities in their own drawing rooms, focusing on an object and then trying to mentally transmit its identity to another member of their party.

Yet this popular, pluralistic science was increasingly at odds with scientific professionalization, under way in earnest since the establishment of the British Association for the Advancement of Science (BAAS) in 1831. The BAAS had, among its goals, the desire to curate scientific activity throughout the country in order to establish unified intellectual and practical standards. The aim was not just to discipline professionals, but also to improve the public understanding of science (as they defined it)—a challenging prospect for any nascent orthodoxy in the first-ever consumer information boom. The midcentury explosion of for-profit media capitalized on this new market for scientific leisure, providing a ready outlet for authors with ideas to sell. This resulted in a tangle of learned and commercial discourses with no clear title of authority. Pundits like John Tyndall and T. H. Huxley stepped into the breach on behalf of educated opinion, but the hard line they took against clerics and mystics was a double-edged sword, ginning up resistance to professional opinion even as they worked to clarify it.[27] Nor did their brand of polemical naturalism sit particularly well with most members of the scientific community, who generally sought a softer stance toward religion that might even include some mutual future.[28] All these voices speaking within and without this new institutional framework complicated the promulgation of a single scientific orthodoxy.

Despite these institutional challenges, by the1870s the BAAS had witnessed over four decades of astounding scientific advancement, its own institutional status rising with this tide. Credentialed scientific practitioners came under more regular professional scrutiny, with their own success increasingly dependent on a certain official compliance. In addition to these official guidelines, scientific watchdogs enforced this hierarchy with attacks of a more personal nature. For those less aligned with the BAAS's institutional goals or poorly positioned within its structures of power, all this oversight and ad hominem hostility could feel quite oppressive. This was especially true for trained professionals occupying the hot zone of scientific spiritualism, but there was also less tolerance for the public's unabated credulity, which seemed to be surging alongside science's own accumulation of discoveries. In 1868, William Howitt, the spiritual enthusiast, gave the number of twenty million adherents worldwide in a letter to the *Dunfermline Press*, while the *Pharmaceutical Journal* (1870) took a dimmer view of this same fact, warning its viewers "that spiritualism was taking up more and more of the public's attention" and that such "faith in the spirit world was steadily increasing."[29] Crookes himself estimated the population of spiritualist adherents in England alone to be "in the millions," justifying his scientific curiosity solely on the basis of that number. This estimate was meant to designate only spiritualism's activist core. Beyond that, there would have been many more people acquainted with these beliefs who did not label themselves as spiritualists. Tag-along friends, curiosity seekers, bereaved clients, and skeptical investigators gave spiritualism a degree of social permeation that would have made first- and secondhand knowledge of the séance experience a fairly standard part of cultural literacy. That spiritualism was a recurring topic in such established journals as *Athenaeum, Critic Literary Gazette, Cornhill Magazine, All Year Round, London Review, Fraser's,* and *Spectator* suggests that, while it remained something of a titillation, the public's interest in it was widely assumed and condoned, and in terms of the commercial press even catered to. This media attention peaked in the mid-1870s, marking the high point of cultural interest in spiritualism, when a scientific investigation like the one Crookes proposed could still expect to take the national stage, and the rising establishment critique against it would only have added fuel to the fire.[30]

In addition to the elite curiosity of celebrity scientists, there was the entrepreneurial culture of mediumship itself keeping the fires kindled under spiritual enthusiasm. These showmen pulled out all the stops for the sake of client recruitment, ever improving upon spiritualism's ritual forms: first there were raps, then there were voices, followed by sounds and smells, then levitating objects, and materializing objects, then ghostly body parts. By the early 1870s, a five-star séance might be expected to boast the full-blown materialization of a walking, talking ghost.[31] The more incredible the claim, the more factual supplementation was required. Spirits of the dead were photographed, measured, weighed, medically examined, and even interviewed at great length in order to validate their authenticity. In addition to the physical evidence supplied by séance investigators, spiritualists could likewise draw on a rash of favorable scientific discoveries that lent theoretical plausibility to this already robust public willingness to believe: cable telegraphy, photography, electricity, magnetism, and phosphorescence all seemed to parallel séance mysteries and even to suggest that science could become the ally of spiritualism. Such theories served as helpful accessories to the established facts of the séance, explaining and responding to them. But they were not called upon to determine their possible validity, which was certified by spiritualists themselves. So while the term "scientific spiritualism" is often used to designate the elite scientific curiosity of transatlantic cable engineer Cromwell Varley, mathematician Augustus De Morgan, and, most famously, William Crookes, it was also a bottom-up phenomenon. Spiritualists consulted science, but they did not defer to it. Rather than being defensive against scientific skeptics like Faraday, spiritualists took the offensive. They impugned credentialed scientists as self-interested and mercenary, applying the same aristocratic criticism used by elites in the first half of the century to tarnish their paid scientific labor.[32] By contrast, the unremunerated spiritual seeker adhered to "the highest ethical law of science [which] was the love of truth," free from the "fanaticism of skepticism" that precluded from observation anything that could not "be carved by the scalpel, seen by the microscope, and analysed in the laboratory."[33] To recruit others to the cause, these spiritualists published "lectures," "treatises," "investigations," and "reports," capturing a tone at once instructional and breathlessly visionary. Here, the empirical encountered the metaphysical with a sense of possibility still intact, encouraged by a literate, scientifically engaged public that had a great stake in the storylines spun about god and nature.

Of course, the true power of spiritualism was that it tapped something much deeper and more enduring than the zeitgeist. At the heart of its widespread popularity was the fathomless power of grief, fueling an insatiable public need to be united with their dead on any terms, at any cost.[34] Bereavement was the bread and butter of the emergent séance industry. The resignation required by traditional Christianity in the face of death left people hungry for a more congenial religious practice that did not place heaven so far out of reach. The séance circle offered a powerful new form of spiritual technology to meet this emotional need. It was also staffed by a more customer service–oriented religious mediary: the séance psychic. This shaman-for-hire could do things with regard to the next world that the ordinary, and less amenable, clergy could not and would not do: foretell the future, narrate past lives, read other people's minds, and, above all, confirm the hoped-for truth of life after death or even summon the soul of a deceased child.[35] The curiosity and emotional desire aroused by such promises helped to enforce spiritualism's legitimization with the public, overrunning the religious and cultural barriers that might otherwise have kept such activity in the margins or underground. Alfred Russel Wallace, as if pitching a new and improved religious product superior to "all that a belief in Christianity could give," touted its unprecedented powers of consolation, giving us a glimpse of the movement's unstoppable zeal: "What use is [spiritualism]? It substitutes a definite, real, and practical conviction, for a vague, theoretical, and unsatisfying faith."[36] Spiritualism was a solace for life's deepest pains and uncertainties, something people would not be readily inclined to deny themselves.

To partake in this spiritual ministry, one did not need to abandon Christianity, just retrofit it with a few practical enhancements. The séance circle had a special appeal for unsatisfied Anglicans, whose theological discipline kept God on high and enthusiasm low. The personal intimacy and spiritual immediacy of the séance circle made it a welcome religious supplement. Part of the attraction of spiritualism was its flexibility: as a practice, it was not ideologically binding, and occasional participation need not impose on one's religious identity. That spiritualism required no denominational surrender, and thus avoided the unseemly drama of a religious conversion, made it appear more conservative than it actually was. A séance circle might even involve members of the clergy. As a support (rather than challenge) to people's existing faith, spiritualism could slide smoothly through the social thoroughfares. Recognition of this flexibility adds to our understanding of the successful transmission of the practices and ideas of spiritualism. It meant the spiritual daring of the séance was not just the preserve of the ardent religious pioneer but could lure mainstream elements as well. Yet its implications for the future of religious practices and beliefs were quite radical in the context of other religious reform movements. Spiritualism not only offered psychic communion with the dead, but in the case of ghostly materializations, it proposed a spiritual reality that one could even reach out and touch. As compared to the more modest aims of the Evangelical and Oxford movements, which sought only to bring more emotion and mystery (respectively) to the Anglican religious experience, spiritualism moved religion forward in the direction of immanence, inviting God back into the clockwork universe and allowing for an intimacy with the dead unknown to modern Christianity. It also restored the exploration of nature to the province of spiritual knowledge, taking back this lost agency of the Anglican priesthood in order to democratize its power.

THE WAR AGAINST THE PEACE BETWEEN SCIENCE AND RELIGION

Seen as part of a grander intellectual cycle, movements like spiritualism and psychical research offered a remedy for the breakup of Anglican natural theology in the earlier part of the century, attempting to restore the link between God and nature in a form more suitable to contemporaries. But natural theology's roots ran far deeper than the eighteenth century. They reached back to the earliest efforts of Roman Catholicism to assimilate classical learning, and to the founding of European intellectual life with the Carolingian renaissance. The assault on this idea, beginning in the early modern period, shook something deep in the cultural psyche, a fault line that widened in the Victorian period as the implications of this growing division entered the public, and personal, sphere.[37] Eighteenth-century Anglican natural theology was the final iteration of this epistemological model, managing to unite mechanical science and reformation theology into a single harmonious worldview in which God had created the universe according to "number, weight and measure."[38]

This post-Newtonian *Summa theologica* steered the ship of state through the turbulent eighteenth century, avoiding the continental extremes of reaction and revolution with its top-down program of forced moderation. Nonetheless, Anglican natural theology began as a fundamentally progressive impulse. It established science as a distinct form of intellectual authority specializing in the natural world, and even allowed for lay curiosity in this burgeoning realm of physical inquiry. This gave rise to a nascent popular science and the protoindustrialization that stemmed from such practical activity.[39] William Paley's *Natural Theology*, published in 1801, powerfully reaffirmed this worldview as it crossed the threshold of the nineteenth century, giving special emphasis to God's biological creation in keeping with the curiosity of the times. The flora and fauna found in God's book of nature now began to populate ladies' sketchbooks at home as well as those of daring colonial explorers abroad, as pious naturalists fanned out over the meadows and across the globe to marvel at the divine perfection of life's many form and functions. All this interest in nature and technology, promoted from the pulpit and transmitted through general education, brought a slow but steady degree of scientific penetration into British popular culture. Yet, these scientific narratives were for the most part understood as unfolding within, rather than apart from, the religious domain, justifying an almost absolute clerical rule over British intellectual life. This control grew alongside the expansion of primary schools and traveled all the way up to the highest levels of university training. By the time Victoria took her throne, this broadening of religious and scientific education had given rise to two conflicting realities. One was the stifling cultural authority Anglican clerics projected from the hallowed dais of their state religion. The other was a growing desire for greater spiritual, personal, and intellectual freedom, at odds with this Anglican hegemony.

The power of religion, both as private experience and cultural authority, enters keenly into the psychical story at the level of personal biography, constituting the milieu in which its founders were raised and lingering on in the ambivalent psychical rebellion against and attraction to religion. Such clericalism was the real "serpent" named by Huxley in his famous tirade, motivated in part from his memories of feeling constrained by the high-handed enforcement of Christianity.[40] By the 1830s and 1840s, High Church authority found itself pressured on many cultural fronts. Emotionally spirited Evangelicals, mystical Catholics, Chartists, republicans, working-class radicals, educational reformers, utopian socialists, and other varieties of social, religious, and political reformers brought a new kind of intellectual and ideological diversity to the scene. The age of "-isms" had begun. As the liberal prerogative of personal conviction shifted more and more authority to the individual, the public's reflexive submission to elite orthodoxies, religious or otherwise, could no longer be assumed. It was not just a matter of asserting the freedom to *choose* one's beliefs; in the great flourishing of pseudosciences and nonconforming religions, we see a vigorous demand to help *create* them as well. Never before had there existed so many substantive alternatives to traditional theism or ontologically diverse sources of moral authority from which to formulate one's personal identity. As the sphere of Anglican influence contracted, it was not clear what new forms of authority would come to occupy the vacancy it left behind. The positivistic outlook of the "warfare thesis" may have assumed that science would be the victor to whom the spoils of the church's lost share of cultural dominion belonged, but such an outcome was far from clear. "Institutional science" as such was still very much in formation, and even if theologians increasingly ceded control over naturalistic narratives, popular ownership of science was not so easily pried away. It continued to assert itself in the intellectual and investigative activity flourishing alongside these sanctioned discourses, and was shaped by independent considerations that tended to reject any overly rigid construction of separate spheres.

As cultural historians have restored these hitherto marginalized narratives to our period portraiture, the view of the intellectual dynamics has become far more nuanced than an assumption of merely "science versus religion."[41] Rather than the neat, elite polemics pitting Huxley's virulent naturalism against Bishop Samuel Wilberforce's theology *in extremis,* (a face-off made famous as "the Oxford evolution debate" in 1860), we see a plurality of discourses offering a mélange of religio-scientific speculation asserted by an increasingly democratized range of players. Spiritualism was just one of several important cultural responses aimed at constraining the widening gap between traditional religiosity and changing naturalistic narratives. What we see in scientific spiritualism and other similar impulses of intellectual reconciliation—be it the Broad Church movement, neurohypnotism, Unitarianism, Neoplatonism, the revival of Bishop Berkley's subjective idealism, mesmeric psychology, or the transcendental physics described in Stewart and Tait's *The Unseen Universe* (1875)—is an effort to herd the discourses of physics and metaphysics back toward some kind of convergence before they could drift outside any possible shared frame of reference. Such would be the inevitable consequences of fully separate modes of speculation. With varying degrees of rigor, these movements attempted to restore the philosophical continuity (and even ontological contiguity) between matter and spirit severed by the two-sphere model of explanation. In these less formal, more experimental narratives, the concepts of natural and supernatural were more porous and pliable. Mind and matter were not parallel, but potentially interpenetrating, as in mesmeric, spiritual, and psychical paradigms.

By challenging the overly strict dualism of matter and spirit, such theories undercut the ideological justification for dividing the two realms into separate scientific and theological areas of specialization, breaching the very boundary that official culture was attempting to put in place. The central anxiety here was not so much about a war between science and religion, as about a deep concern regarding the terms of their peace: the emerging two-sphere orthodoxy. While this new regime of knowledge had clear advantages for the intellectual establishment (theologians would retain their authority in relation to God and academics would impose a new monopoly on scientific knowledge), it had a less obvious appeal for popular and elite seekers left outside the scheme of its authority. If anything, this laity actually lost a degree of inclusion it had enjoyed under Anglican natural theology's looser scientific boundary, without gaining any new religious agency by way of compensation. This elite reformulation likewise affirmed a deeper philosophical chasm, conceding the epistemologies of faith and reason to be irreparably sundered. This left the spiritual questions of most vital interest to modern Victorians unasked and unanswered, and thus limited the potential popular benefit of science while exposing religiosity to uncertain risks.

These popular efforts to reconcile science and religion do not necessarily invalidate the sincerity of warfare rhetoric for disputants or cast this enmity as some kind of institutional contrivance. Surely, the forced theological surrender of naturalistic narratives was accompanied by genuinely felt religious anguish and involved a certain amount of intellectual militancy. However, revisionists have broadened this picture considerably, allowing these subjective narratives of conflict to be seen as part of a larger pattern of institutional realignment.[42] Given the elemental need for both science and religion in the social architecture of Victorian Britain, such conservators as church and academy made unlikely combatants. In this light, the institutional skirmishing is best understood as an intraestablishment rivalry working out new power-sharing arrangements of its ruling class, one eventually embraced by the church as well after its initial resistance in the early nineteenth century. The strategy of separate spheres was ultimately a means of avoiding, not inciting, conflict, eliminating the overlapping scheme of knowledge that had led to so much butting of heads. It allowed the religious beliefs of most Victorian scientists to remain untouched and smoothed the way for a continued clerical presence in higher education. All in all, this epistemological and institutional partition was a highly successful model of secularization for church, science, and society, one that kept to the middle road as much as possible. It held a privileged and protected place for religion while clearing away obstacles to scientific progress.

But it was provisional. The line was not firmly drawn, nor was the desirable degree of separation by any means uniformly determined. The religious outcomes of such a divided scheme remained in question and its long-term intellectual advantages remained in doubt. Narratives like spiritualism provide an insight into the complexity of this secular angst, not just as an elite conflict between two institutions hammering out the terms of modernity, but in the more diffuse setting of personal and popular misgivings directed against this emerging new order. These dialogues of resistance continue to flourish today in various alternative medical, nutritional, and spiritual discourses, but without the great expectations and elite allegiances mustered by the spiritual and protopsychical movements of the 1870s, which still saw science as a somewhat open scheme of knowledge. What remains unique about psychical research is that for a brief generational moment, a group of intellectuals managed to bring this line of metaphysical dissent past the gates of disciplinary discourse, at the very moment science was sealing off its bastions of power. In the 1880s and 1890s, this disciplinary establishment was rapidly gaining control of all recognized forms of scientific power, and it remained an open possibility that psychical science would be codified among them. Psychical concerns tapped broader cultural anxieties that had been percolating throughout the midcentury as spiritualists, naturalists, mystics, and theists all struggled to chart the optimal course for modernity. Researchers leveraged a generalized crisis of faith into a crisis for science itself, forcing consideration of their radical alternative within the heart of science.

GENERATIONAL CRISIS:
THE ONLY THING TO FEAR IS FAITH ITSELF

The contradictions posed by the crisis of faith have proved a fruitful hunting ground for generations of historians looking to find what the fuss was "really" all about. Its specter of atheism seems an odd ghost to haunt the Victorian landscape, given the efforts to hold a place for spirituality. Yet, as Frank Turner points out, it is a story told largely by Victorian intellectuals themselves in distraught biographies, anxious opinion essays, and literary fiction, and only later reified by subsequent generations of secular academics. Its social and cultural dimensions are difficult to evaluate, apart from its primacy as a subject position, yet it also found resonance with the larger public who found some truth to relate to.[43] Multiple intersections, in terms of themes and actors, make the crisis a particularly productive hermeneutic for psychical research and vice versa. Even while religious concerns were excluded from the psychical discipline itself, fears for the personal loss of faith, and for lost faith in general, were an essential part of its context. Many high-profile crisis-complainers, such as Henry Sidgwick, Edmund Gurney, Samuel Butler, Frederic Myers, Leslie Stephen, Arthur Balfour, James Ward, and George Romanes, were also involved, to varying degrees, with psychical research and dynamic psychology.[44] Their writings not only presented a war between science and religion as inevitable, but also saw this coming war as one that religion would inevitably lose. Why did they tell this story, and how did they come to believe it themselves?

If these complainants took their cue from contemporary science, they would have had to concede that most of its leading figures continued untroubled in their religious beliefs and that, at an institutional level, the declared scientific ethos of gentlemanly discourse censored rather than encouraged religious antagonism.[45] Even strident scientific naturalists such as T. H. Huxley, Herbert Spencer, and John Tyndall were not critics of Christ or Christians, but rather critics of clerical worldliness and superstition (making them part of a long tradition of nonconformist insolence toward the established church). Their bellicose language bespoke the relative weakness of scientific institutions, not the *schadenfreude* of an unsporting display of strength. Far from truly being at loggerheads, many religious and scientific leaders came together to advance a popular understanding of biological and social evolution that remained part of a larger divine design, with modern British society sitting triumphantly atop a universal history. This comfortable view was embraced by the majority of Victorians who were so eager to find an accommodation that even extinction was seen as a form of God's benevolent guidance in the direction of progress.[46]

Yet this was not the interpretation favored by those in crisis. Their fatalistic rhetoric insistently doomed all spiritual beliefs to the relentless encroachment of an oncoming, uncompromising scientific materialism. There was some substance to their perception, even if it was personally held. Decades of advances in electromagnetism, atomic chemistry, astrophysics, geology, brain anatomy, and biology had intellectually tamed once mysterious forces and corralled them into a single physical system of energy brought under the mathematical regulation of thermodynamics. By the 1860s, evolution was also integrated into this schema, masterfully consolidating various disciplines into a single, interlocking mechanical explanation, which cast its gloomy shadow over those in crisis. And in truth, the fundamental narrative strategy of Victorian scientific naturalism was to expel God as a causal force in the physical world: all such spiritual energies were to be vacated by mechanism, and the direct hand of God was to be banished from design. It is no accident that Cambridge University, at the front line in assimilating all this new, unfiltered knowledge, was also the center of religious crisis and likewise the intellectual cradle of psychical research. Henry Sidgwick, Samuel Butler, Frederic Myers, James Ward, Edmund Gurney, Leslie Stephen, Walter Leaf, and George Romanes were all present at Cambridge sometime between 1859 (when *The Origin of Species* was published) and the mid-1870s, an era that was also marked by the rising tide of continental materialism in British thought.[47] There they found themselves directly in the path of an influx of aggressively antimetaphysical ideas, pulled in from the continent in the wake of the discovery of natural selection. This new curriculum imported the paradigmatic texts of a mordant continental secularism, many of which had been recently translated through the deliberate efforts of humanists like George Eliot and Harriet Martineau. Auguste Comte's *Course in Positive Philosophy*, David Strauss's *The Life of Jesus*, Ludwig Feuerbach's *The Essence of Christianity*, Arthur Schopenhauer's *Die Welt als Wille und Vorstellung*, and Ernest Renan's *Life of Jesus* unleashed what Frederic Myers called "the first flush of triumphant Darwinism," characterizing the glee with which students paraded these ideas before their more sentimental peers.[48]

In terms of fostering a propensity toward crisis, these competitive social characteristics of a university setting augmented the impact of its intellectual controversies. Such rhetorical debates were not only part of daily life, but also a vital defense of one's intellectual honor and personal beliefs. Pressure to compete enforced familiarity with otherwise unwelcome philosophies and further weaponized particularly sensational texts. Campus combatants took as their model public pit fighters like the scientists Herbert Spencer, T. H. Huxley, and John Tyndall, and cultural critics like the positivist Frederic Harrison, all members of the X-Club (a dining society founded by Huxley in 1864) sworn to defend evolutionary naturalism.[49] These partisans exercised their protection in the press as an attack on church authority, framing a power struggle between science and religion rather than a search for common ground. This story ended with the expulsion of theology from natural explanation and a gloating victory for science. Another factor would be youth itself. Certainly, confronting the pointed atheism of Feueberbach at the age of twenty would have different implications than doing so at the age of forty, when one's identity was fully formed. This was especially true for an older generation of scientists, men like Peter Guthrie Tait, Michael Faraday, and William Thomson, who continued their Christian fellowship untroubled even while they elaborated mechanical theories under an empirical flag. Even those who abjured their religious beliefs did so with equanimity. There was little *sturm und drang* in George Eliot's marvelously stoic renunciation of faith as she calmly endured the demands of her freely assumed humanist philosophy. Crisis was also defused by Herbert Spencer's agnosticism, which saw this admission of uncertainty as a humble testament to the incomprehensibility of the divine, not as an attack on its possible existence.[50] Certainly, the people most immersed in this physical worldview were the rank-and-file scientific practitioners themselves, and yet by and large they tended to avoid subjecting themselves to crisis flagellation. And, of course, there was a sizable college cohort born in the 1840s that shared the susceptibilities of youth and setting with Myers, Gurney, Romanes, and Ward, but who did not succumb to a sense of crisis.

So this emerging institutional and epistemological boundary between science and religion was not uniformly problematic. It allowed for a range of intellectual responses: one could lose one's faith, one could keep one's faith, or one could have a crisis of faith. To understand what substantiated the logic of crisis for its sufferers, we need to look deeper into the interpretations driving its main trope of *conflict* and *catastrophe*. On a personal level, conflict played out as a longing for religion that was inwardly opposed by a fatal attraction toward science. Such an internalized psychological struggle abolished all hope of future social accord between the embodied institutions of both sides, exaggerating narratives of aggression and ignoring evidence of accord. Finally, topping off this sense of imminent religious catastrophe was the inevitable outcome of total societal doom. Henry Sidgwick may have deeply lamented his own personal loss of faith, but his greater concern was that religious doubt would eventually spread to the rest of the social orders, rusting out the moral foundation of society like a corrosive salt. Everything not subject to proof, Sidgwick surmised, would eventually become subject to doubt, withering faith not just in God but in goodness as well. "The ensuing evil," he confided in a letter, "would certainly be very great."[51] Whatever advantages science conveyed, its viability as a replacement worldview was being seriously called into question by the rhetorical response of crisis, a fact put into evidence by its narratives of personal despair. Yet, there was no going back to traditional faith. There was also no going forward without it. Bridges appeared to be burning in every direction.

If we are to understand how the "crisis" operates, in this book and for these subjects, then, it was as an intentional rejection of confidence, of compromise, and of the centrist argument for two separate yet complementary spheres. Instead, crisis rhetoric played up intellectual conflict and spiritual catastrophe. With its pessimistic tone, it effectively blocked the bifurcated path of knowledge marked out for modern secularization, creating an impasse for both science and religion with its prophetic assurances of bidirectional doom. Thus, crisis language was often stuck, the complainer rejecting both religious past and secular future as possible paths out of the modern conundrum of faith versus reason. Yet, by contrast, the emotional and intellectual intensity of crisis rhetoric mobilized a response by ratcheting up the stakes of inaction. The idea of crisis, after all, urgently required intervention. It sounded the alarm. The Victorian crisis complaint, despite its tone of resignation and its affect of malaise, served as a historically dynamic call to action.

Here, it is useful to consider such rhetoric through the lens of William Reddy's theory of emotives, an interpretation of the speech act that is at once obvious and insightful. As Reddy has argued, mainly in the context of the sentimental fervor leading up to the French Revolution, speech does not merely communicate; it is also meant to activate, to arouse, to direct a response.[52] It emotionally conditions the speaker, as well as those spoken to, in ways intended to nfluence behaviors and outcomes. If we look at the crisis of faith as a description of the facts on the ground, it falls short of reporting much of anything beyond the experience of its narrator. Not only was there no pervasive and looming threat of atheism, but also much of the traditional faith that was claimed to be lost was eventually found again in some form or other.[53] Yet, viewed performatively, crisis language makes sense. Such texts strongly activated emotional, psychological, and intellectual states, both for the author in the act of narration and for the reader at the point of reception, heightening the disposition for action. Instead of being strange, historical artifacts, crisis narratives became effectual historical agents, propelling certain kinds of moods and activities that were, in a general sense, resultative, particularly in regard to psychical research.[54]

While I am not advocating heavy-handed overcorrelation of cause and effect, a diachronic study such as this, which spans half a century, lets us approach crisis rhetoric not just as an emotionally reactive narrative, but also as a historically productive one. All this angst about metaphysical faith dying for want of physical proof provided an ideological pretext for the later development of psychical research, as well as the intense and lasting motivation necessary to commit to so ambitious an enterprise. The psychical researchers studied here made lifetime commitments to their research, and saw in it the mature continuation of their earlier intellectual struggles, funded from the deepest sources of personal and intellectual necessity. Observations concerning the utility of crisis rhetoric are in no way meant to deny its authenticity as a deeply felt experience or to see in it some kind of conscious manipulation. These were clearly individuals who had vested religious belief with tremendous value, and that very sincerity raised the stakes on its loss and galvanized much of their powerful agency. Rather, the concept of crisis rhetoric allows us to situate the crisis within a longer arc of historical activity, bringing it meaningfully into this psychical network of personal, ideological, and intellectual relationships, particularly in regard to this group of young intellectuals coming of age in and around the 1860s. This generation occupied an important hinge moment of change and consequence, taking action in a time of heightened possibility, the first step of which was crisis immobility.

The man who became the first president of the SPR in 1882 was also at the center of a major crisis drama unfolding in the late 1860s and early 1870s around the role of religious oaths in academia. Henry Sidgwick resigned his Trinity College fellowship at Cambridge University in 1869 in protest against the religious tests necessary to hold that position.[55] While in public Sidgwick framed his resignation as a matter of principle concerned mainly with academic independence, his private letters also showed his motives to be intensely personal and spiritual as well. He could not in good conscience declare a religious conviction he did not feel. With this highly visible refusal to make a false confession of faith, one of the nation's most prominent moral philosophers had come to the conclusion that belief in the Anglican God was a lie one should not have to tell. Sidgwick's renunciation highlights the ways in which the elite crisis narrative differed markedly from more dilute forms of popular religious anxiety as discussed above. There, metaphysics was never fully forfeited, as a matter of religious faith or as a principle of knowledge. Intellectuals, however, took issue with faith itself as a morally corrupt epistemology, which essentially asserted knowledge of the unknowable. Thus, it was not only the efficacy of faith that the crisis brought into question, but its good character as well. Charles Taylor captures this elite compulsion as follows: "One ought not to believe what one has insufficient evidence for. Two ideals flow together to give force to this principle. The first is that of a self-responsible rational freedom. We have an obligation to make up our minds on the evidence without bowing to any authority. The second is a kind of heroism of unbelief, the deep spiritual satisfaction of knowing that one has confronted the truth of things however bleak and unconsoling."[56]

Thus, as we see in all this intellectual rumination, anxiety about "faith" became particularly agonized when framed as a problem within the philosophy of knowledge: can we in good conscience believe what we believe? The factual rigor of empiricism did not pair well with other forms of knowledge, and if one were to fully adopt these standards, knowledge of God was a necessary casualty. This conflict would be particularly pronounced for rhetorically sophisticated elites, like my subjects, who were trained in the mental procedures of philosophy and thus were more attuned to inconsistencies within the cultural standards of knowledge. The revision calling into question the existence of a crisis, or at least of an ideological one, should not be allowed to discount the moral, intellectual, and epistemological problems the two-sphere model posed to some thinking Victorians. Its simultaneous validation of both faith and reason gave rise to a genuine cognitive dissonance that, for some, compelled the abdication of its core compromise of separate spheres. Thus, the crisis tells a real story of conflict concerning metaphysical versus empirical truth claims and the consequential loss of natural theology's single hierarchy of knowledge. Its language of suffering expressed a mood of intolerable moral and intellectual disorientation. While this sense of personal malaise may have been inscribed within a larger social narrative of reconciliation and progress, it was nonetheless attached to legitimate concerns arising from an intensely thoughtful perspective. What might be at most mentally inconvenient for the majority of Victorians could, for a Cambridge classics student, very well constitute "a horror of a reality that made the world spin before one's eyes a shock of nightmare-panic amid the glaring dreariness of day."[57]

This was a psychological and epistemological crisis running much deeper than the simple narrative of institutional combat put forward by John Draper in *History of the Conflict between Religion and Science* (1874). Although such texts were a factor in its narrative, they were not the source from which the untenable sense of crisis arose. Despite the alarming title, Draper's book had an ultimately optimistic orientation toward intellectual progress and scientific possibility. His attack on Christianity was limited to a historical critique of the church's political worldliness and clerical ignorance, leaving out the issue of authentic piety and the existence of God. W. K. Clifford's *The Ethics of Belief* (1877), however, with its crushing moral obligation of skepticism, is more to the point of the angst of the positivist era. In this epistemological manifesto, Clifford elided Draper's polite distinction between faith and ignorance into a single form of intellectual dereliction, stated in a piercingly succinct slogan for the ages: "To sum up, it is wrong always, everywhere, and for anyone, to believe anything upon insufficient evidence." Those in crisis suffered more than the loss of a longed-for religious belief; theirs was a slough of despond in which faith itself was the torturing sin.[58]

We hear this morally vanquished admission of fault in Myers's own concession of faith: "If there be no God and we perish for ever it may be right to say so and to face the facts as best one can, but one must indeed be optimistic to find much to be pleased about."[59] And yet, on closer examination, perhaps there was more "to be pleased about" than meets the eye. The "rising flood tide of materialism, agnosticism and mechanical theory" may have been a toxic narrative for those in crisis, but it was likewise lethal to the clerics, dons, and Evangelical moralists whose cultural authority freethinking young people like Myers might like to wrest away. No matter how painful Myers's personal loss of faith was (and he nearly killed himself over it), this challenge to the establishment put power into his hands and the hands of a new generation of secular intellectuals.[60] As such, the crisis complaint can be viewed as potential leverage in a complex intellectual and generational revolt that took aim at the clerical and Evangelical power complex into which the generation of the 1830s and 1840s was born. Many psychical researchers' fathers had themselves been clergymen, and one cannot help but see, in this fantasy of a deicidal science dooming all religious belief, the secret longings for a paternal coup. This was, after all, the generation of Samuel Butler and his badgered fictional alter ego, Ernest Pontifex, who likewise came of age in the spiritual prison of Evangelical piety. Scientific naturalism's presumed dispatch of this punishing parental theology may have been on some level a welcome consequence of its purported atheistic threat.[61] This hegemonic religious establishment was hardly going to stumble from a mere kick in the knee by a few depressed atheists. What was required was the advent of a truly monumental enemy: a science that left "extinguished theologians" lying about its cradle "as the strangled snakes beside that of Hercules."[62] Since such a clerical extinction was not a likely outcome at the hands of British science, this mortal enemy had to be summoned from across the Channel: an incantation performed by reading Schopenhauer, Renan, Comte, and Strauss, and then speaking their words aloud to one another in endless discussion. Reluctant agnostics tortured themselves by swimming toward (rather than away from) this undertow of material philosophy.[63] The resulting emotional crisis was then laid out in the press, warning of an oncoming, ineluctable atheism that the British public, lulled into complacency by its historic capacity for compromise, might not have otherwise seen.

In this way, the narrative of the crisis of faith speaks out of two sides of its mouth: bewailing the might of midcentury materialism, on the one hand, while simultaneously reifying the mirage of its power, on the other, through its public narrative of protest.[64] While these expressions of spiritual anguish bemoaned the idea of a predatory naturalism, the underlying allegiance here was not to the faith of their fathers. Nor can the crisis complaint be seen as an ally of those institutional moderates wishing to impose a friendly face on religious relations, even though such an outlook might have relieved some of their purported distress. Instead, this strand of intellectual dejection went further in inferring atheism from physical explanation than any X-Club ideologue might publicly dare, emphasizing an almost lurid materialism. What was the point of taking sides in a war one did not wish to win and rejecting a peace that might lesson one's dread? The answer demands a repositioning of crisis rhetoric outside the ongoing boundary debate of the last thirty years, which looked to maintain common ground, even as science and Christian theology disentangled themselves. In the 1830s, harmony was affirmed through the Bridgewater treatises, eight pamphlets commissioned to shore up God's moral order amid a growing record of cataclysms and extinctions. The point was not to supplant scientific explanation—merely to show the ways in which God could be seen to operate within it. By the 1860s, however, conservation and adaptation put an end to this teleological naturalism. God's guiding hand had no place in this fully automated universe. At best they might make room for free will within the statistical margins of heat dissipation. Even though this mechanism could not be soft-pedaled, it was still the goal of most professional and clerical leaders that scientific knowledge and divine purpose not appear to proceed at odds.[65]

Those in crisis seemed to fire this bridge between God and nature deliberately, even amid their own thwarted longing to get to the other side. If we understand this language as foregrounding a new kind of posture in relation to both science and religion, we can begin to make more sense of it. Crisis rhetoric did not seek to move people to action in the familiar ways of public debate. Rather, such accounts were meant to convey a sense of being stuck, stymied in both directions: reason or faith, reaction or progress, natural or supernatural. They were, as Edmund Gurney later put it, "paralyzed by doubt . . . at the fork of two roads." The point he was making in *Tertium Quid* (1887) was that such binaries were ultimately artificial: "We are told that only two parties are contending, but a tertium quid, a definite position distinct from all other positions, has been overlooked in the argument."[66] *Tertium Quid* was published in two volumes in the wake of five productive years of psychical research. The tone of the essays was still cautious, but the idea of a "third way" was at least publicly staked as a viable possibility.

By contrast, *The Power of Sound*, published in 1880, was fully fenced in, adhering to a strict evolutionary framework in its exploration of musical appreciation. But even as he brought beauty back down to earth, Gurney's theory of music pushed back against Herbert Spencer's utilitarianism, if only a little. Some part of the ecstasy of song belonged wholly to personal experience, irreducible to natural or even sexual selection. This was a rather modest defense of the self, but Gurney could go no further—at least, not in the context of evolutionary determinism. But, informally, Gurney—along with Myers, Sidgwick, Arthur Balfour and other curious agnostics of the Sidgwick Group—was involved in the study of séance phenomena, among other mental anomalies biological materialism could not explain. This was a supplementary curriculum potentially every bit as edgy as the science that had stripped away their faith, but that curiosity had as yet no basis to challenge such sound empirical arguments. The only "legitimate" protests against scientific naturalism to be made at this time were limited to expressions of personal despair.

These were not atheists; they were, in fact, lifelong seekers, many of whom eventually worked out some new relationship to the absolute. The crisis set their belief at liberty to create new convictions, ones more favorable to themselves and less so to clergymen and even to "god," who had hitherto enjoyed a monopoly on metaphysical power. This act of religious rejection, however, still involved tremendous psychological risks for such constitutional idealists, especially in the absence of any new terms or conditions on which to found a compelling alternative to what was surrendered. Crisis rhetoric not only uncoupled a cherished metaphysics from the protection of Christian theology, it did so while calling into being a tyrannical empirical authority. Any speculative idealism launched within this epistemological framework would face the double jeopardy of theological ineptitude, on the one hand, and scientific negation, on the other. The aggressive facticity of scientism may have been required to kill off an older, no longer desirable Anglican metaphysic, but would it come at the cost of any future metaphysic as well? The liability these young intellectuals incurred by linking "proof" to any assertion of "Truth" meant they could not well extricate themselves from the demands of the former without losing face in regard to the latter: all future metaphysical inquiry would likewise have to be conducted along purely evidential lines. This empirical sensibility continued to dog their later days, helping to explain what might appear to be the almost perverse positivism of psychical researchers, continuously undercutting their own longed-for conclusions through the stringency of their investigative measures. Yet, this praxis was morally consistent with their own youthful demands for an evidence-based faith, one free of the mysteries used by their parents' generation (and generations before that) to extort deference. This was a tailor-made solution to a crisis of their own construction. As a disciplinary prescription, then, psychical research can be seen as offering "the hair of the dog that bit you," harnessing the potentially destructive power of empiricism to reenfranchise an orphaned metaphysics under its own protection.

To question the seriousness and probity with which the SPR conducted these researches is to misunderstand its core mission: unbelief could only be resolved using the original criteria that had induced it—the arguments of proof. If members of the SPR were not particular as to this point, they could have joined up with the London Spiritualist Alliance, the Theosophical Society, the Aristotelian Society, or some other modern philosophical endeavor operating outside of academic science. For them, the cleaving of natural theology into separate realms of inquiry put the whole project of knowledge at risk: faith divided from reason left two insufficient halves that did not make a whole. Against this prospect, psychical researchers sought to reconstitute a unitary field of knowledge for the modern age, one in which older authorities would find themselves retired and where a new breed of secular intellectual, such as they themselves, would preside. Despite the seeming extravagance of the psychical conjecture, it was passively permitted by the enigma of consciousness itself, whose structural and operational details had begun to incorporate a "subconscious" region by the mid-1870s.[67] This submerged yet sentient aspect of the mind remained entirely unknown and unfathomed, and in urgent need of a massive research initiative. In these depths of consciousness, there was no real authority to fully disallow psychical powers and endless possibility to imagine them, providing a perfect speculative refuge for metaphysics through the end of the century. Telepathy, retrocognition, communication with the dead, clairvoyance, and second sight all suggested a psychic subjectivity that could not only slip in and out of matter, but also could move backward and forward through time. Psychical researchers thus found a way to sustain "the soul" as part of the ascetic imagination of the secular intellectual, long after spiritualism was forced to the margins and all ghostly apparitions had faded from view.

CHAPTER TWO

William Crookes in Wonderland

Scientific Spiritualism and the Physics of the Impossible

Sir William Crookes is centrally featured in the institutional history of the SPR as the pioneering investigator of a "psychic force" and not as the strange circus master of the Katie King séance. It was a part of his history concerning which he himself preferred to remain silent. Yet, the pictures that come down to us now make it hard to expunge these events from historical memory: photographs taken by a career scientist in order to assert the factual and *physical* existence of ghosts. In one photo, Crookes gazes unabashedly at the camera as he strolls arm in arm with the lissome ghost of a pirate lass. In another (fig. 1), he stands beside her with his hand on his hip, jacket pulled open as if to expose his puffed-up chest, posing like a hunter with his prize (and not like someone with his hand in the cookie jar). When we encounter these images now, the propositions these photos contain appear so incongruous that there is an implied disparagement of its audience: "Who would believe it?" And from there, judgment invariably lands on the photographer himself: "Who in good conscience would make such a claim?" How can we understand an answer that involves Sir William Crookes, the renowned chemist and physicist, discoverer of thallium (1861), Fellow of the Royal Society (1863), winner of the Davy Medal (1888), Knight Bachelor (1897), president of the BAAS (1898), recipient of the Order of Merit (1910), and finally, president of the Royal Society (1913). Yet wedged in among these many scientific honors was this singularly problematic investigation of the medium Florence Cook, begun in mid-December 1873 and concluded in late May 1874, from which these photos survive.

To understand these photos is to finally make sense of Crookes's own scientific career as he journeyed from decorated chemist in the 1860s to

FIGURE 1 William Crookes and Katie King pose arm in arm at a private séance in Crookes's London home in May 1874. This is one of the few surviving images of the original forty-four plates.

celebrated psychic force investigator, disgraced séance impresario, and giant of experimental physics, all in the course of the 1870s. These are not post-cards from the periphery. Such images complete the picture of the complex social and intellectual terrains of science itself, revealing a lost piece of the battleground where various institutional, class, economic, and ideological interests competed for dominance. More than any other scientist of his day, Crookes in his career traversed the spectrum of these localities: he was at once a poster boy of vocational science, an elite ringmaster of scientific spiritualism, a pioneer of the commercial scientific press, the people's scientist, a national celebrity, a professional disgrace, a rebel, a climber, an outsider, an insider, and eventually a colossus who raised himself up to rule the institutions that once held him down. In addition to his prodigious scientific activity (he was variously engaged in photography, journalism, technical chemistry, instrumentation, experimentation, and South American expeditions), Crookes was also a conspicuous scientific performer who used spiritualism to talk his way into the public drama stirred up by the theory of evolution about the place of God in nature and the role of religion in science.

Certainly, the spectacular blowback of the Katie King affair spells out the limits of a unifying project like "scientific spiritualism" in the midst of such intense boundary disputes to ensure their separation. So what made Crookes believe that he could pull this off? This question of risk presents itself again and again when we confront the puzzling facts of these photos. How can we parse the difference between a professional gamble and an act of career suicide from such a far remove? Other scientists besides Crookes had investigated spiritual phenomena and come out in their favor, but few went so far as to supply its artifacts. That was usually left to the mediums, not the scientists studying them. Yet, to the mystification of historians, Crookes produced these photos himself and, in so doing, lost his innocence as a mere bystander or even a victim of fraud, to become a perpetrator of fraud himself. Crookes, as we shall see, well knew the professional risks of taking such photos and, judging by his own spectacular career, well knew how to judge and attain those professional rewards. So why the fatal miscalculation?

While the intellectual and moral integrity of such "proofs" forms a critical part of my historical interest in Andrew Lang, Sir Oliver Lodge, and Frederic Myers, because it speaks directly to subjective issues of truth in epistemology, with Crookes, the fascination (of this particular episode) lies in the deceit. What were the exigencies driving this deception, the strategic understanding with which it was deployed, and the context in which it was expected to succeed (but, in fact, failed)? There is also something improvised and unintended about this whole affair that gives it a certain off-road quality: there is no map for the messy terrain where Crookes's "scientific interest" in Katie King landed him, nearly driving his career off a cliff. This is a glimpse of an episode in an eminent scientist's life that would normally be scrubbed clean in memoirs and admiring biographies. It brings hidden deeds and unconfessed objectives into view. These private spaces, where a ghost and a chemist might meet for a tryst, or two collegial spiritualists might strike a deal to board a prestigious government expedition, open onto an interior view of a culture's larger motivational structure. It tells us what people wanted and how they planned to get it. It identifies the historical moment when things suddenly shift, and someone who thought he was getting ahead finds out he may have gone too far. It is in this back lot in the Potemkin village of official history, where daily life and its unscripted disasters all goes down.

The pursuit of Katie King takes us deeper still into Crookes's sense of professional and personal crisis as his career path collided with personal grief, institutional prejudice, and the larger religious tumult of his day. Crookes's story unfolds very much against this backdrop of belief, as professional, popular, and theological opinion makers worked out their claims to defining the natural order. Conflict between scientific and scriptural narratives were nothing new, part of the intellectual life set in motion by the Enlightenment, but the intensity of this debate in the mid-nineteenth century was something new.[1] It was led by more determined efforts for intellectual independence from above, and driven by more broadly based secular energies from below. As never before, the check on church authority encompassed the economic, industrial, academic, and social capital necessary to transform not just certain intellectual enclaves but the entire culture itself. At the other side of this half-century (1830–80), the modern institutional structures of knowledge would begin to settle into place, scaffolding official science into a network of institutional, disciplinary, economic, and governmental supports, and clarifying its fraught relationship to theology. This new, highly buttressed form of scientific institutionality would coexist in "amicable juxtaposition" with religion as an entirely separate realm of concern.[2] But this final disposition of "non-overlapping magisteria," or NOMA, was still being worked out at the time of Crookes's scientific involvement with spiritualism, and it would take several decades yet to fully sever this attenuated corridor between numina and phenomena kept propped open by Victorian ambivalence.[3]

Most people were quite eager to find a way to keep the blessings of both God and science active in mundane human affairs, which meant tolerating some degree of theological imposition on models of nature. But the militant naturalists, such as "Darwin's bulldog" Thomas Henry Huxley and his evolutionary confrère Herbert Spencer, resolutely held the line against any attempted claw-back by theology. The flair for combat in these quarters gave naturalists a seeming ascendancy in this fight, disproportionate to their actual numbers, substantiating the illusion of an atheistic threat and unsettling a religiously inclined population. Overall, these were potentially favorable conditions for credentialed intermediaries like Crookes who were willing to promote some new theological initiative within science.[4] The urgency of such an intervention increased as evolutionary narratives pushed their way deeper and deeper into the social and human sciences in the 1870s, bringing the biological imperatives of competition and survival to bear on every aspect of human cultural activity.[5]

It was not clear that institutional science could be fully trusted to meliorate this materialistic tendency. Theistic physicists like Gabriel Stokes, James Clerk Maxwell, and William Thomson pushed back against the "machine-ism" of naturalistic science, but their arguments tended to be rather esoteric. Championing free will within a stochastic analysis of entropy was unlikely to produce "the fighting words" needed to prevail in this back-and-forth bluster of the commercial press. One could easily have speculated at the time that the tangible satisfaction of a corporeal ghost might have more popular success. Enter William Crookes. From 1870 to 1874, he offered himself to the nation as its humble yet highly trained scientific investigator, answering the public cry for a rigorously scientific review of spiritualism. The idea of "scientific spiritualism" could mean various things, contingent on class, education, and one's culturally located definition of science. For members of the intellectual elite such as Crookes, the emphasis was on scientific methodology, allegedly forbidding any theoretical bias for or against spiritualism that might cloud objectivity. In reality, this intellectual wing ran the full spectrum of attitudes toward spiritualism, from "all in," like the celebrated British biologist and explorer, Alfred Russel Wallace, to aghast, like the physiologist and natural determinist, William Benjamin Carpenter. Believers such as Wallace, co-discoverer of evolution, hoped for some formal reckoning with theoretical science; Carpenter hoped for a lasting and decisive expulsion of spiritualism from all consideration. There were populist varieties of "scientific spiritualism" as well, which were not so deferential to the theories and authority of professional science. Leading spiritualists, such as Emma Hardinge Britten, referred to a "science of spiritualism" to designate spiritualism as its own disciplinary branch of investigation (as opposed to its object of study). To the extent that spiritualism was a continuation of the investigation of nature, its practitioners saw themselves as part of science broadly defined and found common language with those, like Crookes, working under more elite constraints.

Spiritualism's greatest appeal for the nonpractitioner was, of course, the spirits themselves. These might be encounters at a séance table with the earnestly sought ghosts of departed loved ones, or the titillation derived from reading about such spirits in the popular press. Visionary ideologies were more for the faithful. The mainstream press tended to limit its interest in the subject to noteworthy persons, peculiar practices, and extraordinary claims. Crookes's scientific celebrity and promise of proof was an irresistible combination likely to arouse a broad public interest. At the time Crookes embarked on his program, the public's fascination with spiritualism had grown rather than diminished since spiritualism first arrived on British soil around 1852, advancing alongside an increasingly spectacular parade of phenomena. The knocks, pops, taps, and turns of the kitchen table that constituted early interactions with the spirit world evolved to include far more professional showmanship on the part of both spirits and mediums. Mysterious aromas, floating candlesticks, and ghostly materializations soon began to embellish these proceedings, and verbal communications from the spirit world became much more elaborate as well. By the mid-1860s, writing appeared mysteriously on slates, words tumbled out of gliding *planchettes*, and ghostly voices began making speeches and singing songs. Finally, in the early 1870s, the ghosts showed up tableside and started speaking for themselves, making things very interesting indeed at the exact juncture of Crookes's national debut.[6] By skillfully limiting his championship of spiritualism to a defense of free scientific inquiry, Crookes forced elite scientists to take issue with their own declared values and offered himself (spiritualism and all) as an exemplar of the scientific sprit.

The period of Crookes's researches approached the height of *both* scientific naturalism and spiritualistic investigation, a combustible clash of energies that Crookes hoped to harness in order to propel his scientific career over existing obstacles to reach new, illustrious heights. The scientific order fostered by the BAAS since 1831 might have ended the reign of aristocratic amateurs, but it had imposed new, more rigid forms of exclusion that kept women and laypersons out, and vocational practitioners like Crookes down.[7] Popular resistance to this academic elitism gave Crookes a potentially broad base of support to mount a challenge to both the doctrines and armature of this new professional hierarchy. "Scientific spiritualism" may have incensed

Crookes's professional colleagues, but as a career move, it was not necessarily ill advised, or so it reasonably appeared to Crookes at the time. He continued to occupy this increasingly embattled zone for four years, as the formal opposition to him grew more intense.[8]

Despite this public persecution, Crookes continued to believe that he would prevail in this professional showdown, that he could use his reputational assets to boldly challenge the orthodoxy from which that reputation was on loan. But Crookes was wrong. No sooner had he summoned the corporeal ghost of Katie King in 1874 than he was forced to banish her. He could not resolve what Elana Gomel calls "the inherently paradoxical matrix of the materiality of the spirit."[9] The dissonance that we encounter in the spirit photo is not just a function of our contemporary context; it is rooted likewise in the great consolidating theoretical orthodoxy of Crookes's own time, in which he himself would come to play a significant part. This multistoried edifice of towering explanation had added over the course of the nineteenth century atomic theory, the conservation of force, evolution, and now, rivaling the mathematical achievement of Newton, Maxwell's treatises unifying electricity, magnetism, and light (1873). This last crowning glory had given light a purely mathematical description and confined its once celestial energy to the closed mechanical system of thermodynamics (colonizing a once active region of spiritual speculation concerning "actinic ghosts").[10] There was no forcing the gross matter of Katie King's spirit into this impenetrable intellectual orthodoxy, no matter how hard Crookes tried and no matter how great the public hope that he might do so. (The emphasis here is on *hope* and *might* since, as Crookes would find, spiritualism counted its widest support as a form of longing, not as a convinced belief in such confrontative ghosts.) If the evidence had been more ambiguous or the claimant less eminent, then the dysphoria occasioned by "scientific spiritualism" might have continued as a more passive form of professional discontent. But Crookes had literally entangled himself with this ghost when she placed her hand on his arm. The liabilities of such an assertion were too great to bear, for either Crookes or the Royal Society.

It became fully clear to Crookes, in the wake of the Katie King séance in 1874, that any such ghostly or psychic force wishing to slip past the gatekeepers of science needed better sublimation and a more receptive discipline than theoretical physics. Sadly for Crookes, such a discipline did not yet exist. The boundaries of theoretical physics were too hard, and academic psychology was still too soft. The latter had been struggling to distance itself from philosophy since the early nineteenth century, giving the lead to psychophysics and evolutionary biology in the 1860s in an effort to become an official subdiscipline of physiology.[11] The last thing psychology was likely to tolerate, having rid itself of mesmerism, was bats getting back into its belfry. Even the soft data of mental experience had been purged in favor of neural anatomy, a disciplinary repudiation of "thinking about thinking" that had resulted in a "psychology without consciousness."[12] So when Crookes attempted to exit the spiritualistic framework by substituting the label of "psychic force," there was simply no place for him to go. Physics would not, and psychology could not, have him as part of their intellectual continuum. Unable to pursue his research in orthodox channels, he was forced to proceed through the fraught and conflicted construction of "scientific spiritualism," which eventually led him straight to Cook's door (or rather her to his).

This academic tide would slowly begin to turn by the end of the decade, as the center of psychology began to shift toward the Salpêtrière asylum in Paris and the psychosomatic illnesses of hysteria. With the problematic medium swapped out for the prostrate psychiatric patient under clinical supervision, some limited curiosity in similar trance behaviors found some purchase.[13] The difference between Crookes's study of a psychic force in 1871 and the same topic undertaken by the Society for Psychical Research in 1882 was one of institutional location within the emerging international complex of aberrational psychology. Physics, by contrast, in the 1870s was the most venerable and rule-bound of the all the scientific disciplines.[14] Given its intricate, highly built-up theoretical orthodoxy, even religionists like Tait and Thomson were not going to tolerate a game of disciplinary Jenga with its conceptual framework. They might slip in an infinitesimally small demon (carefully tweaked by Maxwell to explain statistical randomization in entropy), but what they could not abide was the gross ambition of a spirit body, or even the subtler nuisance of a psychic force, wreaking havoc on an edifice painstakingly evolved over the last three centuries. By the 1880s, however, physicists like William Barrett, Oliver Lodge, Balfour Stewart, and even William Crookes found there was room to operate at this physical-psychical border, so long as they made their approach through the basement of aberrational psychology and not through the front door of theoretical physics.

But even in 1871, Crookes would have known, better than anyone else, that such images were a potential flashpoint for public controversy. He was a career scientist, a wary spiritual researcher, and a public expert in matters of photography. He got the picture, but do we? We come upon these photos many decades in the future as images without explanation, which misses out entirely on what Crookes had originally intended them to be:

explanations without images. These icons of Victorian spirit photography have their greatest value as something that was meant to remain unseen and was made to disappear; they are most compelling as a visual record of the politics of invisibility. The act of seeing these photos has itself become a source of misunderstanding. One might reflexively assume that Crookes and his spiritual following were gullible fools, or perhaps that Crookes himself was clumsily deceiving a pack of dull wits. Of course, neither is the case, and the truth, so far as we can make it out, is much more interesting.

PHANTOM PHOTOS

When we as modern viewers confront spirit photos today, we are likely missing what Clifford Geertz called the "thick description" that culturally contextualized these artifacts and gave them plausibility for contemporaries. Without it, we understand only what we ourselves can relate to: the gullibility of grief. A photo then becomes an emotional rather than an intellectual artifact: a timeless *memento mori* documenting the scene of loss and death in the nineteenth century. A grieving mother sits alone in a photographer's studio, the ghostly image of a child framed in the open door behind her. She'll only see him after the plate has been developed, but for now, her face wears a look of unanswerable longing. In another picture, a widower sits with an air of hopeful expectancy surrounded by his children. He'd been told this photographer was a reliable medium, and for a certain fee, he might just see his dead wife complete this family photo as a cloud of light, a ghostly face, or even a hovering human form. Such longing makes sense to us, and provides us with a more convincing basis for Victorian belief than the idea that such belief might additionally, or alternatively, be predicated on intellectual criteria. Both could be true and mutually reinforcing.

Recent scholarship has shown just how complex and multivalent such photographs could be. The iconography of such images was often situational and unpredictable. They were produced and read across a variety of commercial, recreational, cultural, and intellectual terrains, each with its own interpretive scheme for reckoning the merits of a ghost photo.[15] A photo might be variously criminalized, celebrated, remunerated, or ridiculed, depending on which of these rival frames of meaning it traversed, making such an object both a potential liability or an asset. As historian Sarah Willburn points out, Victorians did not just view these images with credulity. Spirit photos were a known part of aesthetic craft photography and also were frequently attributed to deception and craft. This was all part

of the background knowledge against which Crookes's photos had to ne-
gotiate their evidential value.[16] The complex hermeneutics make the robust
belief in spirit photography more interesting because they replace a story
of Victorian gullibility with one of Victorian intellectual agency. Belief in
these photos was not necessarily granted lightly, for it involved some active
formulation as to why they were credible.[17]

This willing suspension of disbelief becomes more interesting still in regard to images of truly arresting flagrancy. In one familiar photo (fig. 2),
Katie King is submitted to a medical examination by Crookes's spiritualist
colleague, Dr. Gully, who clasps his thumb and fingers around her wrist in
order to check her pulse against his timepiece. The scene depicted might
well appear incredible, but final judgment depended not just on the content
of the photo but also on an analysis of its context. Who took the photo, and
what were the circumstances? Were there alternative explanations for these
ghostly effects? What about the moral probity of the medium herself? Such
evaluations would depend upon the expertise and inclinations brought to
bear by the viewers themselves. Most critical would be the extent to which
the viewer held these depicted events to be at least theoretically possible—or
theoretically *impossible*. For spiritualists, categorical denials delimiting the
possible were less authentically scientific than a willingness to allow the
nature of spiritual reality to remain open. This meant *anything* was possible
until proven otherwise, even a scenario where the dead could have vital
statistics for Dr. Gully to monitor. The validity of a spirit photo did not need
to be likely or even probable, but, in the absence of some definite exclusion,
it remained possible. Spiritualists felt they had the duty and the right to
consider this evidence free from the preemptive interdiction of a scientific
theory.

No matter how improbable, such spirit photos have to be understood
in the context of what made them plausible, both for those who produced
them and for those whom they were meant to convince Even the spiritualists who falsified such evidence are likely to have done so in the belief that
the facts they asserted were true, based on the extensive props to spiritualism's authenticity. Fields of force, waves of light, currents of electricity, and
a universal metasubstantive plenum called "the ether" might all be diverted
to the uses of spiritualism with little regard to incompatible elements of
their orthodoxy. More supportive scientific interlocutors, such as Cromwell
Varley, Robert Hare, Dr. A. Butler, Frederick Zöllner, and Alfred Russel Wallace, offered direct encouragement for spiritualism's scientific analogies by
bending their own work in that direction, going so far as to suggest that

FIGURE 2 Dr. James M. Gully checks the pulse of Katie King with his pocket watch. While Crookes describes a similar scene during his famous séance with Cookes, subsequent analysis by Eric Dingwall places the scene at the home of William Henry Harrison, editor of *The Spiritualist*, a year earlier.

spiritualism and science were interlinked and could only advance together on equal footing. In addition to intellectual cachet, there was plenty of social cachet involved. The curiosity indulged by British aristocrats and intellectuals was amplified by the imperial curiosity of Emperor Napoleon III, Kaiser Wilhelm I, and Tsar Alexander II, offering an implicit endorsement to those further down the social scale.

Taken together, these elite sources sympathetic to spiritualism played the central role in manufacturing and sustaining such beliefs, and none more so than William Crookes. The mediums and commercial photographers of this world were, by comparison, merely low-level operatives at the retail end.[18] Crookes did more than just step up to investigate these fraught phenomena of corporeal ghosts; he claimed to have witnessed (and touched and spoken to) one such fully formed spirit himself, the ghost of one Katie King, long-dead daughter of a seventeenth-century Welsh pirate.[19] This stunning assertion was documented in a series of forty-four prints made by Crookes himself in a theatrical display of virtuoso science, using five different synchronized cameras. His photographs documented "one of the most amazing stories in the history of spiritualism" wherein "no such startling demonstration . . . had ever been seen before."[20] And perhaps there was a reason such ghosts had made themselves scarce. Even in the way-out, permissive genre of spirit photography, the appearance of a member of the Royal Society squeezing the arm of a hot young ghost was finally beyond the pale. Instead of pushing the envelope with this risky move, Crookes had accidentally jumped the shark.

Apart from the risks of taking such photos, Crookes failed to consider the potential danger posed by the object of inquiry herself, Florence Cook. This entrepreneurial teenage medium had been running séance tables since she was fifteen and was unlikely to be a passive research subject. Whatever happened during that séance and whatever the nature of Crookes's original intentions and subsequent regrets, he used these photos to publicly back Cook and her phenomena, building an elaborate scheme of evidence on her behalf from which it would prove difficult to extricate himself. This brings us to what is perhaps the most puzzling aspect of this famous photo series (second only to the decision to take the photos in the first place): for all the publicity and detailed knowledge pertaining to the photos (the technicalities of how they were made, the images they contained, the circumstances of their capture), almost no one ever actually saw them. Despite their critical function as visual "proof," Crookes steadfastly refused to offer them up for public display. The pictures managed, nonetheless, to play a compelling role

as testimony *that could be visualized* in Crookes's write-up for the *Spiritualist* reporting the events of the séance. The density of verbal detail provided
a narrative reconstruction of the missing images, while the complex notes
explaining the technical tour de force of their production further implied
their true existence and elevated the quality of the evidential endorsement
they (would have) provided.

> Five complete sets of photographic apparatus were accordingly fitted up for
> the purpose, consisting of five cameras, one of the whole-plate size, one half-
> plate, one quarter-plate, and two binocular stereoscopic cameras, which were
> all brought to bear upon Katie at the same time on each occasion on which
> she stood for her portrait. Five sensitising and five fixing baths were used, and
> plenty of plates were cleaned ready for use in advance, so that there might be
> no hitch or delay during the photographic operations, which were performed
> by myself, aided by one assistant.[21]

Whatever motivated these elaborate efforts to obtain quality images, they
certainly made for good copy, weaving the "thousand words" Crookes told
about his pictures in order to keep them from speaking for themselves. To
ensure this silence, it was not enough that the pictures be kept out of sight.
The objects themselves had to be obliterated, at least in the public imagination. Crookes stemmed the longing to touch and feel these prints by claiming that all forty-four plates were destroyed in a chemical fire shortly after
the séance. (There is no evidence that such an accident actually occurred,
but the story alone was enough to serve Crookes's purpose.) With the photos
safely out of view, Crookes verbally reverse-engineered the images themselves in lengthy descriptions of what they would have looked like had anyone been allowed to see them. He did release a few select prints to close
friends before the "accident," but they were never circulated, thus lending
support, but not visibility, to the claim of the existence of the photos.[22] This
was not sufficient, however, to keep the ghost of Katie King locked up in
the family skeleton closet. Four photos of the posturing ghost resurfaced in
1934, including the one with Dr. Gully measuring her pulse. They had lain
hidden among the papers of Frederick William Hayes, an active member of
Crookes's investigative circle who was present at the Cook séance, and found
only after his death, when his estate was being settled. The pictures of Katie
were immediately (and possibly erroneously) attributed to Crookes's séance
with Florence Cook and ran that year in *Psychic Science* with Crookes's name
emblazoned at the bottom of each print, uniting him firmly to claims of her
existence.[23] These and other images including Crookes eventually migrated

to the internet, where Crookes remains forever yoked to the life-size flesh-and-blood ghost of Katie King. Given her corporeal form, she is less spirit than reanimated corpse, Frankenstein's creature pursuing him as they circle the globe in a digital cloud.

While Crookes could not have anticipated the future threats posed by computer uploads to the internet, as an experienced writer and editor for four different photographic journals variously through the 1850s and 1860s, including the largest, the *Photographic News*, he was uniquely positioned to understand the contemporary liabilities of such a photo, as both a technical and discursive object. From his editorial post, Crookes presided over the debate on photography's evolving professional standards as it aspired to become "the handmaid of science," a discussion that chiefly played out in the trade press. So if anyone would be familiar with what did and did not constitute a respectable scientific photo, it would have been Crookes. Obstacles to such respectability were, predictably, pornography and, somewhat ironically in Crookes's case, spirit photography. The latter in some respects was potentially the worst. In addition to being construed as seedy and commercial, spirit photos turned the trustworthy and objective "handmaid of science" into an agent of deceit. This did not mean that trick photography, in general, was frowned upon by respectable photographers, but there would have been an ethic of transparency governing such images. Aficionados shared their expertise in the interest of the greater scientific good, creating an ever expanding pool of technical knowhow. Crookes must have known that this meant there would be a community of photographic experts on hand with enough detailed knowledge to debunk most spirit photos. Furthermore, such photographers would be inclined to challenge anything of a spurious nature that transgressed the community standards of what Jennifer Tucker calls photography's "fraternal order."[24]

The key was to keep such photos from falling into unsympathetic expert hands, while still using them to leverage the validity of Katie King's phenomena. In spiritualism's formulation of truth, unless a photo could be unequivocally proven to be fraudulent, it necessarily defaulted to the possibility of being true.[25] Even the widespread knowledge of image manipulation could work in the spiritualists' favor. If no satisfactory technical explanation could be supplied, it reinforced the likelihood of that spiritual possibility. Effectively, the spiritualist's burden of proof became the skeptic's burden of disproof, a burden imposed anew with each and every photo since no categorical denials were permitted. By refusing to generalize underlying causes, spiritualists could execute an intellectual end run around any obstacle posed by "assumed knowledge," whether asserted by a university-trained physicist or a highly experienced photographer. Instead, they preferred the authority of their own critical common sense, as upheld by the *popular* positivism of the Victorian culture of proof. Now, when we think of Crookes's missing photos in the light of his highly textured understanding of these competing economies of truth—one based on evidence, the other on expertise—the whole episode becomes less clumsy. By removing the physical artifacts from the equation, Crookes vastly increased the burden on skeptics to prove beyond a reasonable doubt that the pictures were fraudulent, leveraging a state of permanent possibility. While such possibility would be given different weight depending on the degree of a person's spiritual conviction, nonetheless, Crookes would benefit from tilling this doubt wherever it lay.

In a private letter to Captain T. D. William in August 1874, Crookes wrote: "It was at the express wish of Katie King and is also that of Miss Cook and myself that the photographs I took should be kept strictly private and only given to intimate friends."[26] This excuse of spiritual etiquette was not much to offer in the face of what must have been a considerable disappointment. It probably would not have sufficed for the majority of Crookes's readers had they not already received something of greater or equal value: Crookes himself. Given the way such pictures were evaluated by spiritualists, the identity of the photographer in this case mattered as much as, if not more than, the photo itself. Crookes's scientific celebrity and technical expertise were by far the greatest assets that attached to the pictures of Katie King; the photos themselves were, by comparison, just a liability.

While the photos best served Crookes's narrative in absentia, the camera itself remained a highly visible actor in Crookes's investigation. It not only symbolized Crookes's experimental prowess, but also emphasized that these were *physical* researches that involved the world of extended forms. A ghost that could be photographed could be neither apparition nor hallucination, but, rather, was a material reality independent of any state of mind. The camera became thus a mechanical observer of Katie King's real existence, and the photographs were its sworn testimony. Instead of leveraging everything on his personal declaration, Crookes became the blameless stenographer of an unimpeachable witness.[27] Yet, without any photos at all, Crookes's research would have failed to distinguish itself from the countless other séances with materializing ghosts and scientific witnesses. He had ridden into this crowded field (recently trampled by the Dialectical Society) with his armor shining and his pennant flying, and now, he needed something "to show" for it. He could hardly "top" the success of "psychic force" with anything less than spectacular proof. Thus, Crookes did everything in his power to enforce the photos' underlying reality, short of actually exhibiting them in public. This was skillfully done. He described them in vivid detail, gave a highly theatrical account of photographic procedures, distributed select copies to friends who publicly acknowledged their receipt, and commissioned a painting of a plate to run in the *Spiritualist* along with his published research. Thus, these infamous photos, so problematic on display, formed a well-played part of a risky enterprise when rendered invisible.

Nonetheless, it all did go belly up. Why? How? Oliver Lodge alludes to some kind of unanticipated reversal. Though Crookes "entered upon the subject with a light heart, he left it with a heavy one."[28] The abrupt termination of the séance, along with the swift and "accidental" consignment of all plates to the flames, seems to bear out the notion that we are dealing with an unfortunate contingency plan rather than a best-laid one. His lifetime demurral regarding these events showed him to be highly conflicted about his role in them, as well as fearful about their potential consequences. But even if we are to understand Crookes in the context of exigency, it is advisable to first consider what the plan was before it went awry. What had he hoped to accomplish with a scientific séance, and how had he understood the forces at play?

AT THE CROSSROADS OF
REASON AND WONDER

In mapping the logic of Crookes's phantom photos, spiritualism becomes more than a digression on the way from tradesman's son to scientific celebrity; it becomes his chosen path to professional advancement. Along this route, Crookes would find scientific notables like mathematician Augustus de Morgan, evolutionary thinker Alfred Russel Wallace, astrophysicist Friedrich Zöllner, astronomer Samuel Huggins, and transatlantic cable engineer Cromwell Varley, all of whom had managed to secure their career success alongside an active spiritual curiosity. Could he do the same? The research of these scientists was followed with more than the usual interest because of the intriguing spiritual possibilities folded within their curiosity about nature. They asked the questions to which people wanted answers, making their science part of a national conversation. There had already been a great deal of scientific investigation of spiritualism by professional scientists, even skeptical ones, before Crookes threw his hat into the ring. At various points in the 1850s and 1860s, Michael Faraday, W. B. Carpenter, John Tyndall, and

T. H. Huxley had all attended séances as part of an effort to provide the public with a sensible alternative to spiritual explanations. The combined results of these two decades of research, however, amounted to little more than "it was all in their heads." Faraday offered Carpenter's concept of unconscious muscular cerebration," suggesting that the nervous system could execute automatic routines (like turning a table) below the level of rational awareness. Adding to Faraday's physiological emphasis were the "subjective modifications" like hallucination or delusion suggested by psychologists.[29] More popular still was the skeptical consensus that the phenomena were produced by frauds and witnessed by dupes.[30] While this investigative record may have satisfied those already disinclined to believe in spiritualism, many others found it cursory and even arrogant. This null hypothesis failed with others besides those who would not take "no" for an answer. The perceived skeptical bias left many Victorians feeling that science had not fairly adjudicated the great spiritual mystery and that this dereliction was born of hubris. In the words of Hudson Tuttle, these men of science had "signally failed and the magnitude of their failure [had] been in direct proportion to their greatness."[31] Given the compelling testimony of thousands of witnesses over many years and around the world, spiritualism had a hold on the public imagination that demanded its due.

It was in this milieu that Crookes thought to prosper by offering himself as a new, more able, and less biased scientific candidate for the investigation of spiritualism. But a truly scientific spiritualism (one that did not resolve its dilemma through negation) was tricky ground, perhaps far more so than Crookes realized. Such a construct may not have been likely or even possible. Most spiritualists did not wish to cede control of their investigative domain to physicists, any more than physicists welcomed spiritualists into their fold. They wanted intellectual ownership to remain with themselves, seeking alliances with sympathetic scientists, not subordination.[32] As the spokeswoman of spiritualism, Emma Hardinge Britten patiently explained: "the science of Spiritualism can alone explain the Spiritual works."[33] So while the idea of a fully scientific custody for spiritualism had tremendous allure both for Crookes and for a public that continuously remarked, "If only scientific men *would* but examine these things . . . ," such a notion masked a great degree of internal discord and confusion over how that science was to be defined.[34] Crookes meant to annex spiritualism to his scientific practice as a professionally trained chemist, ultimately aiming for some kind of reconciliation within established theory. Britten's science of spiritualism, in contrast, saw spiritualism as a discipline in and of itself, assuming the existence of spirits in the given object of its study. Its practice required exiting professional science's limited theoretical horizons altogether. It was one thing to appropriate the mysterious forces and ethereal plena of physics, and quite another to be held intellectual captive by the conservation of force. William Howitt (a popular historian and genteel leading light of the movement), saw the mechanical causality of science as so intrinsically hostile to the spiritual project that "scientific men are not the men to decide such questions. They have their prejudices and their theories which disqualify them."[35] The phrenologist Charles Grey went so far as to declare "the bigotry of orthodox science equal to the bigotry of orthodox religion," noting, with reference to psychology's material bias against the mental powers of mediumship, "how completely current superstitions [scientific theories] stand in the way of all progress."[36]

While likening academic science to a form of religious bigotry is polemical, this language in some ways correctly mirrored the position of ideological naturalists on the other side of this debate. Faraday's zeal in defending the educated authority of experts reflected a highly deterministic understanding of nature's causal laws and an incontrovertible view of the theoretical orthodoxy describing them. In his lecture on "Mental Training" given before the Royal Institution in 1854, Faraday made Grey's case for him: "Before we proceed to consider any question involving physical principles, we should set out with clear ideas of the naturally possible and impossible."[37] This language was intentionally provocative, and indeed his speech remained an oft-quoted battle cry for spiritualists for decades to come. Faraday offered science not as a reliable arbiter of *likely* causes, but rather as the determining power of what could and could not have happened—period, full stop. It would seem that everything in heaven and earth was dreamt of in his philosophy. T. H. Huxley did not stop at declaring spiritualism *physically* impossible, like Faraday, but went deeper by declaring it *logically* impossible as well: "Psychic force or physical spirit" made no more sense than "a round square, a present past, or two parallel lines that intersect."[38] Such predicates, according to Huxley, had already been preempted by their subjects, and there was no need to investigate further: "Supposing these phenomena to be genuine," he wrote in response to an invitation from the Dialectical Society to join its team of spiritual investigators, "they do not interest me."[39]

What we see in the attitude of Faraday and Huxley is a dispute not just about physical principles; it was also about who would get to comment on those principles. In this conversation about the nature of nature, professional scientists wanted to exclude spiritualists, and spiritualists wanted to include themselves. For Crookes to put himself forward as the public champion of scientific spiritualism, he had to survive the crossfire coming from the activist wings of both sides. One thing these militants could agree on was that "the facts of science are opposed to Spiritualism."[40] These facts were not tangential or parallel or perpendicular to spiritualism, they were diametrically opposed to it. After several millennia of trying, natural philosophy had finally shown metaphysics the door, and now, back came the same idea in a new guise, claiming: "spiritual phenomena are as positive and amenable to law as physical [phenomena] and quite as far removed from the supernatural."[41] Once again, spiritual energies were forcing their way back into the continuum of nature, and naturalists were bent on keeping them out. Crookes had to hold these two opposing forces in tension while playing to the middle: being neither too skeptical, like those orthodox emissaries who had "signally failed," nor too soft, like the scientist whose research had crossed the line to spiritual advocacy. He must claim the right to speak for science with his posture of chaste professionalism, while holding open the possibility that he would abandon its every creed.

Crookes skillfully maneuvered this crisscross of social and intellectual allegiances when he announced his intention to investigate spiritual phenomena, in an article titled "Spiritualism Viewed by the Light of Modern Science," published in the *Quarterly Journal* in July 1870. He played the reluctant hero drafted to the cause, complaining that the *Athenaeum* had outed his private spiritual researchers earlier that month, and it was "a pity any public announcement of a man's investigation should be made until he has shown himself willing to speak out." Yet, he appears to have protested too much. His response to the *Athenaeum* was his manifesto for a bold new scientific approach to spiritual inquiry. But all the while, Crookes kept himself tightly wrapped in the protective cloak of his Royal Society Fellowship, speaking in numbers and distancing himself and his methods from spiritual contamination:

> The pseudo-scientific Spiritualist tells of bodies weighing 50 or 100 lbs. being lifted up into the air without the intervention of any known force; but the scientific chemist is accustomed to use a balance which will render sensible a weight so small that it would take ten thousand of them to weigh one grain. . . . The Spiritualist tells of objects being carried through closed windows, and even solid brick-walls . . . the chemist asks for the 1,000th of a grain of arsenic to be carried through the sides of a glass tube in which pure water is hermetically sealed.[42]

He offered few specifics beyond this, only his sincere intent to carry on where others (the dogmatic skeptics and pseudoscientific spiritualists) had failed. And while he coyly "by no means promised to enter fully into this subject" (his way of signaling he was not going native), Crookes dangled the delicious prospect before his readership that spiritualism had at last found its scientific candidate.

While Crookes appeared to unambiguously declare his allegiance to professional orthodoxy, he was in fact running a more complex game, exploiting tensions between elite and popular science to steamroll the advance of his professional career. He, the spurned son of a tradesman, could make common cause with the layman's sense of elite oppression built into the university system and the governance of science. However, he had to do so from a distance to avoid further class contamination that might injure his own chance of inclusion. Yet, not so great a distance that he could not offer himself as the common man's benefactor and receive his popular acclaim. Crookes's announcement offered his own professional competence as a remedy for the bungling of "pseudo-scientific spiritualism." Thus, he clearly supported the *idea* of an institutional orthodoxy, just not the one currently at hand. He planned to "popularize" orthodoxy not by sharing the power of its practitioners but by making them more responsive (as he was) to the public's interests and outlook.

As a publisher of several photographic journals and *Popular Science Quarterly*, Crookes had already drawn lay support for his professional research from popular commercial sources, but there were dangerous cultural politics here as well concerning the production of knowledge and its distribution of power. As literacy and prosperity spread down to the lower social orders after 1850, more and more people had the leisure and means to pursue their own intellectual interests, both as consumers and as opinion makers. Here the nonspecialist played an active role in shaping narratives of science through the buying, producing, and selling of its lay literature and other educational experiences and equipment.[43] The relative autonomy and bustling activity of the commercial sphere, which gave otherwise dry narratives an info-taining zest, proved highly resistant to the efforts of institutional science to submit the nation to its arid authority. Intoxicating narratives like Crookes's scientific spiritualism made this all the more difficult by splitting the message of orthodoxy and encouraging the science of personal fancy. To a certain extent, in the science of the marketplace, the customer, and not the credentialed scientist, was always right. Institutional science had its champions in this arena as well, who took their share of the public's attention: "popularizers" like Huxley, Tyndall, Carpenter, and E. Ray Lankester. While they succeeded in promulgating scientific theories in ways that made them accessible and widely familiar, their sensational rhetoric was more riveting than appealing. They also went fiercely after their adversaries in the press, dispatching a war party on behalf of science, which might perhaps have been better served by a diplomatic mission. Huxley ridiculed spiritualism as "twaddle" and "a mendacious humbug." Such ridicule fed the perception that academic highhandedness was the problem, and not lay credulity. As W. H. Harrison remarked in defense of Crookes, "the Royal Society worships caste and political power as much as it does science" and "goes about too much on stilts."[44]

As someone who had fluency in both the academic and commercial discourses of science, Crookes may have felt himself in a favorable position to pull off the doublespeak of scientific spiritualism. He knew, or thought he knew, how to say what everyone wanted to hear. As a religious man and spiritual seeker, he understood the resistance people had to surrendering the interpretation of nature to science, especially a naturalistic science free of any theological obligation. Too much that was vitally important to Victorians was vulnerable to how this view of nature was to be constructed. Free will, moral order, and a personal God were all potential contingencies (and casualties) of an overly deterministic understanding of causal laws. Even theistic scientists like James Clerk Maxwell and J. J. Thomson could not be trusted to defend these principles at the expense of established theoretical laws (though they might work around or between such laws to make the necessary room). It was a fairly uniform stance among professionals (across a variety of religious views) that no tolerance was to be extended to speculative metaphysics coming from below, of which spiritualism was a particularly irksome variety. Even ideological foes within the BAAS could agree on that, forming a united modern professional body to protect the meanings and practices of science from outside interference. For a question to be adjudicated scientifically, it had to arise within a delimited theoretical framework and remain within certain knowledge capabilities. Curiosity concerning the nature of God or the disposition of the soul after death violated the most basic ideological commitment of secular knowledge to address problems of the physical world.

Part of this ongoing attraction to religious themes in both elite and popular varieties of science can be seen as the philosophical residue of traditional natural theology. Such a tradition held with Newton that nature told us something about God (and that the church should thus oversee the study of nature). This latency was difficult to get at and dispel. It was in some sense already tucked into the idea of a natural order. This made natural theology less of an artifice than an inference embedded in any notion of a natural design and *Naturphilosophie*: to seek to inform oneself about the physical world was to eventually confront the question of how that world was itself informed. This attitude implicitly held that all knowledge was connected (how the heavens go and how to go to heaven) and that there was really only ever one terrain under investigation. Historian Frederick Gregory labels this schema "common territory," explaining, "There was still one truth to be found. At issue was who had correctly identified the way to get at it."[45] Common territory was a conceptual legacy that continued to dog the debate between science and religion through much of the nineteenth century, long after the collapse of Anglican natural theology, over a prolonged period of flux. (Peter Bowler extends this out to the beginning of the twentieth century with the last throes of a "new natural theology.")[46] The point is that science, even agnostic science, could not be without religious implications because (a) any description of nature automatically seeded clues to a more ultimate or implicate order; and (b) common territory meant that there would be more than one interpretation offered for any given phenomenon. In this contested sphere, one form of knowledge was always asserted at the expense of another form.

Any idea of God implicit in the natural order described by militant evolutionary naturalism would be far removed from traditional Christianity (if He could be said to exist at all). This made mysterious unknowns, like a psychic force or spiritual phenomena, a potentially timely and appealing intervention. Crookes could well capitalize on the elite refusal to take seriously the issue of spiritualism, which he defiantly described as taking a "leading rank among the social questions of the day, and which numbers its adherents by millions."[47] Scientists who were overly dismissive could be shown to the public as acting from self-interest, professional arrogance, or a bigoted allegiance to mechanical orthodoxy. This was all a winning argument for Crookes. With it he could seize the scientific high ground by turning science's two greatest sources of authority—its elaborate theoretical framework built up over generations and its institutional prestige—into seeming liabilities. All the public curiosity surrounding spiritualism gave him a national platform in which to act the trusted intermediary by bringing scientific expertise to a pressing popular concern. And he would speak to the public not in terms of theory or mathematical abstraction but with a common-sense appeal to their own critical judgment of the proofs that he produced. So, though Crookes appeared to be taking on Goliath with his project for a scientific spiritualism, he was in fact drawing on cultural resources that were potentially both deep and wide.[48]

The historian Peter Lamont takes this support even further by suggesting that the potential strength of spiritualism's arguments was not just discursive but extended to the substantive nature of the evidence itself. (He leverages this assertion almost entirely on the phenomena of D. D. Home.) This is not to say that the phenomena were genuine, only that they were genuinely convincing. Enough so that the failure of science to adequately explain such phenomena in the court of public opinion created something of what Lamont calls a "crisis of evidence." It is worth taking a moment to consider that evidence to get an appreciation of how this might be the case. Even if we were to assume this was all a great con (and that would be my starting point), Home still would demand an awed reckoning: his was a superhuman if not supernatural feat of deception. It was twenty-five years of constant, itinerant, and often improvisational phenomena, and it tended to be demonstrated at close quarters before the best educated and most resourceful members of the citizenry. He would go to private homes without prearrangement and still manage to slide their pianos from across the room, topple their bookcases, and pin them to the wall with the séance table. There was a fixed retinue of effects that signaled the phase shift into Home's interregnum reality: a strong cold wind, a violent shock, a crack of sound that sent percussive energy shuddering through the floor. After which things began to happen: levitations, sounds, spectral lights, instruments suddenly being played by hands that weren't there before. To me, that all sounds like the high jinks of admitted frauds, except, on closer look, there are critical distinctions. Those "séance spectaculars" were held in dedicated venues, or the medium arrived equipped with a cabinet, props, and perhaps a confederate. But Home appears always to have worked alone, to have used no means of concealment, and to have performed often in the bright light of several oil lamps and even daylight. Now his seemingly absurd portfolio becomes more difficult to square. This was an age of spiritual enthusiasts, yes, but it also raised a small army of skeptics. The psychical researcher Richard Hodgson went all the way to India to bag his big Blavatsky in a triumph of dogged investigation. Eusapia Palladino and Mrs. Piper both at some point came under indictment by the SPR. And yet Frank Podmore, an investigator who was Home's veritable Javert, never managed to pin a charge on him.[49] Admittedly, to hear casually that Home was "accused" of fraud primes a certain reflex: people must have sensed his deception. But then again, such accusations might be the normal response of that portion of people resistant to having their reality violated. I'm not endorsing the phenomenon. Whoever has been in the presence of a great magician knows that our phenomenological operations can be seriously toyed with. But again, with Home, there were no curtains, props, or performance times to assist with that illusion. At the very least, we must take pause before the written record and try to absorb it. And all this stands before mention of his most legendary phenomenon: the day he flew out the window of one room and entered another in the presence of Lord Adare.

Thus, Lamont productively refocuses the usual question asked regarding séance phenomena: "How could people believe such a thing?" He asks, instead: "How could they not?" At issue here is the unraveling, over the course of the 1870s, of spiritual convictions, which had hitherto rested on a growing bedrock of compelling beliefs. The issue has already been raised that the commercial overreach of spiritualism kept numerous cases of fraud before the public. But Lamont argues that there were other, more deliberate forces also at work, coming from the early scientific "deep state." With its epistemological authority on the line, the defenders of science increased the virulence of their attacks on spiritualism, stooping to slander, criminalization, half-truths, and, in the case of Crookes, even a few lies or at least misrepresentations.[50] This scientific offensive was initiated several years before the Florence Cook affair, but kept ramping up in pressure as its spiritual quarry galloped faster and faster ahead. (Crookes, after all, kept upping his ante until they decisively knocked him down.) This was a pivotal moment in the culture wars where spiritualism and superstitionism came under attack by legal, intellectual, and religious authorities in a campaign of annihilating force (of which Crookes was both the target and immediate cause). Crookes's unprecedented full-court press for a scientific spiritualism, with its celebrated confirmation of a psychic force, had entranced a nation already bewitched by the master wizard and illusionist, D. D. Home. The threat he posed to the Victorian worldview was not focused in the credulity of the public, but lay in the perceived credibility of the evidence itself, potentially elevating Crookes's scientific standing with the people above that of the orthodox obstructionists. Put another way, Crookes threatened to make institutional science look bad, not because he was curious about spiritualism, but because they were not.

The anxiety felt by rationalists concerning spiritualism's runaway evidential claims cuts through E. B. Tylor's privately logged "Notes on Spiritualism," documenting his own sittings with D. D. Home in 1872. As George Stocking suggests, Tylor's own dogmatic commitment to rational progress imposed a duty to reject even evidence he found convincing. He ostensibly undertook this study of Home as part of an ethnographic investigation into surviving primitive beliefs, of which spiritualism was a textbook variety.[51] In truth, however, Crookes's researches of the previous year had aroused Tylor's genuine curiosity. Yet Tylor would not permit himself to be persuaded into any such regressive belief: "even supposing the alleged spiritualistic facts to be all true, and the spiritualistic interpretation of them sound, it might still be wise to reject them." What concerned Tylor most was that spiritualism was able to attract elite thinkers to its primitive mode of reasoning, corrupting the modern mind with its propagation of this "savage survival" of ghostly superstition. Tylor made it clear that if he had to pick sides between "the Red Indian medicine man, the tartar necromancer, the High-land ghost-seer, and the Boston Medium" and "the great intellectual movement of the last two centuries," he would go with the latter, all facts aside.[52]

The considerations raised by Lamont and Stocking are useful here, in that they enrich our portrait of either side of the divide: the spiritualists become less irrational and the scientists more so, as both sides pursue an ideologically and emotionally charged contest over modernity's epistemological values. This is in no way to advance the claims of spiritualism against those of science, but rather to recover some of the contemporary fog of war in which Crookes prosecuted his battle. This ambiguity posed by spiritual phenomena is central to explaining Crookes's calculations and, ultimately, miscalculations about the length and limits of the spiritual support he might find. With Florence Cook, he had badly botched the politics of possibility on which his public support and professional immunity relied. Once Crookes exited the framework of potentialities and began making assertions of unequivocal certainty (touching, seeing, speaking with, and photographing a physical ghost), he had crossed the line of what the public was prepared to believe and what orthodox science was prepared to tolerate. The mainstream press, which had been largely sympathetic to Crookes in the contretemps over psychic force, now skewed in the other direction as the model of scientific spiritualism began to permanently fracture, under the strain of these corporeal ghosts. Instead of making scientific proclamations from a national podium as he had planned, Crookes found himself pinned down in a spiritualist ghetto and having to desperately fight his way out. A closer look at Crookes's personal and professional circumstances leading up to the fateful séance will help us better understand how he got there. How did Crookes fit into the professional world in which he plotted these researches, and what were his constraints as he tried to move through it? How did someone aiming so assiduously for the top of his profession end up so close to the edge instead?

BY HOOK OR BY COOK,

CROOKES MAKES HIS WAY

Until 1867, when he began his involvement in spiritualism, Crookes had labored assiduously along the scientific straight and narrow. His career commenced modestly in 1848, with his enrollment in the newly minted Royal College of Chemistry, where he took his degree two years later. During this time, Crookes applied himself patiently to the mastery of highly complex chemical procedures that later formed the basis of his reputation among spiritualists as a technically skilled investigator of exacting professional standards. Because of the obvious aptitude and industry Crookes displayed, he was selected to continue at the college as senior assistant to Professor August Wilhelm Hofmann until 1854. His duties mainly involved analyzing and isolating chemical substances. This was quantitative, meticulous work that Fournier d'Albe suggests left Crookes fundamentally dissatisfied. In his biography, Fournier d'Albe notes that "the state of chemical science in the fifties may not have been sufficiently inspiring to secure the wholehearted allegiance of a young devotee of science," especially one who "felt fit to be in the front rank, asserted his right to be there . . . maintained it against all competitors to the end of his life."[53] Looking back on Crookes's life, Fournier d'Albe admiringly saw the triumph of spirit and industry; his contemporaries saw more of an upstart. Crookes's romantic and even somewhat grandiose temperament made him ill suited for a life in industrial science. Here he would toil among science's laboring class, just another unsung hero of waste management and better concrete. Such was the chemical trade in 1850, its heady days of atomic discovery behind it, and its toothpaste, solvents, and multisurface cleaners just ahead.

As a professional practice, industrial science offered Crookes few opportunities to partake in the glorious theoretical discoveries that had turned some scientists into household names. Crookes, perhaps rightly, feared that he would be shut out of the scientific limelight, languishing in unremunerated obscurity as a technical drudge. Even though his family was prosperous, the money was from trade and his education was a two-year vocational training. He had not been gently bred on the classics and then sent off to Oxford to contextualize practical science in some grand philosophical tradition or complex theoretical scheme. This university deficit restricted his social as well as intellectual opportunities. Crookes's background afforded him no way of making the professional connections he would need to advance in the class-bound realm of academic science. Only there could his theoretical yearnings take flight, dignifying him as someone who could think about science, not merely apply it. Unlike members of the Sidgwick group, Crookes had no protective social circle to draw around his risky curiosity or insiders opening doors to help him get where he was going.

But in 1861 Crookes managed to make his own future with his discovery of thallium, raising his hopes that he might one day ascend to science's illustrious theoretical class. As Frank James observes, this discovery should in no way be confused with serendipity; it was, rather, the result of Crookes's calculated effort to optimize his time, skills, and resources in such a way as to advance most readily in the world of science.[54] Discovering an element was certainly among the high-profile achievements attainable by someone in Crookes's particular circumstances. When a thin green began to glow one day in the spectral band of a compound he was analyzing, it was the anomaly he had been searching for. Recognition for his discovery was far from instantaneous, but eventually "Crookes the inventor of thallium" joined the parade of nationally renowned scientists leading the march of progress. Crookes used the occasion of his discovery to make a scientific foray into philosophy, contemplating spiritual life in the light of energy conservation for *Popular Science Review* (1861): "Mechanical motion is equally capable of being transformed into heat, light, electricity, or chemical energy . . . these living forces do not die, but become absorbed in that vast reservoir of energy which is the source of all life and light upon this globe . . . If philosophy can thus prove that the latter [physical life] never dies, shall not faith accept the same proof that our own spiritual life is continued after the vital spark is extinguished?"[55] This florid appeal to metaphysics reveals two things about Crookes: his intellectual ambition and his professional naiveté. Such grandstanding about faith and reason might appeal to the readers of *Popular Science Review,* but not to an academic audience trying to tighten its narratives around purely disciplinary subject matter. Yet Crookes takes this moment of national attention to link entropic heat diffusion and human immortality. However, Crookes was not yet important enough, and his musings were still too modest, to give offense or gain notice (a situation he became determined to change).

It took until 1863 for Crookes to finally receive his coveted invitation to join the Royal Society. His invitation was based on the merit of his discovery as well as on the testimony of his mentor, Sir George Gabriel Stokes, who had to vouch for Crookes's experimental *bona fides* and make good his social clout before the eyes of his genteel jurors.[56] The same year he became a Fellow he also assumed the editorship of the new *Quarterly Journal of Science*, which was intended to provide the public with a more learned alternative to the vulgar penny fair. With his own editorial pulpit from which to declaim and the glorious credentials of "FRS" trailing after his name, Crookes braced himself for a meteoric rise. No such miracle occurred. Crookes soon began to realize that, despite his newfound scientific celebrity, the profitably glorious career in science he had planned was far from assured. Complaining to a colleague, Dr. Angus Smith, about his financial distress and intellectual dissatisfaction, Crookes wrote (in a letter dated October 31, 1864):

> No doubt if I were to advertise constantly and give puffing testimonials to tradesmen I could get a connection and make a decent living out of my Laboratory, but as for respectable work as consulting or analytical chemist, I get next to none. I have made possibly £100 in six years at that work . . . my Laboratory and chemical education are only of money value in so far as they enable me to exercise editorial supervision over the *Chemical News* . . . this being an amount of knowledge than any sharp person could get up in six months, nine-tenths of my "brain-force" is lying idle . . . I was scientifically fortunate, but pecuniarily unfortunate some years ago in discovering thallium. This has brought me in abundance of reputation and glory, but it has rendered it necessary for me to spend £100 to £200 a year on its scientific investigation. Of course, it brings more than the value of this ultimately in reputation and position, but whilst the grass grows the horse starves and it is the utter despair that I feel of ever making anything out of my London Laboratory that make me anxious to leave.[57]

At this point in his career, Crookes had a wife and children, and had already suffered a series of rejections for research appointments that fell hard upon him. The idea that Crookes was somehow at his ease in his home laboratory, living off his father's largesse, discounts the genuine anxiety he had about succeeding as a professional scientist, both as a financial matter and, even more important, as a matter of pride. Despite his Fellowship, professionally he was still operating under the dual constraints of class prejudice and a limited education that kept him sidelined from the social and intellectual centers of scientific activity. Yet for all the obstacles in his path, Crookes not only felt "honor bound to make his way in the world" but would somehow find a way to do so.[58]

Stonewalled by academia, Crookes was forced to supplement his income with what he considered to be the most dignified alternative to pure research: science journalism. He wrote for magazines such as *Photographic Journal* and *Photographic News*, and eventually purchased the *Chemical Gazette* in 1859, rebranding it the *Chemical News* and assuming its editorship.[59] But while profitable, his editorial career left him in the company of hobbyists and technical specialists, not illustrious intellectuals. The dismal state of his professional affairs described in his 1864 letter showed no signs of improvement as the decade wore on, until finally, in 1867, tragedy struck—and opportunity knocked. That September, he received the devastating news that his beloved brother Philip had died of typhoid on the ill-fated Cuba & Florida Cable Expedition. Incensed by what he considered to be the scandalous incompetence of the managing company, Crookes publicly denounced its executives in the papers, landing in court with a slander suit against him. The expense and publicity of a court trial only added to his financial woes and raised issues of respectability. Crookes lost the case, but some comfort did come his way in the form of a friendship with the famous transatlantic cable engineer and spiritualist, Cromwell Varley. Varley had read about the case and subsequently sought out Crookes about "contacting" his deceased brother. A nationally celebrated figure in the field of electrical engineering, Varley laced the language of spirit communication with the technomystical jargon of spirit telegraphy in his overture to Crookes, providing attractive bait to both the grieving brother and to the scientist looking for "proof that our own spiritual life is continued."[60] Varley's spirit circle also had a certain promising social luster, with an attitude far more inviting than the aloof "Oxbridge" enclaves of the Royal Society.

Crookes's initial spiritual inquiries with Varley were discreet, intentionally ducking any potential controversy that might come of publicizing such investigations. But despite the potential liabilities of a connection with spiritualism, Crookes realized there were powerful professional and social allies to be made here as well. Scientists like Alfred Russel Wallace, Augustus de Morgan, and William Huggins, as well as the astronomer James Lindsay, FRS, later Earl of Crawford, were active in spiritual circles at the time of Varley's overture. Other cultural celebrities, such as Lord Adare; his father, Lord Dunraven; Lord Tennyson; and Baron Reichenbach, were also active participants, and Varley himself was a considerable spiritual eminence. In this milieu of genteel spiritual curiosity, Crookes found his scientific reputation was finally being given its due. Where academia had treated his technical expertise with seeming disdain, his experimental acuity as a chemist now had particular cachet. He received in 1869 a particularly solicitous invitation from Alfred Russel Wallace to join the London Dialectical Society's national campaign of spiritual investigation, an effort sufficiently meritorious to be looked into by such elite skeptics as Charles Darwin and George Eliot. Crookes, still unsure of the possible career ramifications, held himself aloof, agreeing only to attend an occasional meeting so long as his name was kept out of the press.[61] Of course, word did get out concerning Crookes's possible association with these spiritual researches, but the effects of this publicity were a pleasant surprise. Crookes attracted praise for lending his weighty support to the society's important endeavor, occasioning flattering letters from George Henry Lewes and John Tyndall, who singled out Crookes as an elevating scientific proxy.[62] Tyndall, like Lewes, refused to join the committee, but offered the caveat that "if earnestly invited by Mr. Crookes, the editor of the *Chemical News*, to witness phenomena which in his opinion tend to demonstrate the existence of some power (magnetic or otherwise) which has not yet been recognized by men of science, I should pay due respect to his invitation."[63] (However, this was still 1869, and Crookes was as yet the discover of thallium and not the known frequenter of séance circles or the proponent of a psychic force.)

While Tyndall would soon sing another tune regarding Crookes's spiritual curiosity, for now, Crookes's scientific reputation was already beginning to shine in the light of modern spiritualism a year before he chose to publish "Spiritualism Viewed in the Light of Modern Science." By the time the article came out in July 1870, much of the London Dialectical Society's field testing had already been concluded. No report of the results had yet been issued, however, and expectations were running high. Crookes, outed by the *Athenaeum* anyway and not wanting to miss out on spiritualism's moment of fanfare, thus publicly threw his hat into the ring. And the more willing he was to put himself out in the service of spiritualism, the more attractive his scientific ascendancy was to those interested in advancing the spiritual cause. Just six months later, in December 1870, he found himself aboard the *Urgent*, floating downriver with a raft of scientific celebrities, his national scientific standing at a new height.

Crookes received this appointment to the prestigious Government Eclipse Expedition to Spain and Algiers not from any Royal Society nomination or endorsement form the BAAS, but at the invitation of the American astronomer and soon-to-be president of the American Association for the Advancement of Science, William Huggins, who had become friendly with Crookes through their time spent together in spirit circles. Huggins used his "astronomical cachet" to boost the national reputation of his friend Crookes while likewise tightening spiritualism's connections to this British pantheon of scientific notables. The expedition was a highly publicized event whose participants were sent abroad as icons of the national honor. The *Daily Telegraph* of December 12 contained a leading article unctuous enough to salve any wounded pride:

> The ship *the Urgent* carries a cargo of brain which would be all but irreplaceable for civilization. If we ran through the entire list of the scientific company on board that vessel each name would be a good plea for the favor from Neptune. On to select specimens: there are Professor Tyndall and Dr. Huggins, there are Captain Noble, Mr. Crookes and Mr. Carpenter, there are R. F. Howlett, Mr. Ladd and Admiral Ommaney with others whose loss altogether would be far worse than the sinking of a whole armada of Spanish galleons loaded to the gunwales with silver. We cannot underwrite such a cargo for who is going to put a price upon Professor Tyndall, who to estimate the discoveries of Dr. Huggins during the rest of his natural life, or to put a market value upon the next metal identified by the discoverer of thallium?[64]

Thus, it was his spiritual connections and not his scientific achievements that propelled Crookes out of his professional malaise onto the *Urgent* for a celebrity tour. Such were the strange patterns of interference between science and spiritualism in mid-Victorian society that Crookes found himself having more success in advancing his career in the former by operating through connections with the latter.[65] His public adjacency to both spiritualism and scientific greatness gave Crookes's name a public buzz he had never enjoyed when known solely for his chemical achievements. The spiritual press lit up with his praises, and Crookes found himself living a life "studded with deeds of scientific distinction since his youth," as he basked in the rekindled glow of the green line of thallium.[66] He became a sort of genie's lantern of legitimacy for the spiritual movement, frequently burnished to bring forth more and more of his precious experimental fame—which must have made Crookes feel invulnerable, even while it was making him a target for critics as well.

While this social and intellectual space presented Crookes with surprising advantages, his interest in spiritualism ran far deeper than opportunism. If anything, Crookes's professional conservatism initially interpreted spiritualism as a liability, as evidenced by his attempts at secrecy. But he overcame these reservations in pursuit of a deep attraction to the spiritual program, located partly in grief but also within his own mysticism and intellectual curiosity. Crookes's largely positive early séance experiences nurtured this initial attraction into a sustained and growing interest. Adding to that the warmth and respect of his professional colleagues and the intoxicating adulation in the press allows us to see how spiritualism began to gain irresistible momentum in Crookes life: it was the gathering wind that would blow him out of his personal and professional doldrums (and eventually the gale that would likewise threaten his career).

Disappointingly, however, these winds seemed to flag when the *Urgent* meandered home in a mood of *decrescendo*, with the Government Eclipse Expedition a decided failure due to heavy cloud cover at the time of the of astronomical event ... There were no university overtures awaiting Crookes, no new research opportunities, and even the unsatisfactory report that had finally been issued to the public by the London Dialectical Society took some of the wind out of spiritualism's sails. Crookes felt he was in danger of drifting out of the spotlight, but not for long. He quickly got to work on a mysterious project in his laboratory basement, one that was far more sensational than expeditionary concerns about the concentration of hydrogen in the sun's chromosphere or lackluster results from a séance-as-usual. On July 1, 1871, Crookes announced in the *Quarterly Journal of Science* (his own publication) that he had experimentally confirmed the existence of a "psychic force"—an astonishing finding that would transform existing models of nature and most likely explain the phenomena of spiritualism as well. He had taken as his subject the most prominent magus of the time, D. D. Home, whose well-attested phenomena included levitating the body of Lord Adare in the presence of witnesses and popping in and out of third-story windows. Home was also widely known to assert a powerful kinetic force through the application of mental will alone, a fact anecdotally attested to by numerous witnesses across the globe. But only now had this internationally rumored power been empirically confirmed. Crookes modestly claimed that his research was inspired by "the eminent men exercising great influence on the thought of the country," thereby positioning himself as the heir apparent of the London Dialectical Society, firmly legitimating his research as a continuation of this sanctioned investigation, while totally upstaging it as well.[67] He thus managed both to exploit the publicity and prestige amassed by the London Dialectical Society during its campaign and at the same time to take advantage of its failure.

Crookes's revolutionary protocols developed specifically for the Home investigation offered a dramatically different approach to these troublesome phenomena, which had hitherto proved both ubiquitous and elusive. The only real novelty provided by the London Dialectical Society's research project lay in the quantity and quality of its investigative teams, but the setting remained the same—the séance circle over which the medium presided—making it difficult to swear to the integrity of the data. While some results were promising, they only added to the questions already raised. And thus the issue remained sunk in the same investigative quagmire. Crookes, however, was crystal clear. The program he laid out in the *Quarterly Journal of Science* was meticulously scientific. He described his investigative apparatus in great detail, overwhelming his readers with a wealth of metric information. The accordion cage into which Home inserted his arm to test his alleged ability to play singlehanded was "formed of two wooden hoops, respectively 1ft. 10ins. and 2ft. diameter, connected together by 12 narrow laths, each 1ft. 10ins. long and rounded by 50 yards of insulated copper wire." (This was intended as a sort of Faraday cage to contain Home's psychic energies.) Crookes likewise rigged a device to test Home's ability to alter the physical weight of an object, carefully recording the oscillations on its register when Home gently came into contact with it. All this mechanically observed, quantitative data mooted the skeptical hypothesis that the witnessed spiritual effects were due to hallucination or malobservation (that is, that people did not understand that they were being deceived).

Crookes, the molecular chemist and man of science, had promised a year before to bring his exacting standards of data analysis to the haphazard and impressionistic claims of spiritualism. His experiments on Home made good this vow, supplementing mere visual observations with elaborately devised scientific instruments that could precisely measure and record the force phenomena in question.[68] He also delivered something even more extraordinary: robust scientific proof of the phenomenon in question. In a series of published reports, Crookes confirmed "the now almost undisputed fact" that, "from the bodies of certain persons having a special nerve organization, a Force operates by which, without muscular contact or connection, action at a distance is caused."[69] Crookes named this "psychic force," stating that it could "be traced back to the Soul or Mind of Man as its source," a sound inference based on observing how Home produced these effects through mental concentration alone.[70] Crookes skillfully fortified this seemingly outrageous claim by asserting it through a series of minute calculations, which all added up to the irresistible fact of its existence. Unlike his skeptical colleagues who let their "ideas of the naturally possible and impossible" interpret the facts and govern their thinking about nature's laws, Crookes held that "our only knowledge of the laws of nature must be based on an extensive observation of facts."

Part of the popular appeal of psychic force was that it was so inchoate and poorly understood that academics had no privileged intuitions regarding what it was or how it worked. Even while it remained within the idiom of science, there was no theoretical handle on its concept, no barrier of intellectual exclusion blocking meaningful access to its discourse. The links had been cut dividing the possible from the impossible, the natural from the supernatural, the knowledgeable from the ignorant, and it had been an inside job. Although this is not what Crookes himself intended, it was what his orthodox compatriots had known and feared. The "psychical" designation meant to elude spiritual commitments had neither deterred believers nor placated skeptics. Crookes had, after all, just confirmed that the most famous medium of the spiritual movement, on whose powers of fascination so many people's beliefs relied, did indeed wield preternatural powers. In so doing, he elevated spiritualism's stock to new heights and sent a shiver of excitement through the wider public as to what it all might mean. "Psychic force" found its way into the major outlets of the press, drawing support from radical intellectuals like Charles Bray, a member of the Coventry intelligentsia with interests in phrenology and social reform, whose idealism was offended by academic psychology and its impoverishment of the human mind. As Bray described this unfortunate physiological preoccupation among professionals, "such psychologists don't even know if they have a mind." It even provided a safe speculative outlet for skeptics whose curiosity had been put off by séance theatrics.[71] Crookes's brightly lit laboratory bypassed the medium and the haunted lair altogether and asserted a unilateral scientific authority, placing the majestic Home with his arm in a cage and fingertips monitored by a mechanical graph.

But academia was slow to take the bait. Instead of fanfare, Crookes received a verbal lashing for which he was ill prepared, having thought his scientific formulation would somehow immunize him against the militancy of his critics. (The prominent physicist Peter Guthrie Tait, though a devout Christian, lumped Crookes in with "perpetual-motionists, [and] believers that the earth is flat and that the moon has no rotation," while Allen Thomson went for the jugular, explaining that such psychical credulity resulted when "scientific information and training have been of a partial kind.")[72] Crookes's collapse of science and spiritualism into the single idiom of a confirmed psychic force had exploded the tension building in that relationship over the last twenty years, nearly taking him out in the blast. Yet the intensity of this official response also shows that Crookes somehow got it right, that scientific spiritualism as an undertaking was potentially both serious and significant. While he had meant to disarm the resistance of orthodox science, he had struck its Achilles heel instead. As suggested by Lamont and Stocking, the Royal Society had refused Crookes's paper not so much because his experiment on Home was procedurally or intellectually flawed (Crookes made no theoretical claims), but because it was dangerous. They were not just looking for reasons *not* to believe it (Stocking); their refusal to scientifically adjudicate the matter excluded even its nonexistence from all consideration. This makes sense. The conservation of force firmly fixed the amount of energy and matter that inhered to any closed physical system. Nothing got in. Nothing got out. It thus categorically forbade the kind of mysterious work that might cause tables to levitate, accordions to play, and objects to gain mass or even to appear and disappear from the scene willy-nilly at some medium's behest. Such a finding would have proven too disruptive, not just to the intellectual authority of orthodox science but to its extended institutional apparatus as well, which had come to power under this very domain of thermodynamics.

And so began a series of ad hominem attacks from the watchdogs of science. Such notions were deemed "characteristic of all inferior races" (Tylor's same thesis) and opposed by "the genuine man of science," which, by implication, Crookes and his ilk were not. Crookes was singled out for punishment, dismissed before a general meeting of the BAAS as a vocational flunky way out of his depth, "regarded among chemists as a specialist of specialists, being totally destitute of any knowledge of chemical philosophy." He was to be nothing more than a hired hand of science, utterly incapable of contributing to its intellectual capital. The same brush was used to tar William Huggins, who sat as a witness of Home's phenomena during Crookes's study. Huggins, despite his celebrity as an astronomer, was now "merely a brewer" and deemed "ignorant of every other department of science" but that "small sub-division of a branch to which he has so meritoriously devoted himself." In its efforts to quarantine the class and occult pollution represented by Huggins and Crookes, official science was not just pushing them out on a limb but, as Crookes complained, onto "a twig of the tree of science," so complete was their attempted exile.

The Home episode left Crookes feeling hopelessly stymied by an elitist, insular orthodoxy, hostile to both new ideas and scientific *parvenus*. Yet there was no backing down before such an outrage. He could not make peace with such an institution on such terms. Even as he was being hurled off the battlements of the BAAS, he was scrambling to find tactical high ground, repudiating all those who "damaged the true interests of science and the cause of truth, by thus throwing low libelous mud upon any and every body who steps at all aside from the beaten paths of ordinary investigation."[73] Crookes, by contrast, was a man who understood that "the true business of science is the discovery of truth, to seek it wherever it may be found, to follow the pursuit through by-ways and high-ways, and, having found it, to proclaim it plainly and fearlessly without regard to authority, fashion or prejudice."[74]

Taking such a hard line was precarious for Crookes in the wake of his research on psychic force. What would make him double down on something as risky as a corporeal materialization from the afterlife? Why leap from mechanical drawings of psychic force curves to spirit photos of a pirate ghost? Perhaps Crookes's humiliation and sense of professional checkmate stoked a reckless, pugilistic mood. There was perhaps a moment of forced reckoning that if he could not join them, then he must beat them. Crookes, who still commanded widespread support for his spiritual and psychic researches outside the official venues of science, had a powerful ally in this lay audience if he could continue, or even intensify, their investment in his future research. Standing at this crossroads between popular acclaim and institutional criticism, Crookes understandably chose to continue his scientific mission of discovery through spiritualism's "by-ways and high-ways." Indeed, Crookes's professional and commercial successes of the past few years had been largely reaped through his association with men like Varley and Huggins. These were self-made men of considerable scientific achievement who still took an active interest in both spiritual and financial enterprises. And they did not glower down at Crookes from an ivory tower.[75] Like him, they were outsiders who had to climb the exterior walls of institutional science until they could breach them by some stunning demonstration of investigative ability. Crookes credited their open minds and financial success to their freedom from academia's limiting scheme. His professional future with thallium appeared finite and mundane compared to the glorious possibilities this psychic force fever had stirred in the populace. Having been snubbed by the Royal Society and verbally assaulted in the press by its membership, Crookes felt the futility of further courting an establishment that had always held him back. Entrepreneurial, resourceful, and energetic, with ideas and ambitions too outsized for the small place prepared for him by his fellow Fellows, it is no wonder Crookes thought to blaze his own trail to the top.

When Florence Cook made an unsolicited appearance at his door one evening in December 1873, Crookes may have seen in her the opportunity to take his spiritual offensive to the next level. As one of the most talked-about physical mediums of the past few years, she presented a very visible venue for Crookes to build upon his experimental researches. According to Crookes, the Home investigation had already incontrovertibly proven the existence of a psychic force. What remained to be done, the next natural escalation in this public drama, was to test specific hypotheses as to its deeper nature. Among the most prominent of these was, of course, the spiritualistic hypothesis. Crookes framed this as scientifically correct and necessary, but it was also sensational. In agreeing to investigate Florence Cook in December 1873, Crookes recaptured the public's interest in the researches he had concluded over a year ago, but this time, he was raising the stakes from a mental to a spiritual agency. While he fired up popular support for his spiritual researches, he was still careful to link them to his own scientific authenticity, pitting the humble daring of free inquiry against Royal Society demagoguery so arrogant that it would "attempt to stop the progress of investigation."[76]

This explanation of the lead-up to the séance only gets us to the point where we can understand why Crookes might pledge to investigate a medium known to manifest a walking, talking ghost. This was a high-profile case involving controversy, extraordinary spiritual phenomena, and guaranteed publicity. It does not, however, explain what happened next: Crookes's own personal descent into confederacy. Here the trail goes cold. We are left in imagination standing at the top of the stairs leading to Crookes's dimly illuminated basement lab, wondering what the devil happened down there. It is clear that Florence Cook's appeal for Crookes was more than just that of a strategic research opportunity. She came to him in the capacity of distressed damsel, having fallen under a cloud of ruinous suspicion. William Volckman, an influential investigator in the London Dialectical Society, had apparently tackled the ghost of Katie King during one of her séance perambulations, partially defrocking Florence's ghostly garb and irresistibly forcing the conclusion that "no ghost, but the medium, Miss Florence Cook herself, was before the circle."[77] The sensationalism of a physical assault brought the incident, and the accusation, before the public at large. And though Florence retained some sympathizers at this "gross outrage," the evidence of imposture clearly threatened her reputation and livelihood. Even her gullible sponsor, a wealthy industrialist named William Blackburn, notified the distraught Cook family that "I shall stop payment."[78] (Florence had been supporting her parents' household with her cottage-industry mediumship since she was fifteen.) Desperate to restore her credibility and income stream, the beguiling "Florrie" went personally to Crookes to plead with him to support her cause: if Crookes, the great man of science, would investigate and confirm her mediumship, the power of his word would drive away all of Volckman's doubts.

Given such a flattering appeal, it is possible that Crookes felt moved to act on Florrie's behalf out of gallantry (or some other less altruistic motive

sublimated as gallantry). In any case, he rose to the occasion. When he described to readers of the *Spiritualist* his reasons for undertaking the study,
chivalry was certainly what he wished to imply: "When a few lines from me
may perhaps assist in removing an unjust suspicion which is cast upon another. And when this other person is a woman—young, sensitive, and innocent—it becomes especially a duty for me to give the weight of my testimony
in favor of her whom I believe to be unjustly accused."[79] No details of this
December meeting between Crookes and Cook survive, but by the end of it,
Crookes had agreed to investigate her mediumship and publicly announced
this plan to the press. Four months later he would pronounce her claims
to be true. There is no reason to believe that Crookes embarked on these
experiments with anything but honest intentions. Diaries from this period
show him to be a convinced spiritualist with many affirmative séance experiences behind him. Research success with Home would only have deepened that conviction. His sustained séance activity between 1871 and 1874
suggests that Florence Cook was the culmination of a building curiosity
that had yet to satisfy itself. In any case, the scientific enterprise planned
by Crookes seems to have taken an unexpected turn. Had Florence gotten
hold of the reins (and perhaps gotten hold of Crookes)? Crookes's sense of
his own moral and scientific infallibility, reinforced by Cook's appearance
of youthful innocence, seems to have caused him to let down his guard,
leading him to the fatal miscalculation that he would be running the show.
But judging by his own published reports, Cook became more like a leading lady, and he, her publicist and props master. Not only did she manage to
extract Crookes's public endorsement in writing and pictures but, according to historian Trevor Hall's analysis of the séance, she even commanded
his assistance in staging her phenomena. Crookes personally undertook the
guardianship of the medium's cabinet to which Florence retired during the
allegedly exhausting ordeal of Katie's manifestation, offering "the evidence
of his own senses" that Florence slumbered within while Katie paraded
without. This obviated the possibility that Katie and Florrie were one and
the same person (the cornerstone of Miss Cook's defense) and likewise excluded the possibility that some confederate was hiding therein.[80]

   Hall also leaves little doubt as to why things took this course: the infatu-
ated Crookes and the enterprising Cook had at some point become sexually involved. Years later, Cook (then Mrs. Elgie Corner) is alleged to have
confessed as much to a young man, Francis G. H. Anderson, with whom
she had likewise became involved when he was staying at her guesthouse
in Monmouthshire in 1893.[81] Anderson did not report the story of the affair

to the SPR until 1922 out of delicacy, waiting until both Cook and Crookes were no longer alive. Cook confessed to Anderson that she and Crookes had been involved in a deception of both a sexual and a "spiritual" nature, using the séance as cover for the affair. It should be remembered that Cook came to Crookes desperate for his help in clearing her name. Had she intended to set her honey trap from the start? The affair was probably for her, at least initially, a means to procure Crookes's scientific testimony, whereas for Crookes, that same scientific testimony could well have been the means for procuring the affair. There was motive and opportunism on both sides.

About a month into the investigation, Florence Cook was installed at Crookes's residence, ostensibly to facilitate Crookes's research. The two of them even traveled alone together to France to demonstrate her phenomena abroad, maintaining this cover of medium and unimpeachable scientific investigator to condone what would otherwise have been a scandalous intimacy between a married man and an unattached young woman. The possible sexual improprieties of this relationship, as obvious as they might seem to us now, did not enter into this historical narrative until Trevor Hall wrote about this episode in the 1960s. Like the ghost herself, this material was considered too sensational to be included in Crookes's career biography, and even inspired indignation on his behalf. If the possibility of a sexual dynamic had raised eyebrows at the time, it never constituted part of the negative press Crookes received regarding his research However, such disapproval may have been tacitly rolled into the general breach of professional decorum this séance represented for his colleagues, factoring into the intensity of their critique if not into its explicit content. Additionally, Cook resided at the Crookes home under the same roof as his wife, which paid the necessary obeisance to cultural propriety. We will never know how Mrs. Crookes privately felt or what she may have known, but there was no hint of a rift. Presumably, she would have been equally invested in Crookes's good appearance and, reluctantly or no, would have lent herself to any necessary cover. Crookes appeared to continue on without alteration to the happiness or stability of his family life and social respectability. In this sense, Crookes more correctly gauged his entitlement as a man interacting with a woman lower down the socioeconomic scale than he did his status as a vocational chemist taking on an elite professional class.[82]

Despite its lack of visibility, this affair remains important to understanding Crookes's continued involvement in the séance, after the scientific merits of the study became increasingly dubious. It is clear that from fairly early on Cook's, not Crookes's, agenda was driving the séance and most likely facilitated its evolution into an unmanageable theatrical extravagance. Crookes himself had little to gain professionally and much to lose from publicizing these researches, beyond whatever he garnered from his personal relationship with Florence (which probably appeared to Crookes as quite a lot). For Florence, however, the professional advantages were clear: she not only regained financial compensation from her sponsor, but also greatly increased her earning potential with the fame and restored faith in her mediumship that Crookes's testimony afforded. (He explicitly rebutted Volckman's damaging report for the London Dialectical Society that had led Florence to his door in the first place.) These reports on Katie have none of the scientific character so elaborately emphasized in Crookes's quantitative approach to Home, but rather treat her more like a starlet in a fanzine. "Words were powerless to describe her charms of manner," or "the perfect beauty" of Katie's face as well as "the brilliant purity of her complexion or the ever-varying expression of her mobile features."[83] Given such testimony we can assume that Crookes's investigative priorities had changed from earnest research to romance.

When the last séance was finally concluded on May 21, 1874, it ended with rather abrupt urgency with the ghostly Katie King making a tearful announcement that she would return no more to the earthly plane. This merciful self-banishment was no doubt part of Crookes's negotiated exit with Florence Cook. After all the public testimony Crookes had given on the ghost's behalf, he could not afford the possibility that she would be defrocked in some future séance scenario and thus prove them both to be liars. This farewell guaranteed him a future free of the haunting specter of her haunting specter. However, in terms of its public staging, the parting of Katie King, Crookes, and Florence Cook was one of mutual esteem and affection. Because the loss from her apparition repertoire of the Katie King identity would have been a significant sacrifice for Florence, we must assume it was done for some consideration. Perhaps a deal was struck to publish this report, with the exile of Katie being the cost of services rendered. Cook got to consolidate her gains with Crookes's testimony in the *Spiritualist*, and Crookes got to cut his losses and get free and clear. In his final letter to the *Spiritualist*, Crookes described the ghost's noble words of farewell: "Mr. Crookes has done very well throughout, and I leave Florrie with the greatest confidence in his hands." As for Florrie herself, Crookes assured his readership that "every test that I have proposed she has at once agreed to submit to with the utmost willingness; she is open and straight-forward in speech. I have never seen anything approaching the slightest symptom of a wish to deceive."[84] That being said, both ghost and girl were politely dropped curbside while Crookes sped away as fast as he could.[85]

While Crookes's romantic affair does not directly inform our understanding of his career ambition, the strategies he used to manage its aftermath are telling. In terms of his "scientific" reports of the séance, they were given a very limited airing with a single and exclusive run in the *Spiritualist*, which was owned by his trusted friend, William Henry Harrison. There they would presumably have done him the least professional harm and done Florence the most professional good, had the reports gone no further. But Crookes overestimated his ability to contain a story that had commenced with such scientific fanfare. No sooner had the ink dried on his last epistolary article to the *Spiritualist* than these and other earlier articles from his Home investigation were bundled into the unauthorized best seller, *Researches in the Phenomena of Spiritualism* (1874). This compilation circulated the story of Florence Cook and Katie King far and wide in the convenient and inexpensive format of a single mass-edition book. Crookes's story thus fell into the surge of what could be called "Victorian pulp nonfiction," driven by the second industrial print revolution of the 1870s, which fed the public appetite for what we would now call textual infotainment. Crookes could not escape his own séance publicity machine, which had first fallen under the orchestrations of Florence Cook and now into the hands of an unscrupulous publisher who took a spiritual tryst to be shared only with like-minded confidantes and injected it into a national narrative. The carefully constructed story lines of Crookes's scientific spiritualism fell apart in the catastrophic blowback of bad press his own words had inspired.

What makes Crookes's involvement with Florence so particularly compelling for historians is its off-road quality: there was no map for the messy terrain where his amorous improvisation landed him, nearly driving his career off a cliff. This is a view of an eminent scientist's life normally scrubbed clean in memoirs and admiring biographies, bringing into view hidden deeds and unconfessed objectives. These private spaces, where a ghost and a chemist might meet for a tryst, or perhaps two spiritualists might strike a deal to board the *Urgent*, open onto an interior view of a culture's larger motivational structure. It tells us what people wanted and how they planned to get it. It identifies the historical moment when things suddenly shift, and someone finds that they may have gone too far. It is in this backlot in the Potemkin village of official history that daily life and its unscripted disasters all go down.

On the subject of Katie King, Crookes remained ever terse and evasive, refusing all interviews and all rights of publication, and jealously locking down all images.[86] This lifetime demurral showed him to be highly conflicted about his role in these researches as well as fearful about their potential consequences. When Tyndall raised the question of spiritualism to Crookes at a chance encounter at the Royal Institution many years later, Crookes "was silent and it seemed to give so much pain that he [Tyndall] concluded never to mention the subject again."[87] This was the approach taken by others among Crookes's friends and colleagues, who likewise allowed these highly remarkable events to remain unmentioned (albeit *silently* deplored).[88]

But who was Crookes really (or who was he *also*), behind this display of scientific penitence and remorse? His biographer, E. E. Fournier d'Albe, suggests that Crookes's subsequent scientific researches were entirely independent of "this unfortunate chapter" of séance spiritualism, which he had "done with forever," making "amends by an unparalleled devotion to pure science." But this "unfortunate chapter" of spiritualism was rather longer than the four months spent investigating Florence Cook, having begun in 1867, and much more entangled with Crookes's "devotion to pure science" than d'Albe's marginalization of it might suggest. Crookes never mentioned the ghostly Katie King again, but neither did he ever renounce his testimony. And this was not because, as Fournier d'Albe suggests, he did not want to appear the fool by admitting to such egregious credulity. Nor was it because of the somewhat more obvious reason that he did not wish to acknowledge that he had lied. Crookes's failure to recant goes deeper. He himself needed to believe that the facts to which he had falsely testified could quite possibly be true.

The experimental research program on which Crookes now embarked was far more ambitious than any of his previous investigations into chemical elements, venturing into the deepest structures of matter itself. The next few years were the most astonishingly productive period of Crookes's lifetime and laid the foundation for the accumulation of future honor. But the career ambition and imaginative power that now allowed Crookes to capture percussive light in an evacuated globe and luminous matter in an electrified Crookes tube were intimately related to his drive to summon a spirit on a collodion wet plate. It is quite possible that without the precipitating factors of the one, there would not have been the other, for science and spiritualism were never quite so separate in Crookes's mind as they were made to appear at the crossroads of his professional crisis. Stripping back the palimpsest of both science and spiritualism as it is written and written over and over in Crookes's life story, we see how the integration, not the separation, of these elements pervades the whole. The possibilities Crookes now set about reinscribing into modern physics subtly renewed a questioning stance about its metaphysical exclusion, albeit with far more deference and discretion to the orthodoxy he once defied.

FROM PHYSICAL GHOSTS
TO GHOSTLY PHYSICS

In April 1874, soon after Florence Cook's departure, Crookes began to re-invent himself and his science with a public demonstration of his brand-new invention, the radiometer. This was a windmill-like object consisting of four vanes, each one silvered on one side and blackened on the other, with the entire structure suspended in an evacuated globe. With this ingenious mechanism, Crookes undertook the first step in a series of path-breaking experimental initiatives that would win lasting fame. In so doing, he slipped the hold of his femme fatality, Florence Cook/Katie King, and escaped the yoke of a career in routine chemical analysis. Inside the exotic space of the radiometer, the ordinary became extraordinary. When focused light passed through its rarified atmosphere, it appeared to behave as something other than electromagnetic radiation, pushing the mill round and round as if by some contact force light was not known to possess. Crookes called this phe-nomenon "radiant action."[89] (In reality, the motion Crookes observed re-sulted from the acceleration of air molecules near the black surfaces of the vanes, which absorbed more heat, making the ambient air more kinetic. Thus, on display was only the percussive force of ordinary matter acting in accordance with established laws of gases.)

However, the idea that light could act as a percussive force was intrigu-ing. Did light possess a sort of mass, enabling it to transfer its directional motion to the paddles of the radiometer? Did energy and matter stand in some mysterious relation unknown to science? The elusive mystery of light had finally been packed into the settled theoretical framework of electro-magnetism, but now the tiny circular motions of Crookes's light mill could potentially unwind all that. Challengers immediately set to work testing Crookes's theory; unlike their attitude toward psychic force, the physicists of Section A did not dismiss it out of hand. The seemingly spontaneous mo-tion of the mill conjured out of the stillness inside the glass was as magical a scene as any child's snow globe. Crookes finally had captured the attention of his colleagues and left them amazed.

Crookes had been careful to put his description of radiant force before his fellow physicists in the plainest terms, suppressing any spiritualistic in-ferences or appearance of showmanship. His official report offered only that the radiometer revolved "under the influence of radiation," nothing more, suggesting the very spare hypothesis that light might have a residual force akin to mechanical action.[90] But the wonder the radiometer aroused in "the admiring gaze" of those summoned to judge it at the Royal Society soi-rée of April 7, 1875, was its own kind of threat. Once again, Crookes was found tugging at the keystone of mechanical theory, but this time without the same unified and strenuous objection. His nemesis, William Benjamin Carpenter, raised the alarm about radiant matter and did his best to rekindle doubts about Crookes himself. Was this another attempt at mystification by Crookes, one even more insidious than the scientific pretense of psy-chic force? In truth, when Crookes privately demonstrated the radiometer to trusted friends, the mystical entwinement between the light mill's sponta-neous motion and a psychic force did manage to assert itself. After witness-ing the phenomena firsthand, Francis Galton, whose interest in spiritualism had somewhat cooled at this point, wrote to his cousin Charles Darwin with those hopes somewhat revived on the strength of radiant action: "What will interest you very much is that Crookes had needles of some material not yet divulged which he hangs *in vacuo* in little bulbs of glass . . . Now different people have power over the needle and Miss F (medium) has extraordinary power. I moved it myself and saw Crookes move it . . . Crookes believes he has hold of quite a grand discovery."[91]

The spiritual curiosity of Galton, however, was not Carpenter's principal concern in regard to this enchanting little artifact. What had disturbed Car-penter most was the willingness of the very "physicists to whom we outsid-ers looked for guidance" to throw over established force laws and eagerly take the bait. When "radiant force" was finally put to rest by Arthur Schuster in 1877, Carpenter wrote his gloating "I told you so" letter in *Nineteenth Century* (April 1877), putting what he felt to be the scientific disgrace of "radiant force" before a national audience. Carpenter's rebuke was directed toward Crookes, whom he compared to the judges of the Salem witch trials who likewise "under the influence of a theological prepossession allowed themselves to be sadly deluded and deceived," but he also took aim at the credulity of Crookes's fellow physicists, "in whose judgment the greatest confidence was placed."[92] These "eminent physicists" had seemingly gone weak in the knees at the sight of Crookes's radiometer and its mysterious motor power. Carpenter, a physiologist, hailed notably from the life sciences, a stronghold of ideological naturalists who felt it their duty to combat the aroma of mysticism still wafting about science, often detected in the lofty direction of section A. These "physicists on whom we all so much rely" were stalwart against the overt threat of Crookes's spiritual obscurantism and even his psychic force temptation, but could they resist the *sotto voce* siren's call of a mysterious radiant force? It was one thing for Crookes to threaten the public reputation of science with support for spiritualism, but Carpenter now feared radiant force might do worse still: threaten the scientific world-view of scientists themselves.[93]

But even though the Crookesian mirage of radiant force had been dispelled, Carpenter's article celebrating that fact did not produce the anticipated result. It was he, not Crookes, who found himself on the outs. The perceived ungentlemanly animosity of Carpenter toward Crookes earned him a public reprimand from the president of Section A himself during the next meeting of the BAAS, which was deeply mortifying to Carpenter, who was not present at the time. There was nothing particularly new about Carpenter's article. It ran the same old invective previously directed at Crookes about his "lack of discipline and training" and "failure to cultivate scientific habits of thought," only this time to little or no effect. Crookes had managed to present and defend radiant force in a way that at last engaged the sympathy and support of his fellow physicists. No less a person than his section president defended him as an intellectual colleague. Crookes had finally found his location as an institutional insider in the quarantined space of the radiometer; it amounted to no more than a cubic foot and was bound in impermeable glass, but nonetheless it was his own legitimate scientific terrain. Here, Crookes could safely prosecute the mysteries at the border of atoms and energy, matter and spirit, body and mind. Inside this globe that could fit in the palm of a hand, the potential threat of Crookes's impossible physics was intellectually and physically contained, safely isolated in a display case for the mature consideration of disciplined, professional minds that could handle such stimulus, without risk of titillating the popular rabble.

Now began a fascinating series of studies with another original apparatus of evacuated glass, the Crookes tube, except that, instead of investigating a radiant force, Crookes turned his attention to "radiant matter," probing deep into the structures of the atom itself. The basic experimental setup involved a glass cone wired into an electrical circuit at both ends of the tube. At one end, the wire was actually inserted into the tube, streaming electricity into the vacuum. (This was the cathode or negative terminal.) The current traveled through the vacuum to the anode, or positive terminal, at the other end, where it exited to complete the electrical circuit. Different substances were then introduced into the tube to see how they behaved in the exotic environs of an electrified vacuum The climax of Crookes's career came in 1879 at the annual meeting of the BAAS at Sheffield, when he unveiled his apparatus before the astonished assembly, revealing an otherworldly aurora inside its evacuated chamber. Neon gases glowed green, purple, red, and orange, fluctuating with a strange inner life as they danced about in the serene darkness of the tube's rarified atmosphere. In one famous test, Crookes demonstrated a "molecular ray" (cathode ray) that streamed across the length of the tube, bending to magnetic influences and causing the tube's glass walls to phosphoresce an electric green. Crookes thought he was observing gas molecules traveling in a straight path in the manner of light, a behavioral phenomenon unknown to the ordinary laws of matter. (He was actually observing the activity of electrons, a fact that would not be known for nearly twenty years.)[94] To test this linear pathway, Crookes raised an iron cross in the center of the glass cone, blocking the "molecular ray" from striking the far wall. Everywhere the ray struck the glass, it glowed an eerie green, but directly behind the cross its image was mirrored in darkness, much like a shadow cast in ordinary sunlight. Another strange phenomenon was the inky globes of darkness that formed within the field of glowing gas. Crookes surmised that these black formations were composed of invisible matter, so rarified that there was no collision of excited matter in their midst and thus no emission of light. The more rarified the gas, the larger and more energized the globes of "dark matter" became. The implications were irresistible. Were there some exotic states in which matter might be present but still invisible, not unlike the unseen spirits occasionally revealed to the human eye (or even human touch)?

To his audience of astonished scientific spectators gathered at Sheffield in 1879, Crookes claimed to have bottled and put on display nothing less than "the fourth state of matter," predicted, but not proven, in 1819 by his great experimental hero Michael Faraday. Faraday had proposed that the hierarchical progression of matter moved not only from solid to liquid, from liquid to gas, but was capable of yet higher states of transformation that "were as yet undiscovered by man" until now. Crookes, by rarifying matter *in vacuo* and then subjecting it to electrical stimulation, claimed to have created the trigger conditions that propelled matter to this heretofore mythical higher state. He offered into evidence not just a theory but the stuff itself, the fourth state of matter held in scientific captivity within his Crookes tube. "Radiant matter" had proven, through its vivid exhibition of strange behaviors, that it was indeed qualitatively different from solid, liquid, or gaseous matter. Crookes dazzled the scientific assembly with effects so extraordinary that they nearly rivaled the spectacles produced by spiritualistic mediums, expanding the physical horizons of this world so that they converged upon the next. Crookes subtly proclaimed this fact in his speech to the BAAS:

> We have seen that in some of its properties Radiant Matter is as material as this table, whilst in other properties it almost assumes the character of Radiant Energy. We have actually touched the borderland where Matter and Force seem to merge into one another, the shadowy realm between Known and Unknown which for me has always had peculiar temptations. I venture to think that the greatest scientific problems of the future will find their solution in this Border Land, and even beyond, where it seems to me, lie ultimate Realities, subtle, far-preaching, wonderful.[95]

While Crookes kept clear of any explicit spiritualistic analogies in his speech to the assembly, he did affirm radiant matter's proximity to "ultimate realities," positioning it ontologically at the threshold of metaphysics and intellectually as a departure point for theological speculation. But spiritual analogies are not difficult to tease out. This fourth state of matter was a highly exalted state, with a very resistant threshold of transformation. For ordinary matter to convert to this sublimation, it required exotic conditions, such as an electrified vacuum or, in the case of a spirit, for instance, being undead. This made the spectral gases of radiant matter (or specters of a séance) so rare that they were seemingly "supernatural." Likewise, the laws of nature were differently applied to matter in this rarified state, causing it to behave in apparently "miraculous" ways. Radiant matter operated outside the regulatory physics that governed its other three structural modes, taking on characteristics and capacities more appropriate to light: triggering phosphorescent effects, organizing into a "molecular ray," displaying magnetic characteristics and thermal spectrography. Again, ghostly forms were also known to manifest a similar array of actinic phenomena. This plenipotential quality of radiant and spiritual substance extended beyond the imitation of light. Each seemed to exist in a highly generalized state out of which many different forms or behaviors might be conjured. Radiant matter and spiritual forms could both fluctuate between luminosity and darkness, materialization and dematerialization, substance and energy, and back again, mirroring each other's strange physical properties and perhaps providing a physical model for explaining all the miraculously manifested scents, substances, forms, and sounds emanating from mediums. Given this profound mutational ability packed inside ordinary matter, shouldn't physics expect rather than theoretically preempt events that were also extraordinary? Was the notion that a ghost could be photographed really so very far outside the realm of possibility, given his demonstrated physics of the impossible? What about levitation, physical transmutation, a ghostly pulse, and perhaps even a little spectral flirtation? While ghosts might lurk only in the dim lamplight of the séance parlor, radiant matter had been proven under the blazing lights of an exhibition hall and offered its own clear testimony of other realities hidden from view, so why not an "other side" as well? Once again, Crookes had compelled skepticism about skepticism, doubt about doubt, this time not with the legerdemain of a disappearing photo but with an astonishing spectacle that everyone gathered to see.

The Crookes tube founded an experimental outpost at the brink of reality, pressed up against its outer limits, a millimeter shy of exclusion: supernormal not supernatural (a distinction psychical researchers would leverage to their advantage). As such, instead of being an area of interdiction for scientific pursuits, like a psychic force laboratory or a séance cabinet, the Crookes tube became coveted real estate along the experimental frontier of physics. Radiant matter haunted physics as an unresolved phenomenon within its disciplinary structure for the next two decades, lending encouragement to the suppressed idealistic tendencies in science and putting wind in the unfurling sails of psychical research. It did not, however, lead to any immediate theoretical breakthrough. As the historian Robert K. DeKosky observes, "William Crookes is a puzzle to historians of late nineteenth century physical science. Despite his achievements we are forced to ask, 'Why he did not accomplish more?' Why was it not Crookes, say, who discovered x-rays or identified the electron as early as the 1880's?"[96] DeKosky explains that it was Crookes's commitment to the fourth state of matter that kept him so conceptually rigid, but we could add to that argument that this theoretical loyalty was itself compelled by Crookes's deeper, underlying commitment to spiritualism. Radiant matter was a metaphor too tempting to resist, offering a theoretical gateway in academic physics through which spirits and psychics might slip discreetly. And this theory of a "fourth material state" pointed toward not only a philosophically satisfying future science but one that could be applied retroactively as well: such ghostly physics could redeem the lies Crookes told on behalf of Florence Cook by proposing the scheme of a world in which such lies might be true. That Crookes no doubt knew Florence's phenomena were false was secondary to the fact that he ultimately believed, and very much wanted to believe, such things were potentially true. And he needed others to do so as well. This was true for the sake of his public reputation, his personal sense of honor, and his sense of mission before the world. He needed a science that would make this so.

In the wake of the Sheffield meeting of 1879, Crookes rode this rising tide of scientific fame into the 1880s and 1890s as he garnered more and more honors and offices, all culminating in the highest distinctions to which he could ever have hoped to aspire: the Order of Merit (1910) and the presidency of the Royal Society (1913). By mutual necessity, the world of polite science, to which Crookes now firmly belonged, buried his past indiscretions under a conspiracy of denial. No one asked, no one told, in order to protect science's own institutional investment. And yet, ironically, the higher Crookes rose, the more difficult this silence became to enforce, as spiritualists tried to catch hold of his coattails and ride his soaring scientific prestige. As Fournier d'Albe notes, "Hardly a week passes but [Crookes's] name is flourished in the face of a skeptical world, often in support of the grossest fraud."[97] But despite the attempts of Victorian spiritualists to recapture Crookes and regain their foothold in the discourse of science after1874, the movement was steadily dwindling. Spiritualism found itself increasingly diverted to the fringes of Victorian culture, to be taken up by Theosophists, occultists, and seekers of departed souls. Crookes's photo with the ghostly Katie King may continue to make the rounds of the internet as a form of coerced testimony, but what is missing from this snapshot is how swiftly he let go her arm when she began to drag him down, when it became clear that their philosophical and professional trajectories were dramatically divergent. The earlier and somewhat pervasive intellectual and cultural confusion at the boundary dividing his world as a scientific practitioner and her world as an evidentiary proposition of spiritual faith had been resolved.

But while the corporeal spirit of Katie King had no place in the physics of the 1870s, the possibility of a psychic force was not so easily expelled from scientific curiosity, especially in the field of psychology, culminating in the establishment of the discipline of psychical research in 1882. And indeed, the hidden promise of the unconscious fostered an open, speculative stance about the nature of mind through the end of the century, bringing diverse disciplinary traffic to the hub of the SPR and a more tolerant attitude toward its brand of metaphysics through the end of the century. So much so that when Crookes assumed the presidency of the BAAS in 1898, he dared to reclaim his discovery of twenty-seven years earlier, from the very podium upon which psychic force had been previously denounced:

> No incident in my scientific career is more widely known than the part I took many years ago in certain psychic researches . . . To ignore the subject would be an act of cowardice—an act of cowardice I feel no temptation to commit. I

have nothing to retract. I adhere to my already published statements. Indeed, I might add much thereto. . . . I think I see a little farther now. I have glimpses of something like coherence among the strange elusive phenomena; of something like continuity between those unexplained forces and laws already known. This advance is largely due to the labors of another Association of which I have also this year the honor to be President—the Society for Psychical Research. . . . Steadily, unflinchingly, we strive to pierce the inmost heart of Nature, from what she is to reconstruct what she has been, and to prophesy what she yet shall be. Veil after veil we have lifted, and her face grows more beautiful, august, and wonderful, with every barrier that is withdrawn.[98]

Crookes must have felt a great sense of triumph in standing proud upon his record before the congregation that had rebuffed him, and indeed, there must have been a great sense of personal vindication in rejecting their imposition of silence and its implicit judgment of shame. But, as with radiant matter, interest in a psychic force never actually found theoretical consummation within the framework of nineteenth-century physical science despite its three decades of trying (nor did it compel a new paradigm through the overwhelming force of its evidence and argument). And when particle physics and relativity did come along to revolutionize this cosmic worldview, "radiant matter" and "psychic force" were not taken up and validated as theories ahead of their time, but left behind as Victorian relics.

In the case of radiant matter, its obsolescence was not just a matter of failing to recognize the stream of electrons that made up Crookes's "molecular ray." It was Crookes's fundamental attachment to the concrete, atomic concept itself. Matter in the fourth state possessed the properties of light without being dispossessed of its physical anatomy, simultaneously reifying matter at the deepest levels of cosmic ontology while affirming the ability of science to obtain positive knowledge regarding it. (All evidence was visibly on display in the Crookes tube.) But in the realm of twentieth-century particle physics, once again "all that's solid has melted into air": the stuff that makes up the universe turned out to have little substance, and direct knowledge of it was difficult to substantiate. Matter was rather a ripple in a field, a statistical probability, an information wave, the summary of every possible pathway instead of just the one . . . Psychic force too suffered from the Victorian tendency toward the concrete, placing its emphasis on physicalizing a mental force (or mechanizing light or corporealizing the spirit), rather than proposing that matter itself was "full of ideas." (That is, photons that "know" where to go and partner particles that "sense" the change of directional spin without contact.) This seeming insertion of a "subjective" element into the determination of physical processes (an oblique analogy at best) inspired plenty of speculation about a collective unconscious or holographic universe. Mental influence or "psychic force" did not readily translate into the mystical idiom of the new-age statistical and quantum models that came to dominate parapsychology.

For all this, Crookes still provided the launch site for the modern physics that left him behind and so upheld his pride of place in the history of science. The Crookes tube, and to a lesser extent the radiometer and his psychic force apparatus, were the imaginative acts of a gifted intellect, arising perhaps from the necessity of a "limited education" that directed his genius toward the experimental rather than theoretical and mathematical realms of physics. However, this let Crookes get just clear enough of the "clear ideas of the naturally possible and impossible" to physically manifest the wonderland within nature, even if he could not fully understand it. This quantum universe turned out to be far stranger than anything permitted by those who deprecated Crookes's curiosity and far closer to the "borderland where Mater and Force seem to merge into one another" than their classical physics would allow.[99] Perhaps without Crookes's "peculiar temptations" the reality of a subatomic world would have remained as impossible for scientists to imagine as it was for them to believe in a corporeal ghost.

CHAPTER THREE

Romancing the Crone

Frederic Myers, Spiritualism, and the
"Enchanted Portal to the World"

On the banks of the Niagara River, the falls roaring just above him, a young man of twenty-three regarded the raging waters beneath his feet, and then jumped in. His name was Frederic Myers, and though he could not know it at the time, he would survive this jump to become, by the century's end, a man of great contemporary renown, one of England's leading psychologists, and the visionary pioneer of psychical research. He had recently graduated from Cambridge University with great distinction, had wealth and connections, and was already a highly regarded poet for whom considerable acclaim was expected in the future. So why jump? Looking back on that evening in August 1865, Myers explained his frame of mind. His was not a "leap of faith" but rather a leap precipitated by the absence of any faith whatsoever:

> As I stood on a rock, choosing my place to plunge into the boiling whiteness, I asked myself with urgency, "What if I die?" For once the answer was blank of emotion. As I plunged in the cliffs, the cataract, the moon herself, were hidden in a tower of whirling spray; in the foamy rush I struck at air; waves from all sides beat me to and fro; I seemed immersed in a thundering chaos, alone amid the roar of doom.[1]

The "roar of doom" Myers heard in his ears came not only from the crush of waves without but from a great confusion within. Myers's jump was neither theatrical display (for which he had been known as an undergraduate), nor noncommittal suicide. Myers jumped, in fact, to save himself. This curate's boy turned Byronic princeling, who strutted his way into the fashionable world of Cambridge bloods and boatmen, a poetic prodigy who thought himself as blazing and conspicuous as "the rising sun," now, curiously, did not care if he lived or died.[2] Just the year before, Myers had swum the purple seas of the Greek isles, abandoned to a youth that was "drunkenness without wine" and a hedonism unchecked by the reach of scandal. But now, upon his return, Myers felt strangely emptied in the wake of that lost rush of experience. Rather, at that moment, Myers remembered feeling "numb to all thought of past and future."[3] This condition was so alien and unnerving to this poet, who consecrated feeling as the highest human condition, that he had resolved to risk his life in order that he might exalt feeling again.

Myers looked back on this early period of depression as "my only subjective key to the indifference which I observed in so many of mankind."[4] Though this might be a presumptuous judgment to make about other people (perhaps they were merely less demonstrative than he), the equivalence Myers drew between failing to feel and failing to live arose from deep philosophical commitments that gave continuity to the various phases of his life. The youth standing at the foot of the falls felt that it was passion alone that authenticated one's life, propelling one above the listless mediocrity in which so much of mankind was content to eke out their existence. Passion: this was not the nicety of feeling cultivated by Victorians as "sensibility," nor the rustic sensuality of the lower orders, but the roaring fires of a poet's ardent emotionality, the bodily sensation of being spiritually alive. Myers declared himself in word and deed on behalf of a Romantic mode of being.

It is only Myers's thought, however, not his "mode of being" that scholars today generally classify as Romantic (though both could be said to be true). Frank Turner describes Myers's posthumous opus, *Human Personality And Its Survival of Bodily Death* (1901) as less a scientific treatise than the "last great manifesto of English romanticism."[5] Robert Goldstein sees Myers's psychology as drawn from what he terms "romantic sources," and likewise, Alan Gauld sees in Myers's psychical science, "the late nineteenth century persistence of romantic thought."[6] I would like to give deeper consideration to those "romantic sources" by exploring the full resonance of Romanticism in Myers's life and work. Historians tend to separate the different strands of Myers's life, isolating his science from the religious, personal, literary, philosophical, and even geographical aspects of his biography that give this Romantic valuation its meaning and texture. Without this context, the category "Romantic" itself is problematic, evaporating the boundaries of its own significance as it moves away from delineating a fixed literary movement (initiated by

Schlegel's manifesto in 1798 and concluded with the death of Byron in 1824) to absorbing a variety of "tendencies" downstream of what has been labeled the counter-Enlightenment.[7] In Myers we see not just "tendencies" but powerful intentions and traits of personality that tie him to the original literary initiative as well as to its larger cultural project, which was above all a cognitive revolution. To make these connections clearer, I propose a more organic approach to Myers's thought, demonstrating his deeply personalized biographical and philosophical roots in the early nineteenth-century project of literary Romanticism, and showing how his own poetic style of consciousness gave rise to the Romantic architecture of the subliminal self.

If we accept that Myers's theory of the subliminal self is a romanticized psychology, to what limited extent can we attempt to correlate that to Myers's own Romantic "psychological makeup"? First, it is necessary to recover the specific ways the term "Romantic" has been defined and how it might operate here. To a surprising extent, Myers, as both a historical and historicized subject, occupies a complex conjunction of Romantic modalities. In its broadest sense (as applied by Turner, Gauld, and Goldstein), Romanticism is a metahistorical handle used by scholars to describe those vectors of modern secularization that carry forward the impulses of the Romantic period. Of the important early theorists who abstracted these Romantic values from "the Romantic period," Jacques Barzun characterized its central aspect as a "flight from reason" and a "reaction against the mechanistic, rationalistic world view of the eighteenth century," while René Wellek emphasized a shift in epistemology toward mythopoetic structures of knowledge, giving primacy to the imaginative versus cognitive aspects of mind.[8] The Enlightenment critic Edmund Burke cued this Romantic reorientation early on by reimaging nature as an object of aesthetic terror rather than of practical utility, repudiating the harmonious rationalism of the *philosophes'* nature concept for the *sturm und drang* of the sublime. But perhaps the most insightful summary of the Romantic enterprise, and most relevant in regard to psychical research, is M. H. Abrams's contention that it was, at its core, a secular restatement of religious values within the framework of the human mind: "Much of what distinguished writers which I call romantic derives from the fact that they undertook, whatever their religious creed or lack of creed, to save traditional concepts, schemes, and values which had been based on the relation of the creator to his creature and creation [the Christian triangulation of God, man, and nature] and to reformulate them within the prevailing two-term system of subject and object, ego and non ego, the human mind (or consciousness) and its transaction with nature."[9] Divinity and its ideals were not deleted by Abrams's version of this subjective schema, but rather were reabsorbed within a humanistic, "self-centered" metaphysics, the hallowed center of which was consciousness itself (Schlegel's inner mental life, or *geistige Leben*). It awaited Myers, who was born a poet but died a philosopher, to abstract these romantic ideals of consciousness from their aesthetic embodiment in literary romanticism, painting, and the plastic arts, and reconstitute them in the intellectual format of a Romantic science of mind.

Much of this early conceptualization still informs how we understand the term "Romantic." Despite the difficulty of delimiting its meanings as it drifts into its late, post- and neo-Romantic iterations over the course of the nineteenth century, the overarching idea of a unified Romanticism persists.[10] It encompasses the exaltation of emotion over reason, sensuality over intellect, genius over learning, ecstasy over sobriety, and the monstrous over the normative, all the while striving to access an immanent metaphysics through the body, the senses, the occult, the mysterious, the beautiful, and the strange. The Romantic project was originally worked out in the realm of aesthetics, as the artistic gaze turned inward, away from religious subject matter and nature's outer forms and toward the *noumin* hidden within ourselves (to be later disclosed in the psychical science of mind). The once mimetic object of art was replaced by the subjective process of creation itself, capturing an instant of "the ever creating primal energy of the world which begets and actively produces all things from itself," surging upward to take artistic form.[11] Such were the sources Myers laid open to the subliminal self as it traveled the physicopsychical continuum beyond the limits of rational awareness and onward toward the World Soul.

To varying degrees, all these elements of the Romantic narrative find resonance in Myers's personal life and professional pursuits, and even achieve a kind of ultimate consummation in the monstrous mysticism of his aberrational psychology. Given the strong subjective record Myers has left us in his poetry, letters, diaries, and other "self"-revealing texts, we can move beyond analyzing the intellectual object of his philosophy (the subliminal self) and seek out instead a more complex paradigm, one that relates Myers's strongly affective, self-fashioning nature to the formal outcomes of his thought. As a young man, Myers had a way of being in the world that both consciously enacted and viscerally embodied Romantic tropes, in defiance of both religious and class conventions. In these chosen signs to mark himself as poet, artist, sensualist, genius, and nonconformist, we can see how deeply Myers was "romantically" self-identified and how systematically he tried to correlate this public persona and his personal sensibility. Thus Myers's Romantic selfhood is both a social and a psychic formation, staging the process of modern self-development from the outside in, by drawing on the inherited cultural conventions of Romanticism and from the inside out, by "recognizing" and internalizing these conventions as an original part of himself, drawing these Romantic sources deep into the psychic framework of his evolving point of view.

With this bidirectional self-constitution, Myers fully instantiates Marcel Mauss's *le moi*: the modern, self-positing selfhood emerging around 1800, but given its full latitude in the privileged, poetic life of Frederic Myers. Mauss saw the more traditional category of the person as being derived mainly from external, societal sources (a role one is given to play by society), while the revolutionary, modern *le moi* derives from mainly internal mental processes (the self one posits for society that is ultimately rooted in one's own being). Myers was born in 1843, coming of age in the mid-nineteenth century, when the mental landscape of Romanticism and the civic landscape of liberal individualism came together fruitfully to synchronize the subjective and objective autonomy of this self-determining selfhood. Thus, despite the obvious tension between liberalism's empirical cognitive orientation and Romanticism's lyrical one, together they made possible the expanding spheres of civilian and psychic life within which Myers operated.

The solitude that marked Myers's boyhood likewise furnished the new mental and ideological space of privacy that was itself, according to anthropologist Michael Carrithers, fundamental to the formation of Mauss's *le moi*. This subjectivity was largely enabled by "the relatively autonomous, intellectual traditions characterized by romantic, contemplative, and solitary conceits of selfhood."[12] From this praxis of solitude and self-absorption, seeded by Jean-Jacques Rousseau in the 1760s, arose the quintessentially modern attribute of reflexivity, which came to full bloom in the nineteenth century. Myers often found himself alone in exactly this sort of literary quietude, disappearing with a book for hours on end. He was homeschooled on a large estate with his widowed mother, and here, the logistics of domesticity offered Myers not just privacy but also freedom from peer pressure. It is doubtful that Myers would have felt quite so free to enjoy his excesses of individuality had he not evaded the brute, homogenizing force of male socialization at boarding school. But with only a doting mother and two younger brothers as his primary feedback loop, the precocious Myers was left free to dream himself from textual sources.

Thus, as a young poet and philosopher moodily absorbed in his reading, writing, and reflecting, Myers must have been engaged in a very historically specific kind of mental activity: the cultivation of inner life. This was, throughout the nineteenth century, an ever-growing expanse, first tracking through the discourses of mesmerism and literary Romanticism and then coming to dominate the landscape of late nineteenth-century psychiatry, which was preoccupied with the subconscious mind (what the Romantic poet Coleridge had earlier labeled the "under-consciousness"). It is no coincidence that paralleling the cultural development of *le moi*, a rash of psychological, mesmeric, and literary narratives arose to try to label and give voice to an inchoate inner self arising out of the new facts of reflexivity. (Just compare the explicit, externalized self of the eighteenth-century epistolary novel to the vast interiority of the Victorian novel from Jane Austen to Henry James, books rife with private musings and moral anguish as the modern subjects become increasingly trapped in endless conversation with their once silent partner.)[13] With the "mad rush" of psychiatry into the subconscious mind in the last quarter of the nineteenth century, modern culture began scrambling in earnest to explain its self to itself, for the first time, formally directing disciplinary science to console, classify, and, if necessary, contain the *moi* of the innermost mind. It is fitting that Myers would initiate his own attempt within psychology to codify this terrain in his own Romantic self-image, propelling his "self"-determination into the cultural and historical arena through the theory of the subliminal self.

There is already an abundant scholarship linking modern subject formation to the historical processes of Romanticism, but the rousing novelty of so self-confessed a figure as Frederic Myers is that he consciously enacted this linkage in his life *and* further reified this self in his scientific work, literature, and letters. He is what Jerrold Seigel might call a "concrete" subject, attaining a degree of autonomy, agency, and even originality through the robust "ontogenetic complexity" (Seigel's term for multidimensionality) of the Victorian self-concept. Seigel identifies three main traditions for conceptualizing the self in Western intellectual history: the social-relational self (Mauss's persona), the phenomenological or self-positing self (Mauss's *le moi*), and the biological self.[14] Only a cultural model that integrates the historical activity of all three obtains a real existence and a measure of autonomy—that is, one needs a *moi* that is multiply determined if one hopes to have a *moi* at all. A notion of the self that is conceived with only one of these dimensions—fully instructed by society (relational), fully programmed by language (positional), or genetically determined by biology (material)— lacks the dialectical capacity to stimulate new unpredictable, irreducible psychological formations. (As such, it would devolve into social, solipsistic, or mechanical varieties of determinism, respectively.) Seigel uses this schema to contrast varieties of French and British subjectivities, rooting modern episodes of French determinism (in both its eighteenth-century mechanistic and twentieth-century linguistic varieties) as the cyclical resurgence of an authoritarian political tradition. Thus, despite the appearance of a revolutionary *frisson* in Julien Offray de la Mettrie's *L'homme machine* or Jacques Derrida's deconstructed self, the submission of the self to a single authority marks the eternal return to absolutism. *(Plus ça change, plus c'est la même chose.)* In contrast, the relative freedom of the English political, religious, and social experience supported more complex or more concrete notions of selfhood, historically resisting such fashionable determinisms in their various political and intellectual guises.[15] Psychical research, despite its seeming departure from empirical norms, affirms Britain's more philosophically moderate tradition by holding onto, and even amplifying, the ontogenetic complexity of the subject (and thus its sources of self-determination). The British program offered a collaborative counterpoint to Charcot's strictly neurophysiological interpretation of trance phenomena, sustaining possibilities for the unconscious otherwise redacted by the materialism of continental medical psychiatry.

British Romanticism modified the rational determinism of the French Enlightenment and returned a more complex, concretized modern subject mainly through the augmentation of Seigel's third category: the material or bodily self. While Enlightenment utilitarianism gave its own positive evaluation of the body, making pleasure a source of moral authority and not original sin, the lyrical poetry of British Romanticism further elevated the body's importance by endowing emotional and physical sensation with aesthetic intuitions that amounted to a somatic spiritual intelligence. Thus, the Romantic sense likewise trumped carefully attuned sensory empiricism. John Locke's notion of reflection merely allowed the senses to passively inform the intellect about the outer form of nature, not its inner life. (William Wordsworth's "Boat upon the Thames" is at once immersive and transcendent, penetrating the fleeting, physical moment to find nature's indwelling spirit.) This embodied self likewise strengthened the other dimensions of selfhood, enhancing the relational aspect of individuality by putting greater emphasis on the self-dedication of love and friendship (as opposed to social and matrimonial alliances driven by family or business interests). And personalized and intuitive forms of knowledge amplified the positional dimension by rooting truth in one's gut or one's being (not just in the brain and its universal reason). All this moved the modern *moi* forward along the path of secular humanism in important ways, promoting the cultivation of subjectivity as much as of objectivity in the effort to understand nature and its implicate order. Reality was to be discovered as much in self-contemplation as in outer observation. Attention shifted inward to understand what lay beyond. Whether it was Coleridge's transcendental self or Wordsworth's more sensual "egotistical sublime," self-consciousness, anti-self-consciousness, or visionary solipsism, the leading edge of civilization had become self-absorbed in its effort to attain self-transcending knowledge.[16] In order to escape the limits of one's psychobiology, one first had to tunnel through it, "for in that dark deep thoroughfare had nature lodged the soul, the imagination of the whole."[17]

Myers's own tendency for gloomy introspection was itself a recognizably Romantic form of self-involvement and knowledge-seeking, sending him sailing off the ledge of the Niagara to force a deeper revelation and restore life's meaning. That episode of personal despair plays out against the wider struggle of this modern, self-positing self to adequately provide the source of its own spiritual, intellectual, and ideological authenticity and the need to disclose the hidden means within oneself to do so, whatever that might be: true grit, moral character, rational discrimination, a religious soul? Or perhaps it was an immanent metaphysic sunk in the depths of consciousness in the form of the subliminal self. The heterogeneity of modernity keeps acquiring new options.

In both the occupational self-consciousness of Myers the poet, and the occupational consciousness of the self of Myers the psychologist, we see the ongoing excavation of this inner self of modernity and the opening up of its depths and diversity, its struggles to adequately provide the source of its own authenticity. This makes a consideration of our "selves" also part of the process of this historical and methodological inquiry, embedded as it is in these Victorian processes of "self"-fashioning and schematized by Seigel's productive constructionist conceits. But might we go beyond Seigel's essentially ideological notions of the self, determined by various narratives of self-representation, to something perhaps more basic? Something that concerns more than just ideas about the body, but that is biological in its own right? Recent developments in literary studies have taken just such a biological turn by drawing on cognitive science to "hardwire" structures of narrativity into the brain. These bio-epistemological theories claim a certain material solidity for the narrative self by suggesting that innate neurological schematics pattern the ways in which the mind can interpret and represent "reality." This "wet brain" model of cognitive literary theory offers some basic support for the notion of an elemental, inborn narrative agency in Frederic Myers, one that acts upon the world, rather than being merely acted upon.[18] Such a notion confers "concreteness" upon the historical subject in the form of Seigel's substantive narratives *about* the body (to which Romanticism was an essential contributor), and also extends "ontogenetic complexity" to include a narrative role for the body itself. In the case of Myers, then, can we extend the category "Romantic" beyond the social and intellectual, beyond the cultural and contingent, as a matter of temperament? In other words, to what extent were there Romantics before there was Romanticism? This is not to biologically reify the cultural concept "Romantic," but merely to correlate its significance to something prior to that term.[19] This "something prior" does not signify a life outside of history for some "real, flesh-and-blood" Myers, but it does give us a way to ground his participation in it. His intensity of feeling is particularly important to this study because of its power to mobilize other cultural and intellectual initiatives. Emotions are in part cultural processes that interpret the data of social experience, but they are mediated through the body and directly invest the world with a significance that belongs to the self alone. Through feeling, we find that the world is no longer an intellectual abstraction, but enters into the body and becomes real. Even with a phenomenological understanding of this "reality," where everything experienced as the external world is essentially synthesized within the mind, one's conviction in this illusion requires a mysterious, multivalent process of subjectification that goes well beyond intellect. This is an understanding of consciousness perhaps far better observed by the Romantics than the *philosophes*.

But that still leaves the question of exactly how (and by "whom") all these physical, social, and mental data are subjectified as "individual experience." Why did Myers have these particular feelings, impressions, and passions, but not others? Given the endless complexity, randomness, and indeterminacy of potential psychic formations, how does the dynamic core of consciousness, as described by neurobiologist Gerald Edelman, render something that is highly specific? Here is the heart of human agency broken down by Edelman to its most basic operation: the "active reduction of information" by which the brain takes a formless field of data (that is not us) and focalizes it into a gestalt point of view (that is us). In other words, Myers's mental impressions of his world (Edelman's "qualia"), were not just made, but taken.[20] The interpretive function of the brain or mind splices a shiver of indeterminacy between the acquisition of information and the synthesis of meaning: to choose, to prefer, to prioritize (what Charles Taylor calls the ability to impose a hierarchy of significance) is how experience becomes personalized (and how the phenomenon of personhood is experienced). It is the essence of "subjectivity." In this manner, Edelman's dynamic core reconstitutes this subject like a continuous cat's cradle pulling on millions of balls of string, all with millions of potential configurations. Such are the "functional clusters" of neurons that scaffold the open-ended array of unique conscious states. (Interestingly, Myers's own view of a self in constant flux, centering and recentering upon different neuroanatomical foci, somewhat anticipates this idea.) Taken individually, they structure the experience of existence; taken serially they become the narrative of identity. Whatever one calls this mental process of subjectification, it is far too complex, spontaneous, and open-ended to be rendered deterministically; rather, it requires a moment-by-moment "will to consciousness" initiated from within our own biology and sense of our own existence. The "qualia" we experience, according to Edelman, "rest in our own embodiment and our own phenotype," and operate beyond the programmable limits of either culture or biology.[21] Thus, there is a deeply personal, spontaneous, and provisional aspect to the generic processes of consciousness that enable the "self" to bring itself into existence, reviving the ghost of individuality that guides the transformation of "matter into imagination."[22]

This active, self-constituting subjectivity remains fundamentally unique and fundamentally underexamined, and yet to some extent it drives the historical processes from which the self has generally been excluded. In the case of Myers, a very definite disposition asserted itself in childhood and remained a directing, purposive force with historical reach, claiming new ideological and cultural space on behalf of his own cast of consciousness. This agency roots itself in the unique act of being as well as in the cultural sphere of thinking and feeling. To historicize Myers is to take into account the complex sources of his volition, which are autonomous and contingent, biological and interpretative, belonging to Myers though taken from the world around him. Between his private thoughts and publicly held convictions, as well as in the anecdotal glimpses drawn from friends and foes, or in the meaning of a locket, a poem, a letter, or a drawing, we find modest clues to a self that yet survives the annihilation of history and a spark of life added to our heuristic models of the subject. This allows for a more embodied approach to intellectual history, which accounts for the contributions of "inner life" (the title of Myers's memoirs) and the burden it imposes on ideology to effect what Clifford Geertz called the "symbolic fusion of ethos and worldview."[23] This attunement between "life as it is and life as it ought to be" correlates social conventions and moral codes into a single, seamless logic, but it does not just happen at the level of the cultural system. There is considerable fine-tuning exerted within that system by individuals particularly driven to advance their own version of what is and what ought to be, explaining the unprecedented dynamism of the modern age of *le moi*. Myers, as an endogenous element of his own cultural self-constitution, was such a one. This dynamic does not presume to settle the ontology of selfhood as either mainly biological or psychic, or mainly social or cultural, nor does it make any final claims about the nature of individual autonomy, but it does find some connection between conative drives and cultural formations (and vice versa). Myers, by his character and by the circumstances of his birth, was able to fully occupy Mauss's expanded nineteenth-century notion of the self, excavating its inner depths and exploiting its social latitude. This self has constituted itself as an object at the center of its own world rather than as a "place-holder" in the social pyramid, and has claimed on its own behalf authentic powers of self-determination.[24] For Myers, this extended beyond personal or social identity to include a reconfiguration (and further expansion) of the modern self-concept.

This is not to say that Myers's theory of the subliminal self factored heavily into the future of psychology. It did not, having been phased out in favor of more biological and pathological models of the unconscious (Freudian psychoanalysis), and more deterministic models of human behavior (Watson's behaviorism). But this expulsion from medical psychiatry in the early twentieth century does not necessarily signal failure. It was not the mentally ill person whom Myers's psychology aimed to cure, but rather the philosophically degenerate self. Compared to the hysterical, neurasthenic, neurotic, and dispirited late nineteenth-century psychiatric patient on offer by his theoretical contemporaries, Myers's psychical self had spectacularly enhanced mental faculties. These included the ability to alter one's anatomy, transmute foreign objects, perceive another's thoughts, foretell the future, access our evolutionary past and even guide its forward development. Even if only at the level of suggestion (operating mainly at the level of possibility rather than convinced belief), the vast inner space of the unconscious mind, with its hidden potentialities, continues to enlarge what Charles Taylor called "the sources of the self" for the modern imagination. This unknown, all-knowing psychical subject instantiated itself as a free-floating asset to be taken up in the modern *bricolage* of personal identity in the late nineteenth century, supporting religious and secular intuitions of inner selves, higher selves, and collective selves that still populate our contemporary speech about ourselves. Myers's psychical subject enhanced modern "ontogenetic complexity" by enlarging dimensionality in multiple directions: such an assumed self is endowed with sentient inner depths, greater interpersonal connectedness (telepathy), a physioplastic body, and even a purely vertical fourth dimension, that "immortal" aspect of mind capable of fully transcending time, space, and matter. It resists reduction to neurochemistry by modern psychiatry and the threat of eclipse by artificial intelligence through technology, and thus, even as an illusion, it makes for more concrete, more human subjects even amid the dehumanizing scale of modern life with its megacities, exploding populations, mass markets, and the ever-diminishing "inner life" claimed by screen technology (the new object of our contemplative gaze).

Myers returns us to a potent moment in modern subject formation, when the secular, self-determining and *sui generis* subject, in which the massive enterprise of the twenty-first-century global society has its basis, comes into its own as the dominant cultural actor in art, medicine, literature, politics, science, commerce, and so on. The story unfolds not just in emerging urban settings and the new scale of commercial and intellectual cooperation, but also in the quiet of an idyllic country estate, in new forms of textuality and contemplation. Deeper consideration of Myers's boyhood and early character formation puts flesh upon his ideas by way of biographical "background," showing how the subliminal self proceeds from this particular past and this particular "personality." Any such study of Victorian selfhood (and subliminal selfhood) also returns us to the source elements of our own received notions of personhood, relating who we are to who we were, in a way that illuminates how we are changing.

"FREDDY"

In *Providence and Love,* John Beer maps Myers's childhood onto the cultural and geographic landscape of early English literary Romanticism, the fertile Lakeland country of Myers's birth and imaginative beginnings.[25] As Beers points out, the connections between Myers and English Romanticism were not just "ideas and influences" limited to a boyhood curriculum of Wordsworth, Keats, and Coleridge, but were as substantial as the land itself. Myers grew up in Keswick, Cumberland, not far from Wordsworth's own haunt of Rydal Mount in Grasmere, and his mother's family had close personal ties to the poet, described by Myers as "being amongst his most appreciative friends."[26] Myers was born in 1843 at the Keswick parsonage, the first of three boys, an indulged and precocious child, whose idyllic youth is mistily recollected in his autobiography:

> It was in the garden of that fair Parsonage that my conscious life began. *Ver illud erat* . . . And even with my earliest gaze is mingled the memory of that vast background of lake and mountain; where [Mt.] Skiddaw . . . guarded the winding avenue into things unknown, as if it were the limitary parapet and enchanted portal of the world. I can recall the days when that prospect was still one of mysterious glory; when gleaming lake and wooded lands showed a broad radiance bossed with gloom, and purple Borrowdale wore a visionary majesty on which I dared scarcely look too long.[27]

A hillside beckoning with mysterious glory, a mountain peak guarding an enchanted portal winding into things unknown, a visionary majesty on which he dared scarcely look too long: these phrases wrapped Myers's memories of childhood in a prophetic view of nature. "Nature for Wordsworth," Geoffrey Hartman writes, "is not an object but a presence and a power; a motion and a spirit, not something to be worshipped and consumed but always a guide leading beyond the self."[28] Wordsworth attends in nature a deeper prophecy, a revealing truth, a guide for the impulse to go beyond—an attitude that Myers reproduces in relation to Keswick. Myers sensed that the material world bordered a deeper spiritual reality, penetrated in moments of climactic natural beauty and grasped only by poet or contemplative saint. This was the voice of Romantic naturalism that had whispered in his cradle. Myers extended the identification between himself and the Lakeland to include the corporal and imaginative being of Wordsworth himself, writing that "he [Wordsworth] was a kind of mystical embodiment of the lakes and mountains round him, a presence without which they would not have been what they were."[29] Myers credited this intimate sense of Wordsworth to his "hereditary friendship." It is unclear if Frederic himself met Wordsworth personally as a boy, but Beer suggests it is likely.[30] Myers's maternal grandfather was married to Jane Pollard, a close friend of Dorothy Wordsworth who visited often, and his father, who met Wordsworth through Mrs. Myers's connection, became one of Wordsworth's personal friends.

Myers's mother shared his strong "love of poetry and natural scenery," but the central focus of his upbringing under her care was religion. Her family, the Marshalls, were a prominent Church of England family with strong evangelical traditions that she instilled in her boys. Myers described his father as "a clergyman who both in active philanthropy and in speculative freedom was in advance of his generation."[31] Rev. Frederic Myers was also a close friend of Dr. Benjamin Jowett, a leading Platonist and church reformer who would later publish a famous challenge to church orthodoxy in his book *Essays and Reviews* (1859). Intellectually speaking, Reverend Myers kept fast company, so we can assume that the Myerses did not confound piety with rigidity. Given the reforming, progressive attitude of both his parents, the evangelism practiced in the Myers household was aroused more by the charisma of Christ than the wrath of God. Instead of lectures on fire and brimstone, Myers's moral and religious instruction included Wordsworth's "Tintern Abbey" and nature walks.

Myers's mother fed the poetic and somewhat ecstatic inclinations of her son, though there were early signs that the young Myers's ego would soon outsize his piety. Mrs. Myers, writing in her journal on October 12, 1847, observed ruefully that "he is not grateful or humble as I should like to see him." But what he lacked in gratitude and humility, he atoned for in aesthetic appreciation. She continued, "It is difficult to refrain from giving pleasure to a being so joyous and so easily delighted as he is. Very simple things are matters of almost ecstasy to him—a new flower, a new walk, a bright moon unexpectedly caught sight of—his lessons if skillfully managed—a new book always—a bit of poetry which takes his fancy when he come to it—these he likes 'so very very much—I cannot tell how much Mamma.' And along with this delight is a proportionate pain when he is disappointed in anything he has set his heart upon—or when his will is crossed or his fancy, in anything . . . We should be quite happy with much less if only his moral development proceed as we should wish."[32] Despite this protestation, there is no doubt that Myers, even as a young boy, was encouraged to have a very high self-opinion, and this, no doubt, amplified the willfulness that Mrs. Myers feared was impeding his "moral development." That same willfulness would later bear fruit in Myers's origination of his own idealized psychology. As Freud famously observed, "a man who has been the indisputable favorite of his mother keeps for life the feeling of a conqueror."[33] Mrs. Myers, who had ample means to board Freddy at any school she liked, preferred to keep him home with her.

Her maternal devotion was made all the more complete by the early death of the family patriarch when young Myers was just eight years old, leaving him, as eldest male, the head of the household and uncontested center of her world. The end result of this upbringing was an almost missionary entitlement to save the world. Myers's religious instruction began before he was three years old. As a child, every day, every action, every deed was guided by an immediate and loving Jesus ordering his daily routine—with an easily disgruntled God featuring in the background. Among Myers's papers, there is a mission statement written in the careful hand of the six-year-old Freddy:

June 18 1849
1. To strive more to please God and to be his true servant.
2. To follow Christ and follow dear Papa as he followed Christ.
3. To have more of that love which St Paul describes in I Cor. XIII
4. To be self-denying.
5. To remember always that God sees me and that I shall have to render an account of all my works to HIM
6. To keep more watch over myself—and remember that God will help me to over come my badness

. . .

12. To remember that NOW is the pointed time and that it may be too late to repent even today.[34]

This seems a rather rigorous list of chores for a six-year-old: emulating the Savior, practicing asceticism, maintaining an ongoing Judgment Day dossier to ready himself for an imminent death. According to his mother, Freddy, in a move that would characterize his intellectual practices throughout his life, merely framed a better alternative. He was able to fashion a kind of fairytale religion and leave out the more draconian aspects of his tutelage. "Freddy delights much in dwelling on the idea of Heaven—'all people good, none naughty'–& as happy as if it were always Christmas Day . . . Indeed he seems to have a sort of realizing vision of angels which makes them quite familiar to him—'I had such a beautiful dream Mamma, I saw such a bright angel standing near me.'"[35]

Yet, nonetheless, God was a morally demanding and vigilant presence in young Frederic's life; the evangelical preoccupation with spirit, salvation, and the afterlife created in him a collateral obsession with death.[36] Though the catechism that shaped Myers's upbringing did not emphasize a fallen world and a roiling hell, experience could not long sustain Frederic's childhood attachment to a uniformly benign universe overseen by a loving God in a wonderful place called "Heaven."

When he was six years old, Myers came upon "a sad little mole," left crushed and lifeless in the tracks of a cart. Seeking consolation, he turned to his mother: "I do think that little mole's soul is gone to heaven—don't you?" She then advised him, with overly frank regard for the teachings of the church, that "the mole had not a soul at all."[37] This revelation not only deeply shattered the boy, but scarred the man as well. "To this day I remember my rush of tears at the thought of that furry innocent creature, crushed by a danger which I fancied it too blind to see, and losing all joy for ever by that unmerited stroke. The pity of it! The pity of it! And the first horror of a death without resurrection rose in my bursting heart."[38] And so it would remain, this *idée fixe* of death present from his earliest youth. Myers would have preferred eternal damnation for the mole if it preserved the self from extinction. Whether the predominating sentiment here is one of genuine empathy for the mole or a more selfishly inclined dread of personal extinction, it speaks to a highly imaginative, emotional boy fully engaged with the human condition in a way most children are not. Yet, despite this sensitivity that made the world so pressing for Myers, he was not one to retreat or retire from life. Rather, he would fully dig in, push back, and impress himself upon the world. That death everlasting should sweep off the mole was bad enough, but that this terrible fate might apply to humanity as well was intolerable. That there could possibly be a final and absolute cessation of the self, was such a profound breach of the divine contract that Myers's Christian faith suffered an injury that would eventually prove fatal.

Even as "belief in God fell away," Myers carried forward the indispensable notion of a personal salvation, and this surviving "self" would resurface as a central feature of his psychology.[39] When considering the origins of this "self" infatuation, we cannot dismiss the psychodynamics of having a God acting as a sort of third omnipotent parent, taking a keen interest in all one does. The degree of Jesus's involvement with the daily life of the infant evangelical Freddy could not help but impart some of his cosmic significance to the boy's own self-conception. Add to that the besotted love he received from his mother, and one can see how the little Freddy outgrew the humble devotion of a country parson's son and became a seeker of his own religion. With the attendance of his own "domestic angel," Frederic Myers, like many men of similar upbringing, must have felt a little bit like a God himself, given to theologizing, moralizing, and eternalizing. In his theory of the subliminal self, Myers customized the idea of God to fit more in accordance with his worldview, retaining cherished sentiments of his evangelical childhood (immortal souls, seeking selves, a loving God) while unburdening himself of its interfering dogmas.

When his father died in 1851, Myers found that the consolations of Heaven were unsatisfactory. This was especially true for his mother, whom Myers observed sitting alone for hours in the kitchen, always in tears, repeating "no more happy times." This would overburden the heart of any son, let alone one as quixotically devoted as Myers. Myers began to question the desirability of a Heaven whose gates so firmly parted the bereaved from the beloved. If as a child, and the eldest son, a helpless Myers watched his beloved mother suffer inconsolably, is it any wonder that he should spend much of his adult life trying to reach beyond death to reunite those so parted? Myers wanted a better deal for himself and for humanity.

With his father removed as head of household, Myers, already a "willful" boy by his mother's account, received perhaps greater latitude in his own self-development than was typical of his class and age. He took to writing poetry and reading Plato and Virgil, while his father's core curriculum of clerical classicism fell by the wayside. (Myers appears to have more than fulfilled Mrs. Myers's prescient fears for his moral development.) At the age of fourteen, he placed a prodigious first, second, and fourth in the Chettleham College English verse competition. Upon these laurels he rapidly acquired a sweeping grandiosity. At sixteen the highly original Myers experienced what he called a "conversion to Hellenism," effecting, in his words, "a Copernican Revolution" that displaced the Christian worldview at the center of his universe with a classical one.[40]

That Plato should become young Myers's moral, spiritual, and intellectual guide made both personal and intellectual sense. A pagan philosophy allowed him to bypass the overwhelming moral authority of his clergyman father and evangelical mother in favor of self-rule by literary proxy. He went from being God's chore boy to a Promethean poet, leaving little wonder as to why "Plato [was] held to be little better than a misleader of youth" by his parents' generation.[41] But to overidentify Myers's classical posturing with some reflexive rebellion is to miss its authenticity. This was a precocious claim (in terms of both his chronological age and evangelical era) to his right of personal authenticity. "Classicism," Myers wrote, "was at that time the intensification of my very being," arising from his inner delight and nurtured by his profoundly gifted facility for languages and literature. Myers's Platonism arose out of, not against, his complex familial and educational history at Keswick, upholding through Plato's idealism the essential Christian teaching of spiritual transcendence. Yet, pagan metaphysics could more easily accommodate Myers's Romantic embrace of the senses and the seeking of mysticism in the here and now. What it rejected was Christian self-abnegation, a point of both religious and scientific philosophy that Myers would pointedly exclude from his account of psychical subjectivity.

Plato offered Myers a humanistic revelation irresistible to the increasing scale of his self-devotion. In the disapproving words of Macaulay, "The aim of Platonic philosophy was to exalt man into a God."[42] But perhaps even more important, considering Myers's acute sensitivity, Platonism was a way to overcome the remoteness of Heaven, offering instead the notion that the unity of nature concealed a hidden ideality all around us. This higher order was not apart from, but eternally present in, the mystical-material structures of the universe, healing Myers's divided world. Myers's uses of Platonism narrowed the space between God and man, Heaven and earth, and—of greatest urgency for Myers—the living and the dead. His youthful Platonism was the first step in a lifelong resistance to the troubling duality catechized in the Aristotelian roots of church doctrine, which had too firmly cast apart the terrestrial and the celestial realms. This existential chasm runs deep in Western intellectual history, with Cartesian dualism somewhat restating the salient aspect of this spiritual divide. But with the new scientific preserve of the Newtonian universe, there was hope, for thinkers like Myers, that all existential potentialities—even metaphysical ones—might be folded into a single domain and made subject to scientific consideration.

Plato also gave formal articulation to Myers's burgeoning poetic intuition, as nurtured by a Wordsworthian view of mind in nature. "Nature for Wordsworth," writes Geoffrey Hartman, "is not an object but a presence and a power; a motion and a spirit, not something to be worshipped and consumed but always a guide leading beyond the self."[43] Wordsworth attends in nature a deeper prophecy, a revealing truth, a guide for the impulse to go beyond. In this sense, Myers embraced not his father's classicism, but a specifically Romantic Hellenism. Here the classical discourse shifts away from its earlier political emphasis on Republican Rome and its religious emphasis on biblical exegesis to a purely modern aesthetic basis.[44] Myers's philosophical appreciation of Plato operated as an artistic as much as a philosophical principle. In his biography of Wordsworth, Myers linked the personal influences of the sage and the poet in his own life through their shared insight into a mystical, underlying unity binding man, spirit, and all creation: "To express what is characteristic in Wordsworth we must recur to a more generalized conception of the relations between the natural and the spiritual worlds. We must say with Plato that the unknown realities around us, which the philosopher apprehends by the contemplation of abstract truth, come in various ways obscurely perceptible to me under the influence of divine madness, of an enthusiasm which is in fact inspiration."[45] This confluence of madness, genius, beauty, and inspiration/revelation was already a familiar constellation in Romantic literature, a ready resource for Myers's adolescent aesthetics in the mid- to late 1850s. By the end of the nineteenth century, Myers would fully formulate these early poetic and Platonic impressions into an original cosmology, envisioning a psycho-biological, psychological, and metaphysical schema for a cosmic mind, as interpreted within the framework of World Soul.

FREDERIC MYERS, SATYR AND SCHOLAR

Myers went up to Trinity College, Cambridge, in October 1860 at the age of seventeen, a year earlier than most. He was already gaining some national renown for his poetry by that time. (He had placed second in the Robert Burns national poetry contest the previous year.) According to his own testimony, Myers "wasted no time in throwing his weight about," having already had some practice at home and at Chettleham. At Trinity, Sappho and Pindar, hitherto ostracized from his boyhood curriculum, "brought an access of intoxicating joy" in their sensual appraisal of the beautiful.[46] Myers's Hellenism took a decidedly hedonic turn. With his convergent identities as Romantic poet, Greek aesthete, and mystical seeker, Myers took things a bit beyond the comfort zone of his fellow classical scholars. He gave himself great plumage, fancying himself an Adonis as well as an intellect, and was given to dress and gestures that had a provocative grandiosity. His speech was overburdened by emotional weight, as if each moment, each idea, each feeling, had a particular significance grasped only by Myers—as if, in the words of Alan Gauld, "he could hardly help regarding himself as singled out by Fate for some high destiny."[47] This sense of his own "cosmic import" made Myers nearly universally disliked by his classmates, except for a few athletes and aristocrats not likely to be offended by pomposity. But, in truth, no one could touch him for excellence in classical scholarship, and he was not alone in noting his own exceptional genius. And eventually, that sense of "high destiny," passively displayed by Myers like some adorning attribute of character, would become character in action as he matured and stepped into the world.

The libidinal excesses of the first-generation Romantics had come under considerable fire from Carlyle, Mill, and the Evangelicals, so Myers could not have anticipated any easy time with the dons. Romantics, as historian Julia Wright explains, had become pathologized as layabouts, hedonists, and egotists by Victorians out to set a sterner moral tone for society in both literature and philosophy.[48] Nonetheless, he went about flogging his genius and his passion, dressed in attention-getting shades of red and velvet fabrics, letting loose his inner Byron in an energetic display of cultural protest. But a protest against what, regarding his cherished childhood, his celebrity, his privilege? What motivated this highly signed self-representation sure to make trouble (as it eventually did)? Myers had almost everything his society had to offer, and yet was there something fundamentally missing here that he felt he had to create for himself?

Myers's Romantic affectations in the scheme of his life, amount to considerably more than teenage attention-seeking or some reflexive assault on authority. But Myers did not have a problem with authority—rather, his concern was the prosaic shortcomings of authority. Myers's sartorial semiotics was a socially performed protest of a deeply felt philosophical objection. He aimed his ire at "the new man" of the nineteenth century: practical, industrious, earnest, and modestly attired, a man of efficiency and efficacy, who directed his energies outward to the management of the world, not inward, as did Myers, to the endless reflection upon feelings and philosophy. For those, like Myers, who wanted to express their dissent against utilitarian morality and positivist philosophy, the generalized aesthetics of English Romanticism hung like a ready-made garment. Any young man could so drape himself to express a host of contrarian positions coming under the Romantic rubric: political reaction, excess sensibility, individualism, sensualism, noncategorical artistic longings, or all of these. But even though Myers's Byronicism could be seen to fit roughly in this order, the moral and aesthetic commitments of his lifetime made these gestures something more than those of an adolescent spinning out clichés of rebellion. It may have been theatrical, but it was not just for show.

Even as Myers's aggrandized persona reveled in the self as social object, it was countered by a deeper, introspective turn intent upon experiencing the self as artistic subject. It was both his extroverted and introverted tendency that combined to bring about Myers's downfall in his final year at Cambridge. The pain and disgrace he brought upon himself would prove an important turning point. It all began auspiciously enough. By his senior year, Myers had already won the Chancellor's medal twice for English poems and the Camden medal once for Latin verse, in addition to the Bell and Craven university scholarships. He went on to gain the Camden medal a second time, but was forced to resign the prize because of a plagiarism charge. This really amounted to more of a prank, or, according to Myers, a "literary innovation." The brouhaha arose when one of the unsuccessful contestants pointed out that nearly a quarter of Myers's entry had been lifted from an Oxford book of prize poems. Myers had originally purchased the poems merely to "gloat over their inferiority to his own," but then had the unfortunate idea of culling the best lines of each for his own verse. He laughingly explained to his friends that, like Virgil, he was "collecting gold from Enniu's dung heap."[49]

Apart from its being an obvious dig at the droppings of the rival university, there was literary method to this madness. The explanation he gave to his uncle William Whewell (an uncle through marriage to his mother's sister and at that time Master of Trinity College) in a letter dated May 1864 reveals Myers's familiarity with Coleridge's theory of literary criticism:

> I had thought a good deal on the theory of poetry and had come to the conclusion, of course very probably erroneous but at any rate sincerely held, that the essence and value of a poem consists entirely in the impression which it makes as a whole and that the special parts have no separate function, as it were but are only valuable as building up and subserving the symmetry of the whole.[50]

This strongly suggests that Myers, at twenty-one, was already thinking about the nature of consciousness and its creative activity, though in aesthetic Coleridgian terms versus strictly psychological ones. Myers's "literary prank" had at its core a serious intention to test the nature of imagination and artistic identity. Coleridge famously contended that the essence of artistic power was the ability to take variegated parts and absorb them into a single whole, that the poet "diffuses a tone and spirit of unity that blends and, as it were, fuses, each to each by that synthetic and magical power, imagination."[51] Though Myers neglected to credit other sources for "his own literary theory" (ironically plagiarizing a philosophy to deny a plagiarism charge), the reproduction of Coleridge's central thesis is fairly plain. Myers concurred that the essence of creativity lay not in the ornamentation of lesser parts, but in the ability to *originate* the whole. The thrust was toward totality, unity, or, in the words of Coleridge himself, "something great, something one and indivisible" and not "a mass of little things" for "parts are necessarily little."[52] Myers's letter continued revealingly:

> I considered that the works of others might fairly be useful to him [the poet] where through some casual fitness they were capable of amalgamation with his own peculiar and essential style. It is natural in me to be desirous of immediately reducing to practice any theory which I may have formed and I seized on the opportunity of doing so. So to act, and from such a motive, may have been foolish, reckless, and headstrong and I am painfully aware of my danger of acting on crude and sometimes presumptuous opinions without considering or regarding what others will think.[53]

Myers's idea that the poet might freely draw on "the mass of little things" written by others anticipates in interesting ways the "open-source" authorship of the stream of consciousness that would feature prominently in his psychology. (But instead of having these sources inconveniently externalized on scraps of paper, Myers's "subliminal self" would bring a potential flood of consciousness right through one's psychological core.) That Myers would "practice his theory" in a competition using the work of Oxford prize poets seems to indicate an almost naive eccentricity and may even suggest, according to Beer, "a somewhat unreal state of mind."[54] In any event, the Camden medal affair seemed to seriously puncture Myers's fantasy life. His disgrace was relished by classmates irked by Myers's insouciant pomposity, and they laid it on with satire and scandal-mongering. The University Board even voted as to whether or not Myers should be allowed to take his degree. (The case was dismissed in his favor by a vote of twenty-five to three.) The mortification, and particularly his shame at disappointing his mother, was something of a loss of innocence for Myers. One cannot help but feel sympathy for the Shakespearean scale of his self-destruction: "My name has been held up to scorn in a great number of newspapers. They have written about me in *Cambridge Chronicle, Leeds Mercury, Clerical Journal* and eventually even the *Guardian* and *Athenaeum*. Many of my acquaintances have had their opinion of me greatly shaken and my Mother has been and is seriously distressed. Had I for one moment conceived that discovery would subject me to the least chance of one tenth of these evils, I would not have run the risk for all the prize poems in the world."[55]

What checks most people far earlier along their path to maturity spared Myers until he was nearly a man. This was Myers's first taste of what it really meant to go against society, and his first experience of strong rebuke by the power of convention. He was allowed to take the classics exam, after which he left for Greece, in June 1864, escaping for a few months the dreariness of his chastened life. It appears by his own account that Myers tried to shake off the sackcloth and ashes prescribed by the university, by entering the final, manic phase of his Romantic Hellenism. He traveled alone, and lost himself to the hedonism of the classical world, living "the life of about the sixth century before Christ on the isle of the Aegean."[56] Myers gave a somewhat blushing and evasive account of his trip—mumbling apologetically about lusty, sensual, and thoughtless youth—burying the ripping details in metaphor and mythology. What salacious particulars inspired the following travelogue we will never know: "Few men can have drunk that departed loveliness with a more impassioned spirit ... that intensest and most unconscious bloom of the Hellenic spirit ... Then it was that Praxilla's cry rang out across the narrow seas, that call to fellowship, reckless and lovely with stirring joy. 'Drink with me!' she cried 'be young along with me! Love with me! Wear with me the garland crown! Mad be thou with my madness; be wise when I am wise!'"[57]

Gauld and Beer, among, others suggest that this Greek adventure may have been the culmination of a homosexual phase in Myers's life.[58] (This rumor is mostly traceable to comments made by John Addington Symonds, a Cambridge contemporary, that Myers and his friend, Arthur Sidgwick, Henry's younger brother and a fellow student of classics at Cambridge, "were struck by the same disease. Myers's own evasive apologies concerning the misguided nature of his Hellenism added additional fuel to the myth.) Be that as it may, Myers had, by his own admission, spent that summer out of the box, operating on a moral code conceived centuries before Christ and far afield of the evangelical advisory on chastity and humility. Whatever did happen, the episode was veiled in regret. In the end, looking back on his undergraduate days and Greek interlude, Myers concluded that the Hellenic ideal that guided him, though inspiring "intellectual freedom and emotional vividness, exercised no check upon either sensuality or pride."[59]

Myers seemed to have overreached the limits of a personality perhaps more timid with regard to convention than he cared to admit. He was, after all, a snob in all his tastes and acquaintances, and that sort of elitism is rarely a socially autonomous activity. We may speculate that Myers's escapade in Greece was a case of "be careful what you wish for . . ." Perhaps his inner Victorian realized he did not care for more than a flirtation (as opposed to consummation) with Hellenism. In this last absolute defiance of convention, he clearly overplayed the intentions of his own bravado.

It is this quality of balancing convention with invention that makes Myers such a productive historical subject, positioning him at the social and intellectual hub of Victorian culture rather than on its periphery, where a less cautious subject might have found himself exiled. Even while cavorting about campus in silks and velvets, Myers always gave a somewhat controlled performance, managing to walk away with most of the university's prizes in classics, along with a teaching fellowship after he graduated. The plagiarism charge and the subsequent freefall in Greece were an important part of young Myers's learning to calibrate himself according to the world's measure, to conform and control the self as social object. A lesson learned rather late, but perhaps the delayed onset of such repression allowed him to more fully develop himself as a desiring subject. This was not an easy task in mid-Victorian times. Though the manner in which Myers expressed his passion and dissent became radically transformed by the end of his career, he always remained firmly rooted in his own personal longing, and felt urgently entitled to both articulate and actualize his message in the world. As a man of some social and intellectual consequence, moving in a small sphere at the pinnacle of Oxbridge notables, Myers took considerable risks balanced with strategic cautions. (Myers wanted to be in the position of rejecting elite social convention, not being rejected by it.) It was, arguably, this learned dynamic that enabled him to bring the program of psychical research into the fold of orthodox consideration.

When Myers finally departed Greece, it was amid "a longing which could not be allayed, and the insatiability which attends all unnatural passions as their inevitable punishment." Myers was not a man who wanted more, but rather one who had had too much and was still left wanting. It is clear that when he returned to England, Myers felt overextended, almost panicked, and at an utter loss of direction. This sense of longing that drove him into and then out of the arms of Plato and Praxiteles would shadow him through successive self-iterations well into his middle years. Before there was a crisis of faith, there was this crisis of identity, imposed by a world that could not fully accommodate what he longed to be, or even help him fully imagine it (not unlike the Christian faith that likewise could not accommodate what he longed to believe). When he bade a last farewell to Greece "and turned to go with eyes tear-brimming, a bitter sweet passion of regret," he knew that the sanctuary of this particular fantasy was over. The lights were on, the party was over; it was time to slink away home.

Myers returned to Cambridge that fall to finish up his examinations, getting another first in the moral sciences, and he even determined to spend the remainder of the year preparing for the natural science tripos (the final honors examination at Cambridge). If "the vanishing of the Hellenic ideal left [him] cold and lonely," what warmth might he find in the likes of Darwin, Lyell, and W. B. Carpenter? But he had decided to turn away from Greek philosophy, and that resolution was served in part by this elective submission to scientific fact. This preparation in the life sciences, while adding grist to the mill of his already troubling ruminations on materialism, laid down an important stratum in Myers's intellectual *bildung*, allowing him to survey the world from an alternate lookout. He immersed himself in biochemistry, neurobiology, physiology, psychology, and evolutionary theory, putting himself center stream of the intellectual activity of his era, which he could now occupy both from the standpoint of a first in moral philosophy and a student (if not disciple) of naturalism. He could not marginalize inconvenient ideas now that they were made persuasive and familiar; he would have to take them head on. This curriculum put in place the skeleton of orthodoxy into which he would later pack findings from less permitted fields, giving them purchase in a conversation that they could not otherwise have penetrated. Despite his long months of study, the college did not allow him to sit for the exam, given the unorthodox method of his preparation, crammed into a single year and much of it independent study. Come June, Myers was again at loose ends, disappointed about the tripos and disaffected from philosophy. He left the country, this time for Canada, "as gloomy as [he] had ever been," sounding the depths of his emptiness with that fateful plunge into the Niagara River, "like Byron, the young old man who'd drunk dry the cup of life before his time."[60]

THE HEART OF FIRE

Myers survived the rapids and upon his return from Canada was offered and accepted a position as classical lecturer at Trinity College, beginning in the fall of 1865. But a sense of purpose eluded him, and he began the year in that same mood of indifference he could not seem to shake. His despair did not last long. He found himself soon taken up by the beautiful wife of the vice principal of Chettleham College, Josephine Butler. The gloomy, brooding, and handsome young Myers must have proven irresistible to Mrs. Butler, who was known for performing religious interventions on the lost souls of desperate (-ly good-looking) young men. She reenergized the morally listless Myers with what one biographer called "a highly spiced version of Christianity." In letters exchanged between her and Myers, she explained her doctrine that the most powerful attribute of religious conviction is love, from which flows compassion and ultimately unity with God and one's fellow man.[61] At last the hazy gloom that hovered in what Myers described as the "interspace between religions" burned clean away in the hot fire of a new belief. The ardent Mrs. Butler preached a catechism of the heart, speaking straight to his Romantic longings. Describing his awakening, Myers wrote, "I had been piously brought up but I had never as yet realized that faith in its emotional fullness. I had been converted by 'The Phaedo' and not by the Gospel. Christian conversion now came to me in a potent form through the agency of an ardent and beautiful woman much older than myself. She introduced me to Christianity, so to say by an inner door; not to its encumbering forms and dogmas but *to its heart of fire*."[62]

How different this was from the doctrinal preoccupations of his father's faith and teachings: textual, disciplined, interrogative, in search of universal principles. All this now gave way to the spontaneous faith of feeling and intuition, an ecstatic, sensual noesis of Jesus based on individual experience, not on theological study. In the early flush of his enthusiasm for Mrs. Butler and Butlerian religious ecstasy, Myers, who had hitherto been guided by the higher rationality of Plato or the humanistic sensuality of Sappho, now responded to the innermost urgings of the Holy Spirit, as purely reducible to Love. The inner transformation brought on an outer transformation as well. The self-centered Myers was now all solicitude for the souls of others. One friend observed that "Myers devotes himself to self discipline. He never goes anywhere. He gets up at 6:30 and goes to bed at 10:00. His days are spent in reading *Ecce Homo* and in thinking."[63]

Myers's profligacy, apparently, had been utterly reformed. Much of the emotion and certain aspects of the drama remained, but now Myers seemed more serious and socially committed. Myers had moved from the Eros of early literary Romanticism to the *agape* of Christian humanism. In short, he had acquired moral purpose to add to his aesthetic and philosophical convictions. In a letter written on May 5, 1866, he urged his dear friend Arthur Sidgwick, who remained standoffish with regard to what he perceived as the unsavory sentimentality of Butler and company, to see the light of love.[64]

> The more thoroughly you feel that love is everything that matters the more would it agonize you with shame if once you thought that you had possibly been rejecting such love as the gospel tells of. Nothing would rejoice me so much as your conversion. Great heavens what a prospect! Leagued on Earth with all those whose love is best worth having in a bond closer than any free masonry, enrolled among the countless species of one genus
>
> > "all with forehead bearing L O V E R
> > written above the earnest eyes of them"
> > and all this purely for that asking, surely this is God.

In the thriving debate culture of Cambridge life, Myers found himself vainly testifying for Jesus on the basis of affective experience. Such a dubious epistemology was no defense against the newly minted skepticism of friends and colleagues eager to match their positivism against such easy prey. Against Myers's ecstatic testimony, they argued that the Bible must be historicized, the Earth geologically dated, the feasibility of miracles submitted to questioning. In vain Myers attempted to urge this revelation upon equally passionate converts of the *System of Logic* and *Cours de philosophie*. He could not have chosen a less receptive audience to contest the value of critical reasoning in determining the truth of Christianity, yet the power of his conviction overran these obstacles. As he explained to Arthur in his letter, faith was rooted in an intuitive certainty too deep to be touched by external arguments:

I cannot resist writing to say that the moral evidence in favour of Christianity becomes, immediately the will is thoroughly subjected, quite overwhelmingly strong . . . Even I, wretched and half hearted beginner as I am, can almost say already that I know the thing is true. How do I know it? How do I know that Virgil is a great poet? How do you know that Bach was a great musician? . . . The whole mental process is one in which, as you well know, I have cause to take a desperate interest . . . You cannot say that your critical analysis disproves Christianity. It merely fails to prove it on external grounds. The gospel of John, for instance, Renan supposes genuine though untrustworthy, Strauss (if I am not mistaken) a second century compilation . . . Who is right?[65]

Myers's letter struck at the heart of what would remain a matter of great emphasis throughout his life, resolving itself finally in psychical research. "The mental process of which [he] took such a desperate interest" was this seemingly innate human aptitude to know God. This capacity was related to, or perhaps even the very source of, humanity's aesthetic intuition: to understand the poetry of Virgil, the musicality of Bach, was merely an extension of this more fundamental capacity to recognize the divinity of Christ. This was a Romantic metaphysic, wherein beauty mediated between man and a more exalted plane. This was not beauty for beauty's sake, but rather beauty that served in the knowledge of God. (Myers, intriguingly, identified this religious sensibility as a "mental process"; that is, he rooted something mystical in something neurological. This intimates that his theory of the subliminal self had a far longer period of ferment and was of a more generalized interest beyond psychology.)

Despite the passionate blaze kindled by Mrs. Butler, Myers found that by 1869, his heart of fire had been doused by "the rising flood tide of materialism and agnosticism, the mechanical theory of the universe, the reduction of all spiritual facts to physiological phenomena."[66] This had been intermittently checked by Sappho and Josephine Butler over the course of his decade at Cambridge, but it had now finally overtaken him. As his visceral connection to a spiritual existence receded, intellectual alienation remained behind. He felt his thought compelled in directions where he did not want to go, but which he could no longer resist—the intellectual captive of materialism in crisis. As the rhetorical contest between science and religion became increasingly popularized and politicized over the course of the 1860s, the well of belief from which he drew his earliest memories appeared to be running dry at last. In one of the archetypal "crisis of faith" narratives, Myers described this cultural moment of his personal surrender:

A time when not the intellect only, but the moral ideals of men, seemed to have passed into the camp of negation. We were all in the first flush of triumphant Darwinism, when terrene evolution had explained so much that men hardly cared to look beyond. Among my own group, W. K. Clifford was putting forth his series of triumphant proclamations of the nothingness of God, the divinity of man. Swinburne too, in *The pilgrims* had given passionate voice to the same conception. Frederic Harrison, whom I knew well, was still glorifying Humanity as the only divine.[67]

The issues of belief or unbelief, reason versus faith, creation or evolution dogged him in the lecture hall, at his dining club, on the quad, and while weekending in the country. While at Wykehurst in 1873, Myers met with none other than Herbert Spencer, and they exhausted themselves in argument. Myers, with a bit of class surliness, called Spencer "a philosopher who was nothing but a philosopher," having the "aspect of a serious linen draper."[68] Nonetheless, Spencer and other essayists with the popular touch, such as Huxley and Tyndall, had a magnified effect on such debates by virtue of their rough rhetoric and the want of gentility observed by Myers. They stirred things up, hobnobbing with the general public through the press and sensationalizing scientific debates with pugnacity. They took the dry data of scientific evidence, lying inert and harmless in disciplinary journals, and cross-pollinated it with larger social concerns. Darwinian evolution came crashing over the wall of the natural sciences to infect moral, social, and religious discourses in ways Darwin himself had tried to contain.[69] This mélange of social and religious anxieties intermixing with a host of uncongenial theories and facts came to be known, in the words of Victorians themselves, as the "crisis of faith." It was a catastrophe unfolding not just at the personal level of conscience, but, as Myers made clear in his account, at the civilizational level as well: "Lack of evidence led me to an agnosticism or virtual materialism which sometimes was a dull pain born with joyless doggedness, sometimes it flashed into a horror of a reality that made the world swim before one's eyes, a shock of nightmare-panic amid the plain dreariness of day. It was the hope of the whole world that was vanishing, not mine alone."[70]

Historians have undermined the idea of a pervasive ideological crisis, in favor of a more rhetorical reading of the proposed "war between religion and science," one that must be reconsidered within a variety of social, cultural, and economic contexts. But for those who did not have the benefit of historical hindsight, who extrapolated mournfully from these early intimations of the "war between religion and science," a catastrophist reading perhaps better captures the personal experience, if not the reality, of their situation. Admittedly, Myers's disconsolateness has a certain tropological quality.[71] Yet this should not be allowed to discount that for men like Myers there did exist a core philosophical conflict, reasonably articulated as a kind of ideological Armageddon. Myers himself immediately grasped the random implications of evolution, an understanding suppressed by the majority of the population, who preferred to impose the *telos* of progress despite its express absence from the idea of "natural selection." Myers saw that the God-centered world of his youth, around which all contingent moral, social, and cultural beliefs had once orbited, could no longer be quite so systematic with the new science, that the pieces of this world could well come apart.

While historians dispute the cultural megastructure that Frank Turner called Victorian scientific naturalism, preferring a more pluralistic view of science, crisis rhetoric affirmed such an orthodoxy, or at least the specter of it, overwhelming the traditions of Victorian intellectual life with its sheer competency.[72] Though Victorian science is not reducible to any single creed, the dominant tendencies of its formal explanation were materialistic and deterministic, its epistemology positivistic, and its religious outlook staunchly agnostic. What it gained by this intellectual discipline was a cosmic doctrine of realism and lawful regularity. What it lost, according to Myers, was belief: "In spite of, and by reason of, her studied neutrality, the influence of science is every year telling more strongly against a belief in future life. Inevitably so since whatever science does not tend to prove she in some sort tends to disprove. Beliefs die out without formal refutation, if they find no place among the copious store of verified and systematized facts."[73] It is no wonder that a primary strategy of psychical research was to co-opt, not contest, the authority of science.

Myers, whose emphasis had always been on the interior, the subjective, the personal, and the intuitive felt, by the close of the 1860s, the undertow of the powerful knowledge claims of empiricism. Hitherto, the intensity of feeling had provided the indicia of truth-value, but now it was proof. The new objectivism forcefully swept such cherished personal criteria aside.[74] Public consensus trumped private experience. "Facts" were to be the basis of knowledge, and objectivity the determinant of factuality.

THE SPIRIT OF INQUIRY

Up until 1869, Frederic Myers's development was unusually marked by a fluctuating series of intensely *felt* beliefs. From the tenderhearted piety of his boyhood, to his own personal renaissance as a teenage Platonist, moving

then to the more fleshly Hellenism of Sappho and Pindar, and on to the ex-
ultant Christianity of Mrs. Butler, Myers serialized the Romantic impulse of
his nature in a succession of social ecstasies and ideological forms. But these
episodic passions were becoming increasingly abbreviated interludes, burning out the beliefs that fueled them with escalating rapidity. By the dawn of
the 1870s, Myers found himself emotionally numb, defeated in mind and
body unable to muster even the enthusiasm for a plunge in Niagara.

It was at this time that Myers's former Cambridge classics tutor, Henry
Sidgwick, reentered his life. Sidgwick's singularly un-dashing manner had
been previously found wanting by the undergraduate Myers. The young
dandy had looked disdainfully upon his residential advisor at the Trinity
dining hall, who would gaze "coldly through half slit eyes at the freshmen opposite him," gobbling his supper like "an old goat who ate his beard shoveling
it by handfuls into his mouth."[75] Myers recalled the strong desire to "punch
his head." But ten years later, Myers had changed, humbled in his own self-
estimation even as Sidgwick had moved into a kind of public glory with
his refusal to take the religious Tests in 1869 (continuing this rise with the
subsequent publication of *Moral Philosophy* in 1874). The rational calm with
which Sidgwick stoically bore his own trials inspired Myers, modeling a serene path through his own personal religious anguish. Sidgwick himself had
originally come to Cambridge in the late 1850s with the intention of training
for the clergy but found instead an intellectual climate hostile to faith. The
pragmatic and empirical values of Auguste Comte and John Stuart Mill set
the new standard for intellectual honor, edging idealism and metaphysics
out of their privileged place in moral philosophy, and leaching into Sidgwick's own religious outlook. The more Sidgwick interrogated the foundation of his belief in the Anglican God, the more he felt it give way beneath
him until he could no longer continue as a divinity student. Eventually, he
found himself unable to commit to any idea of God whatsoever. But though
he lost his faith, the religious seeker remained as Sidgwick's guiding spirit.

Under Sidgwick's mentorship, Myers began to intellectualize his approach to the problem of belief. He would try and retry every proposition
to find some philosophical foothold, like a man testing stones across a river
of doubt to get to the other side. This steadier methodology was not without dangers. In Myers's own experience, such an application of reason to
religious questions was just as likely to breed concerns about the question-
ability of religion instead. When Myers was an undergraduate, Ernest Renan's fully historical account of the life of Jesus (*La Vie de Jésus*, 1863) had
so disturbed him that he wrote the author a letter disputing the merits of

the argument (lest he find himself convinced by it). Myers posed his central objection as against Renan's historical methodology, feigning intellectual neutrality regarding the thesis: Christ was a great moralist but not a divine miracle worker. In his letter, the fulminating Myers took Renan to task for his lack of supporting documentation. Where were the historical sources upon which he based this secular history? Myers's invective was motivated in part by the damage Renan's "unsupported" argument had already done to the apostolic narrative, which could not be so easily undone as Myers's own eroding confidence in the divinity of Christ made plain.

But Sidgwick's return to Cambridge brought with it a fresh hope, which broke in upon Myers's gloom on the evening of October 31, 1871, "in a star-light walk which I shall not forget." Sidgwick had been among the founding members of the Cambridge Ghost Club in the late 1850s, but had only been sporadically involved since. Inspired by his experiences, it dawned upon Myers that together they might renew that course of inquiry along more definite terms:

> In a star-light walk which I shall not forget . . . I asked him [Sidgwick], almost with trembling, whether he thought that when Tradition, Intuition, Metaphysic, had failed to solve the riddle of the Universe, there was still a chance that from any actual observable phenomena, ghosts, spirits, whatsoever there might be, some valid knowledge might be drawn as to a World unseen. Already, it seemed, he had thought that this was possible; steadily, though in no sanguine fashion, he indicated some last grounds of hope; and from that night onwards I resolved to pursue this quest, if it might be, at his side.[76]

When Myers published his essay, titled merely "Renan," in the wake of this spiritual resolution, he delivered the *coup de grâce* that had eluded him in 1866. Myers not only skewered Renan's historical methodology, but also tried to force him to the sidelines of any future academic inquiry. Historical accuracy was not the real test for evaluating the credibility of the Bible. It was the authenticity of its supernatural claims. Such miracles turned the life of Christ into Christianity, making Myers's own "quest" to evidence spiritualistic phenomena a far better arbiter in these matters of faith: "If the grand difficulty of believing that God became man was got over, the difficulties of [historical] detail would be far from invincible." If any one of the miracles proclaimed by the Bible (prophecy, transmutation, spirit communication, levitation, incarnation) were proven through some scientifically observed demonstration of mediumship, the bedrock of modern unbelief, the Humean opposition to miracles would crumble to dust. (For Hume, "miracles" essentially testified against their own existence by contradicting laws for which there was abundant proof.)

What Christian miracles required was verified proof of their own. The bizarre events alleged in spirit circles seemed to closely parallel these unverifiable biblical claims but in a manner now amenable to empirical confirmation. Even the more controversial idea that "God became a man" might find some basis in the latest phenomena rumored to occur in séance circles: ghostly spirits who returned to Earth in the flesh. Another boost to Myers's soaring confidence in this delayed riposte to Renan was William Crookes's recently published report regarding an experimentally verified "psychic force." (Crookes's report came out in October 1871, and perhaps played a role in steering "the starlight walk" in this spiritual direction.) A psychic force offered a potential framework within which to interpret the "miracles" of Christ and supernatural events of spiritualism, with reference to some more general law of nature amenable to scientific adjudication.

As Myers and Sidgwick embarked on this investigation with like-minded colleagues, it was not just the possibility of "miracles" they were seeking to defend. More profoundly important were the spiritual values upon which those miracles were predicated. Those heretics of sociology, from Ludwig Feuerbach to T. H. Huxley, used the "falsehood" of Christian miracles as an indictment of Christianity's moral codes as well, looking to supplant these spiritual values with some more naturalistic basis for modern conduct and social mores. (This, in its most brutal form, became "social Darwinism," a term that began to haunt moral politics in the 1880s but percolated earlier in the ethical concerns of crisis discourses.) In the jeremiads penned for the public's edification by Myers and Sidgwick regarding threats to faith, this potential failure of societal morality was a central concern; this is what gave the loss of personal faith its sense of looming catastrophe well beyond the personal scale.[77]

While the unsavory nature of séance mediumship made it the means of last resort for such Cambridge intellectuals, it had the virtue of offering an unprecedented chance of success. Under Sidgwick's leadership, Myers and a few close associates officially formed the Cambridge Group in 1874 (also called the Trinity Group or Sidgwick Group), to initiate a sustained and rigorous academic investigation that might improve upon the sporadic and ultimately ineffectual efforts of the earlier Cambridge Ghost Club. Group regulars consisted of Henry Sidgwick, Henry Jackson, Frederic Myers, Arthur and George Balfour, Walter Leaf, Edmund Gurney, Arthur Verrall, F. W. Maitland, Henry Butcher, and George Porthero. They were mainly agnostic in their bent and careful to distinguish themselves from the out-and-out *spiritualists*. Their aim was to evaluate these anomalies of the séance parlor and judge them on their own merits as physical events. If confirmed, the seemingly far-fetched claims of the Bible would have a kind of precedent in the natural world, providing a basis, if not for the re-Christianization of modern thought, at least for a radically reformed scientific worldview. Such a metaphysic would have the advantage of being unencumbered by clerical baggage and all its worldly politics and theological accretions: a brave new theology forced to streamline itself along the general outlines of the physically known world. Of course, this attempt to ground miracles within physical systems was something of a double-edged sword. Some saw this attempt to reframe miracles as rare but regulated events in nature as tantamount to the denial of Christ's divinity. Not only would such proofs undermine the exceptional nature of Christ, they would place Him amid a tawdry cohort of mediums—not to mention the brazenness of holding God to an empirical standard before yielding one's belief.[78]

The inauguration of the Sidgwick Group happened at a time of conflicting auspices: the tide of spiritual belief and discovery appeared to be cresting, even as a wave of academic opposition was beginning to swell. When Myers and Sidgwick had had their starlight walk and laid their plans for an investigative group, they had been drawn in part by the lantern lit by Crookes's spectacular psychic force. They conceived a plan to arrange some joint investigation, but before they could embark upon it, Crookes was swept out into a rough spiritual sea and had to come ashore.[79] Instead, Myers and Sidgwick decamped for the small port community of Newcastle in 1875, lured by rumors of extraordinary goings-on about town. With their eerie levitations and materializing body parts, the entrepreneurial local mediums had made a spiritual mecca out of an otherwise economically inauspicious place with a distinctly working-class flavor. While these séance setups were somewhat suspect, usually involving some sort of private cabinet to which the medium retreated before the spirits emerged, the spreading fame of the phenomena warranted investigation. Sidgwick and Myers set out expressly to engage the two mediums most celebrated for their creative cabinetry, Miss Fairlamb and Miss Fox, anticipating the compliance due such illustrious visitors from across the class divide.[80]

But the investigators met with resistance every step of the way in a clash of wills provoked by the perception of arrogance on the one side and impudence on the other. They had assumed that as investigators, they would be allowed to enforce certain protocols on the proceedings, but the ladies, according to Myers, made it "difficult for laboratory conditions to in truth be set."[81] It was a matter of constant negotiation for Myers and Sidgwick to entice Fox and Fairlamb into accepting these terms, which were truly rather arduous. Measures against chicanery might include lashing Miss Fox to her chair or sealing Miss Fairlamb in her cabinet, seemingly hostile gestures that not only alienated the mediums, but apparently frightened the spirits away as well (as did handcuffs, body searches, cowbells, and electric wires).[82] Thus, sadly for Sidgwick and Myers, any control they gained over the séance that would render results more conclusive tended to suppress the very results they had come to observe.

In a letter to Henry Sidgwick on December 10, 1875, Myers reviewed the situation: "I still secretly am inclined to think her spirits genuine. I feel that if applying tests, we shall run so many risks of quarrelling with two of the most huffy and somehow spiteful women that I am hardly inclined to push the thing at present. I thought of replying in a few days in a short note something in this way. 'I am sorry that I did not make the purpose of my last note sufficiently clear. It was intended to show you how far I was from the position of absolute skepticism to which your previous note had seemed to assign me. Some little evidence as it seemed to me was still lacking to establish these phenomena on a basis of scientific certainty. It was not my intention to slight the testimony of others or to set my own above theirs . . . Only with a large number of independent testimonies will the world be convinced of the reality of the phenomena. I accept with regret your decision against séance on Wednesday next."[83] Myers considered it his duty to stoop to conquer, Sidgwick less so.

It seems from this letter that Fairlamb and Fox claimed the upper hand in negotiating future access to their talents after whatever initial deference allowed their manhandling by science. Under the new accord, there would be no further interference with their preparations in the cabinet before or during the séance. Not even suspicious thoughts would be allowed to poison these proceedings. Any such negativity and skepticism only upset the sensitive apparatus of the medium's mind and prevented further manifestations, a point readily conceded by Myers in his letter. Like a suitor desperate to be taken back, Myers had to "assure the ladies that [he and Sidgwick] were not like other unbelieving investigators" and that, "should [Fox and Fairlamb] at any future time be inclined to admit Mr. Sidgwick or myself to such a meeting there will be nothing in our minds which can be thought likely to interfere with the success of the manifestations." And in truth, Myers was by nature not a skeptic, but rather, admittedly, "deeply sympathetic with the spiritualist cause." As his letter explained, it was not skepticism that drove him to embrace these aggressive research protocols; it was faith that the phenomena were real and must be irrefutably proven to be so. The ultimate goal, which he shared with spiritualists, was to reveal "the majesty of the spiritual universe into whose intimate structure it may thus, and only thus, be possible to project one penetrating ray," so that ray must be rendered as illuminating as possible through strict scientific transparency.[84]

The relationship between the mediums and the scientific investigators was complicated along many lines: each wanted something from the other, and depending on the individuals involved, this could make them allies or enemies. With a little luck and some expert testimony, two small-town celebrities like Fox and Fairlamb might look forward to some better engagements in bigger towns with a steady supply of clients. If they could manage to steer clear of scandal, they might even find themselves sitting around a dining room table clasping hands with a lord or lady, paid but privileged guests in a social order turned temporarily upside down. A bereaved aristocrat remained a supplicant before the powers of the medium who alone could grant a glimpse of the lost beloved. But no matter how ambitious and enterprising, it was not always in their interest to give outside observers full access to their signature phenomena, especially when the primary motive for the séance was a matter of verification. Even well-paying investigators with a clear interest, like the Sidgwick Group, had at the heart of their mission the stark question: true or false? All this clearly put Myers's "huffy and spiteful" mediums on the defensive. Every potential elite endorsement brought with it the risk of exposure and loss of reputation (and the livelihood that went with it). If the danger was not from the genteel Sidgwick or Myers, who preferred to avoid vulgar confrontation, then it was from the more pugnacious skeptics, like the medical doctor Horatio Donkin or the biologists E. Ray Lankester, who were determined to take down celebrity mediums and disprove their sensational claims. But such confrontation was increasingly part of the séance encounter. Skeptical provocateurs had distributed themselves among the investigative committees of the London Dialectical Society since 1869 and continued to be out in force, terrorizing mediums with their agenda of exposure.

Thus, spiritualism's success in the 1860s and early 1870s was also its Achilles heel. The increasingly elaborate special effects, such as percussive sounds, pulsing lights, luminous powders, and mysterious voices, that kept client traffic flowing through the 1860s likewise intensified public interest, which in turn brought more elite oversight and potentially dangerous scrutiny. And now, arriving on the scene in the 1870s just in time to be refereed by the august London Dialectical Society, the Sidgwick Group, and Professor William Crookes, were spirits that routinely appeared in the flesh and spoke for themselves. Sometimes several might appear at a single séance, as mediums developed a revolving cast of players to heighten the drama of their ghostly theater. With so many phantoms on display, spiritualism seemed poised to despoil itself of one of its greatest assets, the ambiguity that kept all but the most resolute skeptics just a little bit curious. Was any of it true?

But as these ghostly madrigals brought spiritualism careening dangerously close to the edge of absurdity, they threatened to decisively snap the tether attaching such claims to some underlying stratum of scientific plausibility. The momentum that Myers and Sidgwick felt in October 1871 began to stall by 1876. What had been the high point of spiritualism's creative output and cultural appeal seemed to have passed its peak. As the "spirit of inquiry" ceased to haunt academic circles, this adventure became less intellectually and socially tenable. The air of noble curiosity with which elites once defended themselves was stripped away, leaving only a humiliating gullibility standing between them and accusations of outright complicity. This unwinding of spiritual cover was no doubt facilitated by the overreaching escapades of mediums themselves. Full-bodied ghosts promenading about in cheesecloth, past lives culled from newspaper obituaries, and theatrical divas showboating their menageries of the dead upset the balance of plausibility, outpacing the capacity of social, intellectual, and evidential narratives to maintain them as part of "educated opinion," and beginning to finally be brought to heel by professional science.

Myers described this period from 1872 to 1876 as "tiresome and distasteful enough," likening his spiritual expedition to Newcastle "to the hardship which a naturalist will undergo to trace the breeding ground of a song-bird." However, that metaphor suggested that extreme trials were met with great rewards, and Myers counted the costs incurred "as nothing had it only yielded some clear result." But this calculus was changing due to the increasingly aggressive tactics devised by skeptics to put people, particularly those of standing, off such curiosity.[85] Hitherto, the skeptical campaign had focused on making spiritual beliefs *unbelievable*, but a shift in focus had taken as its target the believers themselves. This turned the impersonal rhetoric concerned with disproving the phenomena into a highly personalized form of public disapproval. This potential loss of face was particularly galling for classically trained humanists like Gurney, Sidgwick, and Myers, already sensitive to the gains science was making relative to philosophy in terms of intellectual standing.[86] To expose themselves to the public censure of individuals whom they would not have considered to be their social or intellectual peers risked too much for too little gained. More and more, the cost of spiritualism was not worth the passage, especially if it was all heading nowhere.

Before the Crookes affair of the early 1870s, there was no real precedent for this public lack of collegiality, but now shaming was becoming par for the course. The courts got into the act as well. Professor E. Ray Lankester used the vagrancy laws revived in 1876 to bring the slate writer Henry Slade to court, mainly as a publicity stunt aimed at embarrassing Myers and a fellow Sidgwick Group member, Edmund Gurney. Since the law had effectively banned all spiritual trafficking for money, both men could now be considered victims of a crime and forced to give testimony in open court.[87] While these measures resulted in a few incarcerations, success was not necessarily tallied by the curtailment of commercial actors like Slade, who proved rather resilient. This unsavory aspect of criminality found reinforcement from the early discourses of hysteria, which interpreted spiritualism as a kind of nervous derangement brought on by problematic sexual energies, a psychological stigma far worse than that implied by unconscious muscular cerebration. The real target of this public theater was the intellectual respectability of spiritualism itself. It was no longer just an intellectual adventure to be pursued in the free markets as a matter of private conscience or legitimate curiosity; it was becoming, by this confluence of efforts, a career blockade, a potentially criminal activity, or a salacious derangement requiring psychiatric attention.

While the stigma was certainly dissuasive, there was also little incentive to continue. It was increasingly obvious to Myers and Sidgwick that professional mediums were an obstacle to, rather than a means of, meaningful inquiry. In the end, such phenomena did not equate to scientific evidence and could not be made to serve an empirical inquiry. In the midst of their waning spiritual hopes, one last possibility remained adrift, collecting a few bodies from this spiritual wreck. Myers warily kept his berth in Edward Cox's new Psychological Society (1875), but Sidgwick stayed aloof.

THE PSYCHOLOGICAL SOCIETY

There was enough continuity between investigative spiritualism and the Psychological Society to make it easy for Myers to switch ships in midstream, even as he hoped to sail away from ghosts altogether. Cox's aim in founding the Psychological Society was to gather the lost data of psychology's shadow history and augment it with future research, bringing forth a more expansive view of consciousness. (Myers would eventually achieve this with the theory of the subliminal self, but not in this venue.) Cox was himself an active investigator of spiritual mediumship, but he considered its language sensational and digressive. He preferred the more technical analysis that characterized investigative mesmerism, but he had taken his opportunities where he could get them. Because spiritualism was primarily investigative rather than canonical, it was able to accommodate a diverse range of beliefs and levels of commitment, from science to philosophy and religion, depending on how it was accessorized. But the net effect had been to subsume the magnetic circle into the séance. People came for the trance, but stayed for the spirits, so seductive was the mix of heavenly enterprise and theatrical setting.

Traditional mesmerism was also under attack, at the other end, by physicians, eager to purge its animal magnetism and put an end to special faculties. This had been under way since the 1840s, when doctors like Elliotson, Eisdale, and Braid promoted mesmerism as a surgical anesthetic, emphasizing its purely physical character. Only James Braid managed to "entirely separate 'Hypnotism' from Animal Magnetism," turning it into "a simple, speedy, and certain mode of throwing the nervous system."[88] Even so, the eeriness of trance was a tough sell to professional medicine, and Braid's "hypnotism" stalled academically while mesmerism continued to flourish in the spiritual sphere in England and abroad. But the balance began to turn in the 1860s as Braidian hypnotism established roots in France, promoting itself to French physicians as it had in England, as an attack on traditional mesmeric forces. The altered states paradigm made the most progress in England only when it found an even more demystified version than Braid's in the 1870s, with William Benjamin Carpenter's "unconscious muscular cerebration" (1874), essentially an absent-minded, gross-motor reflex used to explain away table turning. There was no real "trance" here, no second sentience, only distracted individuals fallen under the hypnotic sway of their own automatic suggestion, attributing actions to ghosts that they were in fact performing themselves.[89] From where Cox stood, it seemed that mesmerism's esoteric psychology was being squeezed by physiologists, on the one hand, and spiritualists, on the other, with no real means to maintain the middle.

A more interesting line of research had been broached by William Crookes in the early 1870s, but it had just as quickly come under fire. Cox himself had served as one of the original referees in the investigation of D. D. Home back in 1871 and had been on hand to witness the thrilling affirmation of a "psychic force." But by 1874, Crookes had been swept up in the scandalous affair of the ghost of Katie King, threatening to let the most promising lead that mental investigators had had in years vanish alongside her fleeing apparition. It was then that Cox scrambled to found the Psychological Society to give a protective berth for "psychic force" outside the contaminating context of spiritualism. Crookes's laboratory study had allowed for the return of a purely mesmeric or "mental" analysis of phenomena such as trance, table turning, somnambulism, mind reading, magnetism, and so on, otherwise overtaken by spiritual explanation (or denied altogether). In this hope, Myers and Cox were one. The Psychological Society excluded all mention of spiritualism from its investigative program, designating its area of interest to be "psychical phenomena," to be discussed exclusively in terms of the mind. But consciousness, for Cox, remained pointedly open-ended, and "psychical" was more a way of rejecting spiritualistic attitudes and methods, not its principal object, "the soul." Cox also studiously abandoned any reference to "mesmerism" as well, but that was about other people's prejudices, not his own. At the time of the society's founding, Cox was personally involved in both spiritual and mesmeric investigative lines of inquiry, but was wise to the cultural sea change of 1874 (see chapter 2) that forced him to jettison that cargo to make a go of "psychic force."

It is a testament to the distance Cox put between spiritualism and the Psychological Society that he was able to entice both Crookes and Myers, two shell-shocked veterans of the culture wars, to climb aboard in 1875 on its maiden voyage, with their names on the list of founding members. But the Psychological Society remained a fringe organization, making little headway even amid rumored French researches that Cox got early wind of. Hypnotism, hysteria, and automatic suggestion offered far deeper probes into consciousness than Carpenter's stingy "muscular cerebration," but Cox could adapt no plan to fit in. His society was ultimately unable to exploit this scientific trend, shuttering its windows even while the clinic of Salpêtrière was preparing to throw open its doors. This failure to connect to this movement in psychology was more than a matter of timing. Cox's organization lacked the essential elements that would give psychical research its longevity and success, and it would be a mistake to call Cox's Psychological Society a precursor to the Society for Psychical Research, as is often suggested. Despite the seeming correspondence of their "psychical" object of study, there are few real points of contact between the two. Of the regular membership of the Sidgwick Group, only Myers's name appears on the roster for all four years. Sidgwick himself, feeling increasingly compromised by all fringe speculation, refused to join the Psychological Society (though he would energetically assume the presidency of the SPR). As for Myers, his autobiography mentions neither Cox nor his organization, but does describe the time period itself rather dismally. Furthermore, Myers, so prolific as a psychical researcher, never actually contributed anything to Cox's *Proceedings* beyond his name, which appeared every year on its published list of members (a list that may have been easier to get onto than off of).[90] William Barrett, who was at this time pioneering the exact sort of laboratory parapsychology that Cox had hoped to sponsor, only affiliated himself with Cox's venture at the request of Myers, remaining an honorary member only. Like Myers's Barrett's name appears on the rolls, but he too contributed nothing of his own original research. The paper he presented on "Thought Reading" to the Royal Society in 1876 (but was not allowed to publish in its *Transactions*) was never offered to Cox, nor were any of his other follow-up investigations, which appeared only as mentions by Cox, not accounts by Barrett.

The difficulties facing the Psychological Society stemmed in part from the founder himself. Cox simply could not attract the kind of elite membership needed to sponsor the dialogue with science managed by the Society for Psychical Research. Cox was a successful sergeant at law as well as a passionate amateur of the study of mind, but he was no Oxbridge intellectual and lacked any professional scientific pedigree. His past offenses against elite psychology included authorship of a major textbook on phrenology published in 1839 and the dogged pursuit of strange mental phenomena ever since. Given this, he found himself and his society somewhat stiff-armed by those fearing to fall afoul of academic science and its increasingly rigorous intellectual policing. In this somber mood, all such colorful curiosity was in poor taste. The research agenda proposed in 1875 was never actively pursued; instead, contributions to the *Proceedings* were often a rehash of phreno-mesmerism presented by Cox with a slightly more mystical gloss. Even Cox's exclusion of spiritualism backfired because it effectively narrowed his popular audience without adding much to his roster of elite support.[91] (The SPR would manage an investigative branch of such spiritualistic phenomena as "hauntings and apparitions," behind its far more redoubtable respectability, though this was highly psychologized in its presentation.) The Psychological Society thus failed to capture the public imagination, garner elite interest, gain disciplinary standing, or even maintain its internal momentum; it was dissolved in 1879, a few years after its inception, without much change to the domestic narrative of psychology. So rather than serving as a test run for the SPR, if anything, it appears to model a cautionary tale.

Its most dramatic impact on Myers's future psychology came perhaps through the vast inventory of knowledge it added to his larger understanding, sources left out of his preparation for the natural science tripos at Cambridge. This helped build up for Myers a uniquely comprehensive, perhaps schizophrenic, account of consciousness from which his thinking could draw. Cox, if nothing else, was an attentive scholar of this alternative psychology with roots going back to the Count of Cagliostro, Emanuel Swedenborg, Franz Mesmer, the Marquis of Puységur, and all varieties of mentalists and magicians beguiling viewers on fairgrounds with claims of their feats of psychic power. Thought reading, charismatic healing, fortune telling, levitation, possession, and hauntings—all the archetypal attributes of the magical mind—where they were not outright denied, were studied instead, creating a scholarly literature of consciousness that expressed the Enlightenment's secular impulse in conflict with itself. This uniquely Western project to secularize and psychologize magic did not begin with the subliminal self (though it was arguably consummated by it); it rode a longer wave of opposition to the mechanization of the body and the disenchantment of the mind. Its most significant idiom, the alternative consciousness paradigm, started with Mesmer in the 1770s and led directly to Myers along a fairly narrow corridor of hypnotic research, which he retraced as part of his work for the Committee on Mesmerism, drawing inferences from the spiritualistic and magnetic phenomena that crisscrossed this medical history and fed his own personal intellectual interests as well. Mesmer's dissertation "On the Influence of Planets on the Human body" (1766) was a heterogeneous mix of Newton's theory of tides, mainstream medicine with its pumps and fluids, and a little Paracelsian biomagnetism thrown in as a secret ingredient.[92] Mesmer at first practiced conventional medicine for ten years, trying magnets only as a guess, based on the cosmos he described in *Planets*, a world of attractions, effluvia, and energies needing to be balanced. He does not add to electricity, magnetism, and gravity, but in the mixing and balancing it becomes more esoteric, especially when it involves the vital element in humans, "animal magnetism." Mesmer's initial cure was spectacular, progressing through trial and error to the use of Mesmer's own "personal" magnetism as a better, more direct effect. But even though he was practicing resultative medicine and "doing no harm," his Parisian clinic could not survive the threat to medicine itself: doctors did not heal through personal powers, but through fixed procedures, and by grant of professional license. The trailing gowns and dramatic staging did not help either with the ethos of standardization.

The expulsion by the Franklin Commission was decisive, making Mesmer and mesmerism the eternal pariah. Mesmer's protégé, the Marquis de Puységur, had to take this floorshow "inside," retreating to his country estate where he became more and more preoccupied by mesmeric states of consciousness. Where "animal magnetism" used the body to cure the mind (establishing homeostasis), Puységur began to discover that the mind could cure the body as well, through thoughts alone. This was the beginning of talk therapy and hypnotic suggestion, but there was something more here: a moral, intellectual, and even spiritual component at work in mesmeric sentience. Puységur had discovered the "higher self" in the depths of consciousness, whose line of descent would lead only to more esoteric phenomena (thought transference, clairvoyance, rapport, mesmeric possession, and so forth) but would also provide the fountainhead for Romantic psychology. Poets, artists, and psychonauts began to explore what Samuel Taylor Coleridge labeled the "under-consciousness": an undiscovered Atlantis of unimaginable cognitive riches, where the relationship between art and beauty, God and man, was to be worked out in the realm of human creativity. (As historian Alan Richardson has pointed out, Romanticism had its own wing of experimental psychology, which included an interest in psychoactive drugs, Eastern religion, altered states of writing, and even brain anatomy.)[93] This was the rich history catalogued by the Psychological Society, an otherwise dispersed and denied body of knowledge that at last found a remarkable curator in Cox. By joining Cox's venture, Myers specifically addressed himself to this history, not just the mental phenomena intersecting with spiritualism, but also its vocabulary and philosophy, as well as its occasional points of crossover into mainstream science.

Whatever Myers's initial hopes were when he joined the society, by the end of his tenure, those hopes had been rather bitterly disappointed (once again). He described the dreary interregnum between 1876 and 1881, which marked the end of his formal interest in spiritualism, on the one hand, and the beginning of psychical research, on the other, as a period in which he could work "but fitfully," feeling "half-sickened" by "Craft, with a bunch of all-heal in her hand [spiritualism]" and the surrounding "vassal legion of fools," among whom he no longer wished to be numbered.[94] But across the Channel, in the heart of materialistic French neurology, a revolution was already under way, news of which had yet to reach English shores. But it would come eventually with tidal force, catching up Myers in its onrush of experimental discovery concerning the unconscious mind. Starting in 1881, as details of this new, excavated terrain began to emerge more fully, Myers was once again resolved to continue the pursuit of his elusive quarry, not as a visiting spirit at a séance or ghostly writing on a slate, but as the conundrum of consciousness, in the riddle of the brain, whose mystery was the calling of psychology. Perhaps the expectations raised by French psychiatric hypnosis were unreasonably fired by Myers's grasp of the literature (after all, the medical survey in France began and ended with Braid's *Neurypnology*). But it did not matter, because he sensed something new, something truly profound in its discussion of consciousness. It had so turned out that Myers's "limitary parapet and enchanted portal of the world" would prove to be the prison gates of an old gunpowder factory turned women's sanitarium, the Pitié-Salpêtrière hospital in Paris.

CHAPTER FOUR

"The Incandescent Solid beneath Our Line of Sight"

Frederic Myers, the Self, and the Psychiatric Subconscious

PSYCHOLOGY LOSES ITS MIND

The women in the photos wore immaculate white, their poses staged in a serene, clinical setting. Often they were photographed with their beds plainly visible, a nurse present to chaperone those who might not otherwise be able to control themselves. With their thin nightgowns and air of prostrate abandon, the patients gave the medical voyeur a sense of total access, satisfying his longing for unilateral intimacy as his diagnostic gaze peered past these intimate arrangements into the darker disarray of their innermost minds. On display were not just the physical symptoms of hysteria, but the very experience of it: the first-person drama of the patient told through a diagnostic portraiture complete with theatrical lighting, props, and poses. Consciousness itself had become the new object of scientific curiosity, and madness its most promising hunting ground. In that instant of the *attaque d'hystéro-épileptique*, the superficial armature of identity fell away, and the hidden gears of subject formation were laid bare. One moment the patient was in repose, completely "herself"—but turn the page, and her whole body had become convulsed, her face rigid with a telltale lunacy. The onset of such seizures could be quite violent. One woman gave the appearance of having been decapitated, her head under a pillow, stomach arched upward in the agonized *arc-de-cercle*.[1] After this initial crisis came a kind of decompensation or lethargy, followed by the final crisis: the patient entranced within her own delusional reality. She became a Medusa fixing a menacing stare, a fantasist dancing like Salome, a saint praying, a lurching automaton, a maniac in the throes of *erotica delirium*.

Such was Jean-Martin Charcot's *Iconographie photographique de La Salpêtrière* (1876–80), timed for release as a single, elaborately produced volume just before his trip to the London International Medical Congress in 1881. With its narrative use of imagery, as well as the number and quality of photographs, including forty large plates, the volume grabbed the attention of readers. This visual sequencing of the *mise-en-scène* of madness was something of a medical breakthrough, fixing as a photographic object psychiatric ephemera otherwise lost behind institutional walls or domestic shutters. Medical interest was also no doubt raised by the fact that the chosen subjects tended to be young and appealing, with the camera savvy of movie starlets. If not, then the *arc-de-cercle* would do in a pinch. In addition to its illustrations, the text totaled over 350 pages, with clinical notes documenting the patient's verbal stream of consciousness as her sanity flowed away. One reviewer for the *British Journal of Medicine*, scandalized by the general lubricity of the contents, complained of "long pages of the obscene ravings of delirious hysterical girls, and descriptions of events in their sexual history."[2] But perhaps the most curious elements of this lurid photographic essay were the tidy pen-and-ink drawings, placed in a separate section, depicting somber men of medicine clearly in the midst of mesmerizing their patients. The facts implied by this juxtaposition were revolutionary in both theory and practice: here the hypnotic state of trance and the psychotic break of hysteria were part of a single neurological continuum, and thus, tripping the one could trigger the other.

Until then, only limited reports of Charcot's researches into hysteria had circulated in Europe, just enough to send rumors gusting through the halls of science, but the exact nature of his project remained poorly understood. The *Iconographie photographique* was the trumpet blast preparing the medical community for his pending ascendance over their field.[3] He was hardly the first to hunt down the infamous *bête noire* of hysteria. It had been under modern medical investigation since the 1850s as a psychosexual disease of the nerves (as opposed to an affliction of the womb, per Hippocrates), yielding, in the words of Pierre Briquet, a "Proteus which presents itself in a thousand guises and cannot be grasped in any of them."[4] But Charcot's situation was uniquely conducive to success: in 1863 he had been placed in charge of a facility numbering some five thousand patients, over half of whom were afflicted with neurological diseases. He took upon himself to diagnose and classify these diseases, creating a potential cornucopia of case studies. For all his failings assigned in retrospect, Charcot offered a significant step up for the inmates of the Salpêtrière at the time, allowing the institution to flourish as a teaching hospital rather than a custodial facility. This new psychiatric interest he spearheaded within neurology allowed these inmates to be redefined as medical patients to be treated, not social problems to be detained.[5] But his real research advantage came with the recognition that hypnotic trance was itself a hysterical symptom, a case he argued empirically through clinical observation. He noted the similarities, such as physical rigidity, extreme suggestibility, the tendency toward amnesia, the disinhibition of libido and emotion, catatonic lethargy, and even the signature sleepwalking of the deepest states of hypnosis. Ultimately, Charcot found he could use hypnosis to medically induce the hysterical crisis itself, collapsing mesmeric trance and hysterical *somnambulisme* into a unified field of study, with a passkey to altered states of mind.

By 1878, Charcot was using hypnotic induction to dramatically accelerate the pace of his research, conjuring up hysterical symptoms at his medical convenience. The essential point for Charcot was to substantiate hysteria as a real disease with a definite, physiological basis, no more "psychological" in its etiology than gout, paraplegia, goiters, sclerosis, wasting disease, or other anatomical afflictions for which his contributions were already celebrated. The only difference was that the principally afflicted organ for the hysteric was the brain, and the manifest symptomology was mental illness.[6] For all the highly personalized elements of their psychiatric dramas, the patients in the photos all conformed to Charcot's rigorous disease profile, moving in lockstep through a three-stage hysterical nosology: first, catalepsy; second, lethargy; and third, the total disassociation of trance somnambulism. This last phase signaled that the disease had become acute, uncoupling the brain from external reality. Such patients wandered lost in the mirrored hall of their lunatic imaginations, periodically caught and released by powerful urges they could no longer regulate or even understand.

Charcot's neurological approach offered certain advantages for those seeking "a truly scientific psychology." That meant building up knowledge experimentally and working within established physical theories. The path Charcot struck began with the nervous system, and progressed through dissection, experimentation, and clinical observation, correlating the subjective and objective conditions of brain disease. But the ultimate goal was to arrive at a general theory of consciousness. Through painstaking observation of the broken brain, science would eventually grasp its normal function.[7] Without such an initiative, psychology would remain a no man's land, stranded between physiology and philosophy. Given the allure of his project, it did not take long for a line of intellectual pilgrims to beat a path to his door.[8] He had managed to eclipse even the ascendancy of German experimental psychology under Wilhelm Wundt, which approached consciousness as a quantifiable neural process converting stimulus into perception.[9] The goal was to calculate the energies spent in performing this "work," thus bringing the mind fully within the framework of energy conservation laws, and psychology within the realm of physics. But for all the scientific appeal of Wundt's Leipzig laboratory, an international draw since its establishment in 1879, these granular data did not add up to mental experience. The Paris sanitarium was potentially a more productive setting for this enterprise. What Charcot had delivered with his *Iconographie photographique* was more than just the secret selves of women in nightgowns who were off their rockers; here was nothing less than the secrets of the self, that endlessly complex surge of neural activity out of which human identity cohered.[10] And Charcot's state-funded clinic could offer something Leipzig could not: a ready supply of experimental subjects for whatever studies could reasonably be justified. Charcot boasted of "a casting annex and a photographic studio; a well-equipped laboratory of anatomy and of pathological physiology ... the teaching amphitheater ... with all the modern tools of demonstration." The momentum now shifted irresistibly toward this more theatrical psyche of the Salpêtrière, remaining there for the next two decades.[11]

Yet, for all his stagecraft, Charcot made clear that "hypnotism" was to be construed conservatively, as nothing more than a physical shock to the nerves, restricting cerebral energy. These were the auspices by which hypnotists like James Braid and Rudolph Heidenhain had likewise courted science, emphasizing hypnotism's use as a medical instrument and framing it within a psychophysiological paradigm that excluded any special force.[12] But the hysterical crisis of *somnambulisme* was not the respectable trance of midcentury surgical hypnotism, with the patient lying insensible in an anesthetic coma. To the extent such "altered states" had found some respectability within medicine, they were first voided of an active will. But not so the perverse saints and villainesses on display in the *Iconographie photographique*. Here the hypnotic pass did not shut off sentient activity, but rather instigated violent forms of it, setting off strange firestorms of willful, runaway personalities, hot spots of subregional selves flaring up across the brain. Charcot's protocols were devised to aggravate (not anesthetize), pursuing ever greater penetration of his patients' subconscious drama with ever more aggressive techniques: Electric shocks, the Chinese gong, loud noises, a flash of magnesium, or electric light were all used in order to poke and prod the mind in and out of madness.[13] However constrained Charcot's notion of hypnotism, it replaced Braid's "on and off switch," with a spinning dial pointing toward endless varieties of self-initiating sentience, waking the mind from mesmeric sleep to the nightmare of hysterical *somnambulisme*.

There were also the strange physiological symptoms of Charcot's hysterics, suggesting an unsettling degree of plasticity in body as well as mind. A flash of light or a sudden sound might produce hysterical blindness, spontaneous paralysis, a sharp pain or numbness, or fluctuations in body temperature, heart rate, or emotional states, driven by the hysteric's own runaway self-transmogrification. Charcot's "conservative" neurology thus encouraged a far more porous understanding of the mind/body relationship than conventional biology would normally allow.

Packed in between these headlines of hysteria and hypnosis was the obscure clinical practice of "metallotherapy," harking back to the most arcane varieties of mesmeric influence. Charcot first witnessed the phenomenon in 1876, during a demonstration by the magnetist Victor Burq organized by his intern Charles Richet. With a pass of a magnetized metal, Burq was able to induce tremors and paralysis in the bodies of Charcot's inmates, moving their symptoms about from limb to limb or even melting them away. Charcot, intrigued, led a formal study the following year, sponsored by the Société de Biologie, at the conclusion of which metallotherapy became a regular part of his clinical regimen. Thus, Charcot had allowed biomagnetism to gain an early foothold at the Salpêtrière, trailing its possible field theory of remote energetic "influence."[14] (Animal magnetism by any other name . . .) From the start, then, Charcot seemed to be sailing close to the wind, but always, as will become clear, with Charles Richet's hand discreetly on the rudder.[15]

And now, in 1881, the once despised mesmerism, chased out of town at the end of Franklin's pitchfork, was being given a hero's return by a man capable of convincing the world that there was "no man was more opposed to quackery."[16] It took someone like Charcot (head of the Salpêtrière since 1863, chair of neurology at the University of Paris since 1872, and now, the parliamentary appointee to the chair of diseases of the nervous system) to place this forbidden curiosity under his protection. This time the return of the repressed came pulling a new dynamic model of the subconscious in its wake, the depths of which had yet to be sounded. In the audience during Charcot's barnstorm was somone particularly receptive to his preaching: Arthur T. Myers, a practicing physician, brother of Frederic and companion in his many spiritual investigations. Here at the congress, Arthur was moving in his own professional milieu, but he was also a patient, having suffered from temporal lobe epilepsy since he was twenty years old.[17] That condition, along with his medical and personal curiosity, had already made Arthur something of a study in the phenomena of hypnotism. He, of all people, could see its complete geometry: as a resurgent therapy, a scientific frontier, and the outer strand of a deeper mystery entwining mediumship, hauntings, psychic force, and other unknowns of conscious experience (to which he was subject with every seizure). The many-chambered mind Charcot now invited science to explore might not be so readily contained by the brain his neurology described.

In the wake of the congress, Arthur Myers made at least one trip to the Salpêtrière that year to see for himself the truth of these wonders. He met with Charcot, and also made a side excursion to Nancy, where a quiet country doctor named Liébeault had for decades been using hypnotic suggestion as medical therapy. It was Liébeault's theory of suggestion that most inspired Arthur, but nothing matched the scale of Charcot's unfolding research enterprise. Whatever the theoretical limits of his neurology, the scope of Charcot's quest and the technical brilliance of his endeavor created a thrilling sense of possibility. Frederic Myers described the powerfully resurgent mood among the disbanded Sidgwick Group that Charcot's work inspired: "It was to men wearied but not broken, discouraged but not despairing, that at the end of 1881 a fresh call to exertion came."[18]

Far from challenging Charcot's neuropsychiatric approach to the subconscious, the Society for Psychical Research based itself in the penumbra of possibility cast by "hysteria," pitching its outpost in London the same year that Charcot opened the doors of the Salpêtrière to found the world's first neurological research clinic. The SPR's principal areas of interest were, according to Frederic Myers, "thought transference as the primary aim, with hypnotism as its second study, and with many another problem ranged along its dimmer horizon."[19] These investigative priorities placed the SPR well within the extended borders of this psychological *terra nova*, as did their principal identification as an experimental discipline. Psychical research was thus, from the start, conceived and designed as part of the formal research initiative under way in experimental psychology, assuming a degree of Charcot's institutional protection even before it was formally offered. It took until 1885 for it to gain its intended traction at the Salpêtrière, at which point Myers would move into the heart of its intellectual apparatus for a brief but extraordinary tenure. Here he began to forge his alternative model of the dynamic subconscious, one that could assimilate all the abnormal and supernormal data exploding across this field of investigation.

THE SOCIETY FOR PSYCHICAL
RESEARCH AND THE SUPERNORMAL MIND

The Society for Psychical Research was founded in London on February 20, 1882. The sense of mission it renewed in Frederic Myers was not a rekindling of the old Evangelical fires of his youth or a return to "the gropings and the *tâtonnements*, the disappointments" of his recent spiritual researches.[20] Nor should we view this as a kind of redux of the ineffectual amateurism of his most recent venture, the Psychological Society, with which the SPR should never be confused. What Myers and a few dedicated colleagues had in mind was something entirely new and fully disciplinary. They would not repeat the miscalculations of the past. Myers had come to realize that "in beginning inquiry with so-called Spiritualistic phenomena at all, they were somehow beginning it at the wrong end."[21] Any such question must be approached from the gateway of matter, arising organically from the observational data as part of an inquiry judicable by science. To designate this psychical terrain as part of a knowable reality, Myers coined the term "supernormal," a word intentionally "formed on the analogy of abnormal."[22] While unusual, psychical phenomena were not *un*natural. Their possible existence in no way threatened the established facts of known physical laws, but they did claim some space in nature free of the determinations of those laws. As suggested by its prefix, the supernormal was part of nature's upper deck, physically contiguous with, but not necessarily contained by, matter. It "exhibited the action of laws higher, in a psychical aspect, than are discerned in every day life . . . apparently belonging to a more advanced stage of evolution." This was a propositional naturalism of a very particular kind, holding out the power to obviate the dyad "natural and supernatural" altogether by redistricting reality in such a way that it contained both those elements. Thus, regardless of its positivistic practices, the SPR asserted a disciplinary object with the potential to upend the physicalist dogma entrained in its own experimental verification. And now, the opportunity to merge this supernormal initiative with new trends in abnormal psychology was at hand.

In terms of an immediate experimental model, however, the SPR did not look to the medical pathologists at the Salpêtrière. Instead, the society focused on lines of inquiry initiated several years earlier by William Fletcher Barrett (later Sir William), a professor of experimental physics at the Royal College of Science for Ireland. Barrett had been independently investigating thought transference (the society's designated priority) since the mid-1870s, and according to Myers, his results were "already rising within measurable distance of proof."[23] Barrett had even presented his research into thought transference in a paper read before the meeting of the BAAS at Glasgow in 1876, titled "On Some Phenomena Associated with Abnormal Conditions of Mind." He was unable to penetrate disciplinary curiosity any further, facing the head winds stirred up by Crookes a few years earlier.[24] While Barrett had looked somewhat condescendingly on Edward Cox's Psychological

Society, joining only as an honorary member at Myers's behest and submitting nothing for publication, he threw himself wholeheartedly into the psychical endeavor.[25] Here, he was a founding force, helping to structure the SPR's earliest research protocols along the lines of his own professional standards. Even the eminent Henry Sidgwick agreed to risk lending his dignity once again to this promising venture, bringing instant *gravitas* to the society in his role as first president. Myers himself was content to sit on the council, joined by Edmund Gurney, William Fletcher Barrett, Arthur Balfour (future prime minister), Edmund Dawson Rogers (journalist), and Henry Sidgwick's wife, Eleanor. Other councilors of note included Frank Podmore (future founder of the Fabian Society), Hensleigh Wedgewood (eminent etymologist and cousin to Darwin), and Gerald Balfour (brother to Arthur), as well as the physicist Sir Balfour Stewart, a Rumford medalist and celebrated pioneer of heat mechanics, and Lord Rayleigh (future Nobel Prize-winning physicist). There was a consensus among the founders that nothing about the objective nature of these so-called psychical phenomena could be assumed, not even their real existence. The society's mandate was merely "to examine without prejudice or prepossession and in a scientific spirit, those faculties of man, *real or supposed*, which appear to be inexplicable on any generally recognized hypothesis."[26] The term "occult" was deliberately left out of the SPR's operational vocabulary after much debate, because it brought with it an unwelcome air of mystery. This posture of resolute agnosticism might at first appear purely renunciative, an abnegation of metaphysics to achieve the necessary entrée. But such constraints worked in the other direction as well by suspending *all* dogmatic prejudgments, meaning the physicalist dogma that had come to blinker the scientific imagination. Psychical researchers were not commanded to inhabit the cosmos of thermodynamics, composed only of "ether and atoms and where there is no room for ghosts."[27] They could wait and see. This in no way refuted the validity of such scientific theories, which stood on their own efforts and proofs. The point was to question their *totality*.

This somewhat open intellectual border initially accommodated a sizable contingent of genteel spiritualists led by the Rev. Stainton Moses, but we should not let this early composition characterize the society's agenda. Psychical research was not scientific spiritualism by another name.[28] Spiritualism as an explicit object of investigation remained off the roster during this early trial of disciplinary incubation, though some of its phenomena were otherwise represented in mesmeric and apparitional lines of research. The SPR founders agreed that the society could not be, by either fact or reputation, a continuation of that earlier, failed venture and still hope to achieve its disciplinary aspirations. Sidgwick had refused the presidency until he could be assured that the spiritualists did not expect a similar welcome for their spirits and spirit guides in the current line of research. The aim was to work within the professional standard, not alongside it, below it, or above it. In the past, a respected clergyman like Moses or a noted intellectual like Sidgwick might have tried to wade into the middle of modern scientific debates with his extrascientific credentials, but times were changing (faster now because of all the boundary confusion that spiritualism had stirred up).[29] There was only one way for the SPR be a significant part of a scientific conversation, and that was to be a scientific discipline. Certainly, compared to Crookes's antagonistic launch of "a truly scientific spiritualism," the attitude of the SPR toward institutional authority was far more deferential.[30] Where Crookes had faced off against the Royal Society and tried to instruct its scientific agenda from the sidelines, the SPR assiduously courted inclusion within the establishment. The goal was to eventually obtain full disciplinary recognition, with its own berth in Section D of the BAAS (but not prematurely, at the expense of intellectual freedom).[31] While the founding roster was stocked with social and philosophical elites, it also included distinguished scientific practitioners, who were in many ways at the heart of its prestige. This was not a social club or a place for vague curiosity. The society's steering committee sought out productive intellectuals who could fully implement the SPR's research mandate, recruiting anyone trailing an "FRS" after their name, especially those in the "hard sciences" known for their experimental rigor.[32] There is nothing quaint here. Psychical researchers did not label themselves "scientific" as *they* defined the term, but according to the disciplinary standards of the psychology of the day, which was the branch of inquiry to which they had attached themselves. While those disciplinary standards were admittedly nascent, there was nonetheless a definite structure to this legitimacy. At its center, as of 1882, was Charcot's neurological clinic at the Salpêtrière, with which the SPR sought (and eventually attained) a strong mutual alignment, eventually drawing into its sphere high-profile European researchers in psychology, physiology, neurology, psychiatry, and psychophysics, such as Babinski, Beaunis, Bernheim, Binet, Dessoir, Féré, Flournoy, Janet, Liébeault, Liégeois, Lombroso, Ochorowicz, Ribot, Richet, Taine, and von Schrenck-Notzing.[33]

Since one of the aims of this book is to reassert the ties between Victorian psychical research and academic science, it is important to document the extent of its disciplinarity. These ties are not confined to methodological considerations or participation in professional networks. The purported "faculty folded up inside of man" had implications for the basic ideological disposition of science being worked out internally among professionals and fretted over publicly in cultural opinion. The question as to whether mechanical theory would continue to move from nature to body to brain, or whether some mental principle would push back in the opposite direction, came alive with the science of mind in the late nineteenth century. Precisely as a discipline, psychical research was a way to focus these ambient philosophical and epistemological concerns attached to the idea of science as part of its formal research program. Institutionally speaking, science was not set up to have that larger discussion, or even to engage with theories outside its circle of prediction. Consequently, at the organizational level, the SPR did not formally assert a theory of psychical phenomena or even offer official confirmation of their existence, a point worth emphasizing. The members were not propagandists making a point; they were researchers seeing what could be said. Even when leading members, such as Myers, Gurney, and Lodge, felt the preponderance of evidence was pointing in the direction of proof, as officers of the SPR they refrained from ordering an institutional endorsement of that opinion. To do so, they knew, would be to exit the scientific worldview, which did not recognize such a force, and what they really wanted to do was to remain and change it. Nonetheless, critics tarred the SPR with pseudoscientific beliefs that the organization did not officially hold.[34] Historians continue to engage in this sort of characterization today, though in a far less antagonistic spirit, when they position psychical research as part of the continuum of Victorian supernatural curiosities and Gothic entertainments. While movements like scientific spiritualism, mesmerism, and theosophy intersected with psychical research in terms of some individuals and certain points of inquiry, the SPR's methods, along with the epistemological values that informed them, were ultimately antagonistic to these enterprises. This commitment to skepticism sets it apart from the tendency toward "revelation" in the pronouncements of Helena Blavatsky, Emma Hardinge Britten, or Aleister Crowley. Even though Lodge, Myers, and Lang did at times argue positions well beyond the psychical evidence, they did so *outside* of the formal publications, acting as speculative philosophers furthering the science, not chosen sages channeling the divine.

Of course, even to investigate such phenomena was to closely entertain the possibility of their existence, suggesting the existence of a not-so-skeptical bias, but was that inclination necessarily *un*scientific? Given the unexplored nature of consciousness, couldn't the same be said of those discounting such a possibility prior to proper investigation? As a body, the SPR existed mainly to referee and coordinate that missing research effort, which, as of 1882, had yet to be properly undertaken by any branch of science and, given the unanswered claims of spiritualism and the clinical research piling up at the Salpêtrière, seemed urgently overdue. Such an agenda could just as readily accommodate foreclosing on the possibility of such phenomena as endorsing it, so long as something definite had been ascertained and added to the foundations of knowledge. It must also be considered that the ultimate utility of psychical research required a high degree of disciplinary compliance. In order for psychical data to infiltrate other disciplinary discourses, the data had to be obtained and presented in a way that such an audience would find compelling. To that end, the SPR issued both a *Proceedings* and a more abbreviated *Journal*, with strict standards of peer review governing publication, so that readers could formulate a critical opinion regarding what they read. That meant authors had to fully describe their experimental protocols as an essential part of presenting their data. Additionally, any hypothesis ventured in its pages had to be warranted by the evidence at hand and interpreted with reference to (if not necessarily in compliance with) the established laws of science. This often made for dry, laborious reading, as researchers attempted to establish the objectivity of their process for the reader, regardless of what their privately held hopes might be. Clearly, these publication were not pandering to the public, but nor had they captured the attention of psychology.

The *Journal* was added to the flagship *Proceedings* in 1884, "for gratuitous circulation among the Members and to ensure to our Members and friends a speedier knowledge of matters of interest," in the hope of raising the society's profile.[35] The first two years were slow going in terms of expanding academic interest beyond the initial contingent. The American psychologists William James and Stanley Hall were the only high-profile researchers to join in 1883, while Charles Richet and Charles Féré joined the following year, suggesting that some small momentum was building on the continent beyond the anglophone world. While the council itself refrained from offering any endorsement of researchers' conclusions, the obvious rigor it enforced laid the grounds for others to develop their own convictions (or at least curiosity). As an organization, the SPR had to walk a line, relying on lay members to support the work, while aiming to become part of an exclusive professional discourse.[36] Eventually, this restrained, scientific propagation of knowledge paid off, allowing psychical research to fly in formation with the rest of French experimental psychology by the middle of the decade. An extensive academic library kept members abreast of key disciplinary developments, and its publications always included a thorough literature review of new and exciting works. This attention to other disciplinary developments fully embedded psychical researchers in a larger intellectual network, such that they might become purveyors of the latest scientific information and not just consumers of it. As researchers refined their methodologies toward a higher empirical standard, they enriched experimental procedures for all of psychology with special card decks, tests for eidetic (pictographic) as well as verbal telepathy, and the statistical analysis of results (originally worked out by Charles Richet, the first of the Salpêtrière's circle to break ranks to become a corresponding member of the SPR).[37]

One thing becomes very clear in reviewing both the published works and the private correspondence of these early researchers: they held the scientific character of the SPR as essential to its mission and fundamental to the satisfaction of their own intellectual curiosity. In order to scientifically validate these phenomena, it was necessary to practice a valid science. To do otherwise would be a pointless exercise from both an institutional and a personal point of view. Even though its leading researchers, Gurney and Myers, were Cambridge classics majors, they had a highly disciplined intellectual curiosity concerning the physical and biological sciences. Myers prepared to sit for the natural science tripos as an undergraduate, while Gurney actually passed the Cambridge examination for a bachelor's of medicine in 1880. Both read voraciously across psychology, physics, chemistry, and physiology. This made them gifted polymaths versed in the content of science and its habits of mind, rather than two interlopers from the humanities bringing their gauzy speculation, as might be assumed at first blush. William James best summed up Myers's own astounding scientific transformation in his memorial address written on the occasion of Myers's death: "Brought up entirely upon literature and history, and interested at first in poetry and religion chiefly . . . Myers had as it were to re-create his personality before he became the wary critic of evidence, the skillful handler of hypothesis, the learned neurologist and omnivorous reader of biological and cosmological matter, with whom in later years we were acquainted."[38]

This intellectual transformation came with a more personal one as well. The original Christian impetus that had guided Myers's spiritualism in the early 1870s no longer took the lead. His search for the simulacra of the miracles of Jesus had become a far more open-ended inquiry, one that involved rethinking even fundamental assumptions about "theism" and "creation." Christian miracles and holy saints were subsumed in a growing body of psychical and psychiatric explanation that abandoned theological explanation in favor of some new, yet-to-be-named interpretive scheme that could place them in the natural order. This did not mean there was no *potentially* spiritual or religious significance in the psychical project. But for Myers and others, that was a strictly personal matter. Any metaphysical hypothesis in excess of the facts threatened its own evidential basis. Whatever latent spiritualistic desires were being harbored by these early psychical researchers, they were threatened rather than served by the active spiritualists in their midst. Those members who could not fully share the proper scientific attitude were left off committees, marginalized at meetings, and left out of the *Journal* and *Proceedings*. The spiritualists found themselves on the receiving end of this disciplinary policing, causing Stainton Moses to complain on their behalf that "spiritualism was not being properly entertained or fairly treated by the Society for Psychical Research," and in 1886 the spiritualists departed without any overtures urging them to stay.[39] The SPR's enrollment, already at one hundred fifty by the end of the first year, was now over four times that number, with over twelve thousand *Journal* subscriptions distributed to libraries, universities, and private homes throughout Great Britain and America. The SPR was also by then fully networked into the academic circles of aberrational psychology in Europe, Russia, and America.[40]

It is interesting to note that it was only after this purge, when the membership that remained shared a primary commitment to a scientific program, that the subject of séance mediumship began to make its way more freely into the official research agenda. Eleanor Sidgwick herself published "Results of a Personal Investigation into the Physical Phenomena of Spiritualism" in the *Proceedings* around the time of Moses's departure.[41] The society's large-scale and sustained investigations of the famed mediums Leonora Piper and Eusapia Palladino did not ramp up until the coast was fully clear, beginning in 1889 and 1894, respectively. The original committee structure established in 1882 pointedly excluded séance mediumship from the psychical agenda, while creating separate committees for thought transference and mesmerism as distinct branches of psychological inquiry. (The SPR chose the term "mesmerism" over "hypnotism" for the committee name, to designate the inclusion of a wider range of physical and psychical influences interdicted by the "hypnotism" approved for deployment in professional settings and discussions.)[42] Along that "dimmer horizon," four additional committees were formed to cover the more fringe phenomena. These encompassed Reichenbach's Odic force (a vitalistic energy field emitted by all living objects to which certain individuals were sensitive); the personal testimony of reliable individuals regarding apparitions and hauntings; Crookes's psychic force (or, more generally, the causal laws behind the physical forces and effects attributed to spiritualism); and last, a literary committee to gather and review particularly compelling instances of all phenomena under review by the society. By virtue of their inclusion in the psychical program, all these objects of investigation were unified at their point of origin: consciousness. (To be psychical was to be, by definition, "of the mind.") A dedicated committee, however, in no way implied an endorsement of the phenomena under investigation.

To the extent that ghosts were admitted into this lineup with the committee on hauntings, they were treated as purely "subjective" phenomena, assumed to unfold within the mind of the observer. This was not to treat such visions *en masse* as purely delusional. All options were on the table: these experiences could be attributed to indigestion, fever, or fright, on the one hand, or viewed as an actual telepathic communion with the mind of a disembodied spirit, on the other. Either way, the emphasis was taken off vulgar, corporeal spirits like Katie King and retooled as a perceptual phenomenon. The testimony gathered by the Literary Committee or submitted directly to the *Journal* came from ordinary, unremunerated individuals offering their direct, personal recollections of seemingly supernatural experiences. Even so, such accounts were presented to readers as "cases," purposely employing the distancing language of psychiatry to quarantine psychical objectivity from the personal point of view. The committee (Myers and Gurney) placed an advertisement asking for "any good evidence bearing on such phenomena as thought-reading, clairvoyance, presentiments, and dreams, noted at the time of occurrence, and afterwards confirmed, unexplained disturbances in places supposed to be haunted; apparitions at the moment of death or otherwise; and of such other abnormal events as may seem to fall under somewhat the same categories."[43] Hundreds of letters came pouring in, as the committee's request triggered a catharsis of repressed encounters too long held in reserve. To warrant consideration as evidence, however, some kind of verifying, third-party piece of information was required, although one's good character also served as an important element of his own endorsement. These stories were relayed not by spiritual mercenaries or captive madwomen, but by judges, policemen, clergymen, and respectable matrons. These accounts were not "pressed eagerly by vain or imaginative informants; Rather they were for the most part won with difficulty from opposing reserve."[44] Bundling these scattered, underground anecdotes together, attaching names and institutional certification, suddenly raised these half-heard whispers in the dark to a whole new level of scientific persuasion.

ADVENTURES IN CONSCIOUSNESS

For the society's inaugural investigation, Myers and Gurney teamed up with Barrett to continue his highly successful series of investigations into thought transference begun the year before, showcasing the extraordinary abilities of five daughters of the Rev. Thomas Creery, ages ten to seventeen, who would ride their growing fame right past the Fox sisters to become the new darlings of supernatural display. The girls performed their feats of non-sensory communication without benefit of hypnosis or any other alteration of consciousness. This ordinary brain-state was significant. Had the girls been entranced, their communication might possibly have been attributable to some form of sensory hyperacuity that exaggerated ordinary channels of perception. It also confirmed that "thought reading" was a potentially normal capability, exercised here by healthy, wholesome girls, breaking the association with altered, pathological, or even possessed states of mind. This was a well-chosen introduction for the SPR, conveying its aspirational tone while showcasing Barrett's scientific methods. (The typical psychical subject was a cooperative and consenting individual without mercenary motives, who walked a middle ground between the abject psychiatric inmate of the Salpêtrière and the manipulative medium of the séance.) These angelic cler-ical offspring were ideal candidates for Myers's experimental foray after his rough handling by the Newcastle mediums, who lacked all due deference to the demands of science.

A sample of the report prepared by the committee gives a sense of these early trial protocols, which Oliver Lodge would subsequently configure more stringently. The Creerys were carefully policed and not given a pass as a clergyman's daughters. Here, as an extra precaution, only the research-ers knew the object to be guessed, and they communicated this information only after the designated "mind reader" had left the room:

> One of the children was sent into an adjoining room, the door of which I saw was closed. On returning to the sitting-room and closing its door also, I thought of some object in the house, fixed upon at random. We then all [meaning the sisters as well] silently thought of the name of the thing selected. In a few sec-onds the door of the adjoining room was heard to open, and after a very short interval the child would enter the sitting-room, generally speaking, with the ob-ject selected . . . I wrote down, among other things, a *hair-brush*; it was brought: an *orange*; it was also brought: a *wine glass*; it was brought: an *apple*; it was brought: a *toasting-fork* failed on the first attempt, a pair of tongs being brought, but on a second trial it was brought.[45]

In Myers's personal addendum he noted "a far more vivid impression of their genuineness than the bare printed record can possibly convey," and as-serted, on behalf of his team, that "we could find no resemblances between these phenomena and those known as mesmeric; inasmuch as a perfectly normal state on the part of the subject seemed our first pre-requisite."[46] With that last observation Myers emphasized that (a) the phenomena were not produced through the effects of trance and thus could not be catalogued as neuropathological; and (b) no controlling "influences" appeared to be asserted by one party over the other, of either a biological or mental variety. The rapport between the girls was of a more mutual kind, one that left their mental faculties unperturbed: an extraordinary means of quite ordinary communication It was Myers who had suggested substituting the term "thought reading" for the commonly used term "thought transference," to reframe the association between minds on more democratic terms (updating that term later that year, with his neologism "telepathy"). While the hypnotist "transferred" his thoughts to the empty vessel of the hypnotized, in "thought reading" half of that agency was shifted to the percipient, who actively read and received what the other had transmitted.

The vast implications of this study required confirmation from the highest possible authority, preferably from a source outside the society. Edmund Gurney thought immediately of his old friend Oliver Lodge, who was fast becoming one of the most eminent ether scientists of the day. Unlike Myers and Gurney, Lodge was a trained physical experimenter, and unlike Barrett, he had no ties to psychical research, making him a perfect candidate. Additionally, Lodge was currently located in Liverpool, where a traveling conjuror had recently inspired a series of thought-reading tests all over town, uncovering the uncommon capabilities of two otherwise ordinary shop girls at a drapery factory. A furor ensued of unquenchable curiosity. The local excitement kept building until an appeal was finally issued to the SPR to come and investigate. The eyewitness report came from an esteemed local magistrate, one Malcolm Guthrie, who was at once the head of a law firm, a judge, and brother of a respected scientist. Any perceived Liverpudlian deficit of character inhering to this working-class drama was thus overcome, and the situation was deemed worthy of scientific attention. Furthermore, the similarity of these phenomena to those of the Creery sisters presented exactly the opportunity for testing those results sought by the Committee on Thought Transference.

Nonetheless, Lodge declined the request to investigate. Despite his continued warm relations with Gurney, he was inclined to be suspicious of the SPR as a continuation of Gurney's unfortunate spiritualistic hobby.[47] (Lodge himself was born a potter's son and had to steer clear of such elite digressions; he recalled finding Gurney once buried under piles of records attesting to apparitions, clairvoyance and other nonstarters for a conscientious naturalist trying not to so bury his career.) Myers had to convince him to take the investigation seriously, carefully explaining the psychical commitment to disciplinary standards. Far from being asked to compromise those standards, Lodge was now being called upon to guarantee them. Myers was being genuine here. If the point was to bolster the credibility of the already rigorous Creery experiments, then they had to raise the profile of the experimental conditions. The only way to do that was to bring in an outsider with a reputation for integrity and rigor. Perhaps Lodge himself had a certain amount of unconfessed curiosity concerning these strange claims, because he eventually did agree to referee (not conduct) the investigation, keeping his distance by framing it as a favor to a friend. To his surprise, Lodge found himself utterly drawn in by the phenomena, and immediately began planning a second set of experiments that he himself would design.

For the purpose of this follow-up study, Myers and Gurney released Lodge to make all of his own arrangements. Lodge relocated Miss E. and Miss R. to his research lab at the University College, Liverpool, and expanded it to include a headmaster and a biologist unknown to the girls and thus unlikely to collude with them. He tested a variety of telepathic pairings among the four, disrupting any prearranged routine the girls might have relied upon earlier. The communicants were both blindfolded and placed behind a wooden screen, mixing up the pairings of "agent" (the transmitter) and "percipient" (receiver). The results he obtained greatly added to the thrill of possibility suggested by Myers and Gurney's initial report. The Liverpool testimony was particularly apt (or adept, depending on how you look at it): too near the mark to be mere coincidence, sufficiently incorrect so as not to provoke suspicion. The imprecision was also well suited to the diffuse signaling expected from a field theory of transmission. Lodge's report, "An Account of Some Experiments in Thought Transference" appeared in the society's journal for that same year (1884), and was full of such intriguing near misses. This excerpt from the report features Miss E. and the headmaster as agents looking at the three of hearts, while the percipient (Miss R.) is questioned by Lodge:

MISS R: Is it a card?
LODGE: Right.
MISS R: Are there three spots on it? . . . Don't know what they are . . . I don't
think I can get the color . . . They are one above the other, but they seem
three round spots. I think they're red but am not clear.[48]

In another test, two agents stared at the cutout shape of a teapot, made from silver paper. One of the agents, however, kept thinking to himself how like a duck the teapot looked. The percipient announced that she saw a silver duck, but the drawing she made strongly resembled a crude reproduction of the original teapot. So here it was the image (teapot) and the idea (duck) that were cross-wired. Altogether, it suggested a kind of static in the lines of transmission well known in cable telegraphy. Lodge's informal speculation steered the interpretation into the framework of an ether field through which mental energy could repose and propagate:

> That consciousness is located in the brain is what no psychologist ought to assert, for just as the energy of an electric charge, though apparently on the conductor, is not in the conductor, but in all the space round it ... so it may be that the sensory consciousness of a person, though apparently located in his brain, may be conceived of as also existing like a faint echo in space, or in other brains, though these are ordinarily too busy and pre occupied to notice it.

Lodge's ether-field theory gave Myers a way of approaching these psychical phenomena from the standpoint of their physical mechanism, lending theoretical and corporeal "substance" to psychical speculation. This would prove useful to Myers when it came to navigating French neurophysiology, by giving him a mechanical language that was not strictly atomic or biological. Lodge surmised that the mind behaved much like a tuning fork, establishing syntonic resonance more easily with some minds than with others, and only very rarely was such attunement so perfect that it allowed for telepathic communication. This rapport could be ideational, eidetic, musical, or emotional, depending on the sympathy between brain structures and signaling waves.

The general consensus among the leadership of the SPR, after two years of endeavor, was that the evidence they had compiled for telepathy was highly convincing. But at the same time then-president Sidgwick still cautioned against endorsing the phenomenon, noting at the general meeting for May 28, 1884, that "none of our critics appear to me to appreciate the kind and degree of evidence that we have already obtained."[49] He went on to describe a skepticism so wily and adamant that there was no point in trying to intellectually engage with such animus. Instead, the dignified Sidgwick urged, "Our aim, in my opinion, should rather be to consider whether we can learn anything from our critics—even from ignorant and prejudiced critics." He added, "We cannot precisely define the requirements of a fair mind in dealing with matters so unfamiliar; and that we ought to continue patiently piling up facts and varying the observers and conditions, until we actually get the common sense of educated persons clearly on our side."[50]

Scientific recruitment particularly required that membership in the society not imply belief in its object, leaving researchers and readers equally at liberty to decide for themselves.

While the SPR seemed unable to defuse the indignation of British skeptics still incensed by spiritualism, it was equally hard pressed to banish the indifference of French academics, who tended to ignore such metaphysical insurgencies when they came from abroad or below. Rather, the threats that concerned Charcot tended to come from inside the high walls of academia. In the past it had been the introspective dons of philosophy; at present, more and more, it was the professionals of Nancy with their theory of suggestion. This blind spot would later work in the SPR's favor, as Charcot seemed more willing to tolerate a deferential interest in telepathy than a seditious hankering after suggestion. But as of the May 1884 meeting, no psychical curiosity had yet broken French ground.

Part of the issue was its lack of salience to French neuropsychiatry, whose core questions involved consciousness as an embodied process with madness as a guide to mental anatomy. What was the use of this extracranial force, scouted out in the main by British physicists (Crookes, Barrett, Lodge), of little interest or concern to its own discipline?[51] Whatever the lure of its disciplinary prestige, the physicist's approach lacked the nuance of psychological experience. The work of the Committee on Mesmerism bore more directly on this disciplinary terrain, but the categories risked moving from fringe physics to a state of disciplinary infringement. Even though the SPR divided mesmerism and hypnotism at the level of nomenclature, cabining off "hypnotism" as a modern term free of all mystical forces or faculties, that division was polemicized from the start. The psychical discourse treated hypnotism not as something separate but as a subset of mesmerism, the partial recognition of a larger truth yet to be established. That was not the approved semantic strategy of modern science, which was about denial, not subdivision: there was no healing force, no communal connection. The phenomena observed were psychologically driven effects (whether powered by "suggestibility" or originating from some lesion in the brain). This shifted the burden of experience entirely onto the patient: the one with psychiatric symptoms, not a degree in psychiatry.

Mesmeric ideology not only confused the divide between the mental and the physical (the magical and scientific), it did so as an attack on professional integrity, merging the bodies of doctor and patient, and implicating the doctor himself as the source of "animal magnetism." In a series of trials launched by the Committee on Mesmerism between 1882 and 1883,

Gurney, Myers, and Barrett broke the seal on this inquiry, not to confirm some "animal magnetism" or "electro-biology" but rather to consider "the specific rapport that hypnotism fails to account for."[52] Their data seemed to plainly confirm the existence of "the community of sensation" (the sharing of sensory perceptions) and even some merging of minds (though not necessarily transferred from mesmerist to mesmerized). It was the purported success of these trials that earned a note of caution from William James about rushing to publication.[53] But having recently attended several of Charcot's famous lecture-demonstrations (1882), James well understood the politics of mesmeric exclusion. They were not to be trifled with. However convincing the committee's evidence, Gurney would be hard pressed to get that French cohort to make it part of their brand of psychiatry.[54]

Thus, the difficulty in navigating this psychical approach across the Channel was how to how to raise interest in their phenomena without simultaneously raising the alarm. If they were to find space for psychical speculation, it required a proper language better attuned to the concerns of French psychology. From that standpoint, all the psychical sectors of interest were potentially problematic: spiritualism had always been beneath its consideration, while mesmerism had been formally deleted from it; telepathy was too remote a concern, while hypnotism was held too close. And yet, in truth, all the lines of psychical curiosity brought it directly into the realm of psychology's most fundamental concern: the nature of consciousness. This was the question the Sidgwick Group had been circling from the start, excluding nothing in the scope of their inquiry. Whether plumbing the depths of the evolutionary past or contemplating destiny beyond death, all this connected at the hub of human identity, which remained, in terms of modern explanation, a giant question mark. What physical science had done to illuminate nature, it was now incumbent upon psychology to do for the human mind, using every tool at its disposal to answer the question: Who are we?

However Charcot wished to circumscribe that question, he had already raised it. However he hoped to restrict hypnotism, he had already revived it. However he tried to objectify his patients, he probed their depths of experience. The professional lure of his neurology was its promised capture of human personality. It was only when Gurney and Myers began to consider these telepathic and mesmeric questions in the context of consciousness—not as a force but as an extraordinary form of subjectivity—that they began to connect more meaningfully with seekers in psychology. Despite the blinkers Charcot placed on hypnotic research, he could not make everyone blind to its curiosities. All the symptoms he itemized as elements of hysteria were also reminiscent of those who had been "mesmerized": hallucinations, double personality, alienation, loss of sensation, dissociation, aphasia. As more and more attention was turned to exploration of these phenomena, it was hard to keep the tide of magnetic literature from rolling back into France, carrying certain of its influences. Late in 1882, as he was exploring some of the neglected occult literature for his treatise, *L'homme et l'intelligence* (1884), Richet began to actively research automatic writing, thus reviving, for his own scientific purposes, some of the oldest practices of spiritualism and mesmerism.[55] Shortly thereafter, the young Pierre Janet embarked on his doctoral research into dissociation, which would also draw him into similar explorations of automatic intelligence. It would seem that, just as Gurney and Myers were trying to close the distance between French psychiatry and British psychical research, two small boats were being simultaneously launched from the opposing shore, ready to meet them halfway.

THE INNER SEA OF ME

At around this same time, and presumably independently, Gurney first, and then Myers, began to experiment with their own automatic writing protocols. Their novel strategy for accessing the unconscious was well adapted to both present circumstances and past experience. With no fund of hysterics on hand and an interdiction on using paid séance mediums, they turned their inquiry upon themselves, mixing structural elements of the psychiatric case study with some of spiritualism's oldest pieces of paraphernalia. The planchette was a small, circular board supported by two casters and a vertical pencil, introduced around 1855 as a method of spirit communication, meant to improve on "table tapping." (After 1891, the Ouija board supplied the spirit with a ready alphabet, as well as the words "oui!" and "ja!" for even more efficient communication.) In the séance setting, aspiring spiritual communicants would place their hands dumbly on the pencil board, awaiting for some spirit operator to take control and begin to write. Normally the device involved multiple sets of hands, which helped disguise the origins of the secretarial impulse, giving the planchette the appearance of its own life. As the planchette swooped about the paper, participants felt it do so in the utter absence of any intelligent direction on their part. This aligned precisely with how Myers defined trance automatism, as "the action of the body without will or volition of the conscious self." Yet something "took control" of the planchette, operating through the participant's central nervous system. The act of writing itself necessarily made the case for a complex automatism. These "spirits" could spell and tell, and in some cases write poetry and foreign languages.[56] Stranger still, use of the planchette was often a group activity, requiring the coordination of all this "unconscious cerebration" to cohere into writing, something that made little sense outside the spiritual trope of channeling or mesmerism's communion of minds.

Put in the right hands, that same occult instrument became an advanced tool of exploratory psychology, a technique they used with others and directly upon themselves. Gurney was principally interested in what he labeled "hypnotic memory," memories of events he could not consciously recall, but could record with a planchette placed behind a screen.[57] This early study of alienation, where consciousness divides into multiple streams to be experienced serially or simultaneously, drew William James into the psychical milieu in ways Barrett's physics could not. Myers's program was equally ambitious, having "selected automatic writing . . . because of its direct bearing on one of the most interesting of telepathic problems—the relation of consciousness to telepathy and the extent to which the hypothetical impact is consciously present, whether as will, thought, emotion, or sensation in either mind."[58] The study of automatism afforded a less structured approach to telepathy, considering it not as a force or faculty but as part of the pervasive and impressionistic field of human experience that might come in the form of presentiments, emotions, strange kinetic impulses. Such were the mysterious incursions, large and small, rare and common, that compromised the self's sense of mental boundary. In its most extreme form, this intrusive alterity might take on the appearance of "demonic possession" or the more benign séance spirit guide, establishing obvious continuities with the split personality of *dédoublement*. Though spiritualists described an external agent and psychiatrists an inner latency, the alienation was similar if not the same, with the body being operated by a fully contained separate identity, complete with its own intention and chain of memories.[59] Myers in no way meant to reduce mediumship to a mere diagnosis or confuse mental illness with spirit possession. The idea was to recognize the continuities between these experiences and the possible ways they might inform each other, bringing together the overlapping psychical, psychological, and neurological phenomena that "would indicate the true relation to each other of the processes in which our being consists."[60]

Myers's first survey of automatic writing was published in November 1884, under the title "On the Telepathic Explanation of Some Phenomena Normally Classified as Spiritualistic." It featured, as its most intriguing and oft-referenced case, the strange episode of Mr. A, "a sane and waking man holding a colloquy, so to speak, with his own dream." In his own account, Mr. A described sitting for four days, alone and meditative with his hand on a pen, eyes blurring and unfocused in the candlelight, drifting in and out of awareness. His hand would occasionally twitch of its own volition, and once even "moved violently," but the summary after two days was a jumble of letters without meaning. But, finally, on the third day, "Clelia" announced herself in a thin, spidery hand, composing her thoughts in riddles and anagrams so complex that they outstripped Mr. A's conscious capacity to spontaneously produce them. The original text was typed up and included as part of Myers's formal report. Here Mr. A poses a question (Q), to which "Clelia" gives an answer (A):[61]

Q. 1. (rep.) What is man?
A. 1. Tefi Hasl Esble Lies. (This answer was at first written right off.)

Q. 2. Is this an anagram?
A. 2. Yes.

Q. 3. How many words in the answer?
A. 3. V (i.e., 5).

Q. 4. What is the first word?
A. 4. See.

Q. 5. What is the second word?
A. 5. Eeeeee—

Q. 6. See? Must I interpret it myself?
A. 6. Try.

Presently I got out:

"Life is the less able."
Next I tried the anagram given upon the previous day, and at last obtained
"Every life is yes."
But my pen signified that it preferred the following order of words,
"Every life yes is."

By the fourth day Clelia had gone. The pen that had lost all deliberation, becoming "altogether wild, sometimes affirming and sometimes denying the existence of Clelia," before falling silent, had finally performed as follows: "u.c./c.c." The last initials were surmised by Mr. A to signify "unconscious" and "Clelia's consciousness" (whatever that might be). All he could conclude about this experience was that something apart from his conventional "personality" had asserted control over his pen during Clelia's script. The question for Myers and psychical research was to determine what or who that mysterious operator might be: an indwelling ghost, a submerged self, an inner wisdom, or some deeper species of a collective madness? In Mr. A's summary analysis, he offered only two rather conventional possibilities, though his experiment seemed to open the door for far more. Mr. A's first proposal was scientific: Clelia was his own unconscious cerebration directing the planchette in order to fulfill his own expectant desires. (Really, this was Carpenter's thesis.) He cited as supporting evidence his own literary flotsam strewn across the content of Clelia's conversation, gleanings he recognized from recent books he had read. But this did not really explain his *ability* to write such anagrams. Mr. A's second explanation was somewhat obligatory. Clelia was a spirit, "operating upon the cerebral particles," who fled because she was "weary of my unbelief, and I am weary of her coquetry."[62]

It is worth drawing attention to some of the stagecraft involved in Myers's selection and presentation of Mr. A's case. Why not discuss his own autohypnotic experiments? Why rely on Mr. A? In the article, he readily advises that readers should attempt this for themselves in order to fully appreciate the reality of these effects. Otherwise, all this testimony would carry little weight, even anecdotally, with no reference for the sensation of alienation described. Then there is the matter of Mr. A. himself, who remained resolutely anonymous, whereas the other cases discussed in the paper came with either names or significant identifiers. Mr. A's only credential was as "a friend," yet one whom Myers stated "could be trusted entirely." While probably not Myers himself (he later confessed he could never get such anagrams), Mr. A was most likely a fellow psychical researcher, and a prominent one at that.[63] He was conducting autohypnosis in the spring of 1883 in order to channel his subconscious, specifically mentioning subsequent tests with a planchette. These elements were part of Gurney's pioneering experimentation with automatic writing in 1883, skills unlikely to be mastered by people beyond his immediate, trusted circle. The fact that Mr. A, whoever he was, had the discipline and dedication to spend four days staring at a candle during his Easter holiday, only to produce a script devoid of any real bang or embellishment has that signature air of abnegation that makes this early membership so morally enchanting. At the very least, it suggests neither the effort of a novice nor that of a mercenary professional.

There is also something contrived about the way Mr. A discusses the case. The device of the study itself suggests an advanced psychological curiosity, yet his actual analysis is strangely lacking in imagination, forcing the choice between unconscious cerebration, on the one hand, and ghosts, on the other. While those might have been the questions put to the planchette by researchers in the 1870s, one would expect more from the sophisticated trawlers of the 1880s, who were confronting a new, dynamic paradigm for the unconscious. And we note the way Mr. A tees up the ball for Myers: first drilling holes in both his proffered explanations, then wistfully "hoping that some psychologists may clearly explain it." This allows Myers to step in, like Salviati to Mr. A's Sagredo, illuminating a new way to perceive the truth hidden in familiar phenomena. Whoever or whatever this Clelia was, for Myers she represented a distinct subject position, the precise location of which was difficult to determine or restrict. Whether she was a submerged aspect of Mr. A's own intelligence or a telepathic stream of ideas plucked out of the mind of some passerby, her authorship signified a specialized sentience, not just some ordinary activity on the margins of awareness. In some sense, this was where "unconscious cerebration" most fell apart as a thesis because Mr. A was most definitely aware of Clelia, enough so that he engaged her directly in conversation. This was, then, at the very least, a distinct center of sentient initiative rather than "unconscious" or even "subsidiary" neural process. To the description "sub-conscious," Myers added a more expansive term to describe the scope of human intelligence: processes that were "super-conscious," suggesting that the mind had not only a basement but an upper story as well.[64]

This foray into automatic writing allowed Myers to cast his psychical query deeper into the stream of experimental psychology, finding ways to adapt his mesmeric and telepathic questions to its neurohypnotic language and body of concerns. Taken in this light, the Clelia transcript functions like the patient narrative in a psychiatric case study, an intake of data at the point of mental disturbance. Just as a clinician might take the dictation of the hysteric's first-person madness, Myers let Mr. A and Clelia "self-" report, reserving the final analytical prerogative to himself from the pristine distance of "scientific observer." This was by far the most important protection offered by the dual narrative of the case study, which put a firewall between the psychical researcher and the psychical subjectivities he proposed to study. It was not enough to quarantine Mr. A's dissolute automatism within the autohypnotic transcript. Psychical hygiene had to be doubly ensured by scrubbing him clean of any biographical markers that could have tied him to that organization. So even though Myers qualified Mr. A as a friend and impeccable source, once he became the subject of his own investigation, he became a permanent liability of sorts whose identity could never be revealed.

Myers's brother Arthur (another possibility for Mr. A.) turned out to be the mysterious "Dr. Z," an alias buried so deep that it did not come to light for nearly a century.[65] He had been an epileptic patient under the care of Hughlings Jackson since 1877, keeping a diary of the way seizures altered his ordinary consciousness (excerpted by Jackson in 1888 for a research paper). We can understand such anonymity as a straightforward case of doctor/patient confidentiality, especially when the patient was also a doctor. The professional encounter between psychologists and subject depended on expertise, not identification, for its efficacy. That Arthur was "Mr. A" may very well be suggested here, but the main point is to consider the strategy itself.[66] How adaptive it was; how intuitive in its timing with regard to Janet and Richet.[67] There is also the experimental daring of the researchers: their willingness to find a way, go deep, try new things, and try them on themselves. Above all, they make their way into this conversation about automatism ahead of nearly everyone else.

A month after Myers published "On the Telepathic Explanation" in November 1884, a sudden thaw in Ango-French relations came when a flattering reference to psychical research appeared in the prestigious *Revue philosophique*. The article in question was by Charles Richet and featured his latest test in automatic writing, a study whose design he was glad to credit in part to previous published research by the SPR. Richet's test placed two tables in a room divided by a curtain, with group one chatting and tipping a table at random, while group two wrote down the letter to which they were pointing when so cued.[68] The setup was different, but the questions were similar: What were the acuities of the brain's secondary intelligence? Did it include some telepathic aspect? Myers referred to this as Richet's most important study, but the paper included more basic telepathy trials as well, involving guessing at card suits—but with a twist. Richet introduced statistical analysis to better evaluate his results, using a massively expanded subject pool and logging 2,927 independent guesses (claiming a probability for the existence of telepathy of +1/10). The randomized telepathy trial was born, allowing potentially subtler patterns of telepathic phenomena to be detected as a normal variant in human cognition. The SPR's satisfaction regarding Richet's "La suggestion mentale et le calcul des probabilités" was evident in the speed with which it was reproduced in the *Proceedings* a few weeks later (December 1884). Far from being ignored, in this corner of France at least, psychical research was an instigator in its own right, honing psychology's cutting edge.[69] Myers wasted no time in following up with "Automatic Writing II" in the January 1885 issue, mentioning Richet extensively in his footnotes. An important conversation had begun. To expand that conversation, Myers made a point of addressing Richet's wider cohort as well, spelling out the ways in which psychical studies of automatic writing intersected with their own contemporary investigations of diseases of the nerves. He mentioned aphasia, auditory processing, word blindness, agraphia, hand dominance, hemispheric functions, and more in a tour de force of medical literacy across his text and footnotes.[70]

The true turning point came in August of that year, with an invitation to visit the Salpêtrière, no doubt at Richet's behest but presumably also with Charcot's blessing. While there, Myers witnessed what had hitherto been for him a fascinating montage of pictures, studies, records, and rumors concerning the vagaries of trance consciousness. Charcot's stature had only grown in the interim since his international debut at the congress. Visitors to the Salpêtrière began to propagate their own lines of hypnotic research, with centers rapidly arising along what Eugene Taylor called "Charcot's axis." This included Liébeault, Bernheim, and Liégeois in Nancy; Eugène Azam in Bordeaux; Max Dessoir in Germany; Sigmund Freud and Josef Breuer in Austria; Flournoy in Switzerland; Ochorowicz in Poland; and Gurney and Myers in Britain. There was also an offshoot headquartered in Boston, the American Society for Psychical Research, led by William James and James Beard, established in 1884.[71]

By 1885, hypnotic research had a kind of centrifugal energy that required harnessing, lest all this momentum break away from the Salpêtrière and clinical neurology. At Richet's urging, Charcot established the Société de Psychologie Physiologique, an international organization that could hold all these various tributaries in a single conversation, with biweekly meetings and a bulletin for private circulation, in which to publish the best presentations.[72] Its membership was a comprehensive gathering of minds at the forefront of psychological inquiry, incorporating not only much of "the axis," but also German psychophysicists like Wundt and Helmholtz, as well as more traditional philosophers, such as Paul Janet, who was appointed to the vice presidency in an ecumenical gesture. (Everyone knew, however, that they all sat at Charcot's table.) As the society's name implied, this was a physiological psychology, where access to the mind was to be plotted through the nervous system and worked out by means of experiment. And above all, judging from the contents of the *Bulletins de la Société de Psychologie Physiologique*, everyone was on their honor not to directly challenge Charcot's hysterical theory of hypnotism.[73] It seems that the only sector excluded from this ecumenical group was Charcot's nemeses from Nancy (Liégeois, Bernheim, and Liébeault). Psychical research, by comparison, was most welcome. Myers was extended a membership toward the end of that first year, joining two fellow travelers from the American society who were already on its roster (though William James and James Beard were there at the outset, no doubt because of their status within academic psychology). Still, psychical affiliation was *not* disqualifying, which cannot be said of the Nancy school.

The topics under discussion around this time ranged over phenomena like hallucinations, eidetic memory, psychostimulants, and handwriting analysis, sticking to the seam of the mind-brain connection and always couched in the language of quantities and data collection. In work by Gley, Richet, and Rondeau, a potion of hashish was carefully prepared with chemical exactitude, to see if a pigeon could get stoned without a temporal lobe. (No, but a dog did.) The famous Dr. Beaunis offered a seemingly extraordinary account of delayed hypnotic suggestion, made more scientific by the specification of a time period of precisely 172 days in the title of his article. Other investigators measured how psychological factors affected the metrics of both auditory and visual sensory perception (how loud, how far, how fast).[74] This was all very interesting, very impressive, but was it revolutionary? Was there space here to truly advance an open inquiry into consciousness? Sure, they were willing to turn inward to observe the hidden scaffold of identity, but what ultimately did they see? They were as committed to using biological processes to explain mental experience as any British evolutionary materialist. The difference was that where Darwinians like Spencer, Huxley, Tylor, and Lubbock kept their vigil over man's "rational utility," now *psychologie physiologique* was intent on setting that fragile function aside, excavating underneath the executive personality to find all sorts of alternative sentience. For all the investigative energies focused by the Salpêtrière and the society, no line of investigation fanned out toward questions of more ultimate significance. The facility itself, now a teaching hospital, seemed to stand at the crossroads of tragedy and hope: hysteria was a disease to be managed and studied, but not ultimately cured. But the pessimism here went well beyond this patient prognosis. The true casualty at the heart of modern psychiatry for Myers was sanity itself. Even the healthy individual was said to rely on a jury-rigged conglomeration of hereditary parts, ready to buckle at the evolutionary seams under the pressures of modern life.

This was nowhere more in evidence than with *dédoublement*, which not only diminished the personality, but split it entirely in two, illustrating the synthetic nature of the human subject. As such, two selves ultimately meant there was not really "any one" in the first place. Myers's own exploration of the intersections between psychiatry and spiritual trance had been open-ended, leaving room for metaphysical insurgencies that might still conform to the shape of this behavior. But here in the midst of Charcot's clamoring research machinery, all the evidence piling up seemed cut to a certain pattern. Any hoped-for glimpses of an unseen world were called hallucinations. Even savantism in this pessimistic scheme was just further evidence of neural degeneration. As with Clelia's clever anagrams, occasionally a well-placed brain lesion might result in the outpouring of poetry, mathematics, music, or art.[75] Thus, while Charcot placed hypnotism at the heart of academic curiosity, the essential features of its "ill repute" remained. Even without the necromancing mentalists or healing hucksters, hypnotism was a biological attack on the order of the mind, a practice suitable only for those already penned within sanitarium walls. Far from having any therapeutic value, the act of hypnosis itself stressed latent susceptibilities even in the normal brain, creating a hairline fracture in the integrity of consciousness that could become progressively worse with every pass of the hand. For Charcot, who had few qualms about intentionally aggravating a medical condition in order to study it, the immorality belonged entirely to Liébeault, a doctor harming the healthy with a discredited practice. Suggestion was not only bad medicine; it was bad faith, going against standards already set by the profession that forbade this charlatan's cure. Thus, the dispute between Nancy and the Salpêtrière was less petty than it appeared, given the ideological, professional, and theoretical concerns involved.[76]

After departing the Salpêtrière, Gurney and Myers went to Nancy to meet with Liébeault (and, more briefly, Bernheim). By 1885, the relationship between Nancy and the Salpêtrière had become quite acrimonious, as the exclusion from the society suggests. And in truth, no two views of hypnotism could be more antagonistic: the one saw it as a cure, the other, as a disease. Yet both programs offered important experimental insights for Myers and Gurney, who steered clear of any overt ideological battles in order to meet all evidence without restriction. That being said, Liébeault's ideas had more intrinsic appeal for psychical research's maximalist project (and for other researchers closer to home, less free to express it).[77] As a practice, Liébeault's therapeutic hypnotism was true to mesmerism's origins as a physical cure, but his research into suggestion brought him closer and closer to Puységur's early "talk therapy" (without its mystical component), recognizing the fundamentally psychosomatic nature of many health disturbances. All this was safely contextualized within Braid's mechanistic *Neurypnology* (1843), which had guided Liébeault's medical practice since the late 1850s, cleansed of any force or mysterious influence.[78]

Adding to the appeal of therapeutic suggestion for Myers was the daring of Liébeault's evolutionary reasoning explaining hypnotic efficacy: trance states were part of the brain's normal evolutionary programming, an auxiliary system entirely distinct from ordinary sleep and wakefulness. This "somnambulistic" setting allowed the brain to correct kinks in the affective or logical structures of consciousness, and any somatic complaints with which they were associated. At his clinic, Liébeault used the Braidian nervous shock to pacify the executive brain, dropping in vigorous suggestions for self-improvement like a computer coder reprogramming a system. (The familiar "Every day, in every way, you are getting better and better" is a global prescription for self-improvement devised by Liébeault's protégé, Émile Coué.) This diminished capacity, which Charcot saw as evidence for a broken brain, was the key to the "fix" of suggestion, allowing ordinary patterns of consciousness to be overwritten by the doctor's words without resistance. A hypnotic suggestion was more than just something to think about; it became part of one's thinking. To the extent that "personality" was an evolutionary instrument to navigate the world on behalf of the body, the ability to reboot it in a more functional form seemed highly desirable (as the thousands of transformational weekend retreats will attest). This idea of the hidden evolutionary trance utility would prove an intoxicating thesis for Myers, reconciling the rational and irrational rather than opposing them.

But there were deeply troubling aspects of Liébeault's therapeutic narrative as well. Its benefits seemed to come at great cost to moral identity and free will. Even in the most constructive applications of hypnotherapy, the healing mechanism amounted to a substitution of the doctor's curative will for the destructive will of the patient. This might seem insignificant in the context of breaking a bad habit: merely pitting the exogenous compulsion of the hypnotist's positive suggestion against the endogenous compulsion to bite one's nails. But it did not end there. Some of Liébeault's more recent experiments, with which Myers now became acquainted, involved a tampering with the will that was far more sinister. One such demonstration, conducted three years earlier in 1881, involved a lovely young mademoiselle who, when placed under hypnosis, obligingly shot at her mother upon Liébeault's command. Liébeault had naturally left the weapon unloaded, but, nonetheless, it was the thought that counted. These were not mere theatrics. Liébeault had made sure the girl had seen him put bullets in the chamber earlier that day in an unrelated context; she "knew" on some level that the weapon was loaded, but she shot anyway. When asked her reason, she could only describe an irresistible compulsion to act.

Could so vile an act as a matricide be met with no more resistance than a kneecap might offer a ballpeen hammer?[79] In a similar case, Liébeault coaxed a respectable French matron to shoot a government official, but here she was fully conscious, acting on a previous hypnotic suggestion made hours before. This seemed to confirm the thesis of French psychiatrist Théodule Ribot, a Charcot protégé, that all our seemingly deliberate moral choices are actually the result of deterministic processes cascading below the threshold of our conscious awareness.[80] It was also precisely the argument made by the alienist Jules Liégeois, who added to the scandal of Nancy by using Liébeault's experiments to degrade the notion of criminal responsibility in legal arguments. If the mind could be broken into and vandalized by a hypnotist's tricks, then principles like free will, self-control, and personal responsibility were so vulnerable as to be meaningless. In the case of the homicidal French mademoiselle and madame, the moral values cherished at the core of their being had been stunned into submission by the glint of a pocket watch. These were not the shrieking hysterics of the Salpêtrière, who lacked moral will to begin with. This was your mother, your daughter, your sister-cum-cold-blooded-killer.[81] While Myers played some of this off as the moral jeopardy that naturally attached to being French and therefore "sensitive," the problem had been made universal by evolutionary biology, which had already salted away a certain amount of natural determinism into our self-determination. Hypnotism was only now revealing the extent to which this was true, calling into question even the real existence of this self, let alone its self-assertion. In some ways, Liébeault's therapeutic model of hypnosis, which normalized trance, was more problematic than Charcot's pathology because it diseased the moral will of all humanity. For his own part, Myers subscribed to neither view, seeing them both as partially true but fundamentally incomplete. Even if this trip signified admission to their circle of investigation, could psychical inquiry flourish in such a climate? The overall scheme of continental thought seemed to foreclose his most urgent questions of ontology. Why seek to understand the nature of something that is not recognized to exist? Whether the self was being undermined from without by suggestion (the Nancy school) or attacked from within by disease (the Salpêtrière), the thrilling pace of this research was leading to personal extinction.

In a letter to his wife, Myers seemed to celebrate their success, writing that "the way in which we were received by Savants in Paris was most gratifying. We are far better known than we expected and people seem to expect great things from us in the SPR."[82] Yet he returned home in September from this professional triumph with a sense of dejection. The impressive tide of research in which that continental army of researchers was engaged seemed to be moving in the opposite direction. The result of these ruminations was an uncharacteristically bleak paper read before the SPR on October 29, 1885, summarizing what he had learned on his trip. The state of consciousness research on the continent was not pretty. Myers warned his readers in the opening paragraph that "the facts and inferences contained in the present paper will be novel and even startling" because of the "depressing view of man's dignity and destiny which this train of argument implies." This was not the sentimental view of the self in which most Britons had taken unwitting comfort. That quaint notion was based on eighteenth-century common sense and did not hold up well against nineteenth-century science. Myers quoted Thomas Reid's philosophy of the self as "expressing the views of the bulk of [his] readers":

> My personal identity implies the continued existence of that indivisible thing which I call myself. Whatever this self may be, it is something which thinks, and deliberates, and resolves, and acts, and suffers. I am not thought, I am not action, I am not feeling: I am something that thinks, and acts, and suffers. My thoughts and actions and feelings change every moment; they have no continued, but a successive existence; but that *self* or I, to which they belong, is permanent ... Identity, when applied to persons, has no ambiguity, and admits not of degrees, or of more and less. It is the foundation of all rights and obligations, and of all accountableness; and the notion of it is fixed and precise.[83]

On that last notion alone ("it is the foundation of all rights and obligations"), Reid's common-sense self should be upheld, even if proven to be only an illusion. But Myers, with grim scientific resolve, would have none of it. He made it "the task assigned to this paper" to dismantle this fallacy according to the facts: human personality "had been shown through hypnotic experiments ... to be neither definite, permanent, nor stationary; free-will is shifting and illusory, and memory multiplex and discontinuous, and character a function of these two variables, and directly modifiable by purely physiological means." Myers identified experimental psychology as taking the lead on this scientific initiative, making its case not "by merely introspective analysis but by a study as detailed and exact as in any other natural science."[84] He was referring to that great wave of neuropsychiatric research set off by Charcot in the 1870s and now sweeping over Europe, Britain, and across the Atlantic. The momentum of these arguments had carried the day, bringing forth facts to which modern minds must now defer:

We start, then, with the single cell of protoplasm, endowed with reflex irri-
tability. We attempt a more complex organism by dint of mere juxtaposition,
attaining first to what is styled a "colonial consciousness," where the group
of organisms is for locomotive purposes a single complexly acting individual,
though when united action is not required each polyp in the colony is master
of his simple self. Hence we advance to something like a common brain for
the whole aggregate, though intellectual errors will at first occur, and the head
will eat its own tail if it unfortunately comes in its way ... We rise higher; and
the organism is definitely at unity with itself. But the unity is still a unity of co-
ordination, not of creation; it is a unity aggregated from multiplicity, and which
contains no element deeper than the struggle for existence has evolved in it.[85]

Had the once sentimental Myers, yearning for spiritual transcendence
for the past twenty years, now suddenly thrown in with the biological ma-
terialists and fashionably dissolved the self into the primordial ooze? These
were the accepted facts of evolution, and thus the starting point of any naturalistic inquiry concerning the biological structures of human consciousness. The multiple lines of thinking in organic chemistry, psychophysics,
brain biology, evolutionary psychology, and experimental psychiatry had
converged by 1885 into a kind of settled opinion among naturalists: we were
not cut by design from whole cloth, but were assembled haphazardly in a
piecemeal fashion. By this understanding, the singular, synthetic experience
of being a self was an illusion built up entirely by the convergence of subordinate functions and cellular parts. It thus, conceded Myers, "must be analyzed into its constituent elements before the basis of a scientific doctrine of
human personality can safely be laid."[86] Though plenty of English scientists
would have taken exception to such an opinion, its special authority for
Myers came precisely from its grim, uncompromising view of the facts (the
signature virtue for the ethics of belief.)

Nothing seemed to put the case more plainly that the person was "a unity
of co-ordination, not of creation" than the strange condition of *dédoublement
de la personnalité*, which showed this composite selfhood splitting apart at
its seam. This alter appeared to disclose a cruder, more basely biological
variant of the self, not unlike Liébeault's murderous matron and matricidal
mademoiselle. Such a ready unwinding of one's moral character seemed to
suggest that it was among the more superficial aspects of personality, and as
psychiatry delved down to the deepest layers of our sedimentary selves, such
civilities as "thou shalt not kill" quickly began to give way.

But Myers also took note of experimental psychology's cascade of negative results: these scientific findings, while of enormous value, were only

true "so far as they go."[87] Such conclusions had been drawn from patients locked away in mental asylums or people perverted by hypnosis and put up to some dark task. This pool of data was pathologically skewed and fundamentally incomplete, and gave rise to theoretical biases that further inhibited research. While it was clear that hypnotism and hysteria had a physical impact on the nervous centers of the brain, it did not follow that the resulting trance phenomena they produced were likewise of a purely physical nature. There was valid evidence suggesting that there were aspects of brain function that transcended brain biology. "One such discovery," Myers argued, "that of telepathy . . . has, as I hold, been already achieved."[88] While telepathy was not formally recognized by the SPR, it was a main line of study and had been rigorously and repeatedly investigated now for several years. Even more exotic phenomena were also under inquiry at the SPR, such as clairvoyance or telepathic apparitions projected from the minds of the dead or dying. Why interpret the subconscious through the lens of our evolutionary past, urged Myers in the closing arguments of "Human Personality in the Light of Hypnotic Suggestion," when there were also clues intermixed with its phenomena pointing to a higher stage of evolution altogether?

Despite this last, sanguine note, Myers's lecture generated an overall negative buzz, having entangled itself too closely with continental philosophy. Yet, Myers had a duty to report on the progress being made in *psychologie physiologique*, offering the one thing he could to forestall the force of its conclusions: the continuing hope of telepathy. Such facts could not be established as a matter of scientific consensus, but the research thus far was promising. Myers's invocation of telepathy as a defense against biological reductionism was less wistfully rhetorical than it might appear.[89] A few days before he delivered "Human Personality" before the general meeting, Richet paid him a visit in London on October 25 and 26, a fact he noted in his diary without further elaboration.[90] However, hardly a week later, Richet presented a paper by Myers at the November 2 meeting of the French society, which he had presumably collected during this visit. "De certaines formes d'hallucinations" was the first formal introduction of the work of the SPR in a professional setting in psychology, briefly detailing the contents of the forthcoming volume, *Phantasms of the Living* (1886). Richet himself had tentatively raised the issue of ghostly hallucinations in July, referencing certain foreign journals but not by name. He only meant to affirm the reality of the psychological experience, not the ghosts themselves. But Myers's paper delivered a considerable twist by suggesting that such hallucinations were also "veridical." This term classified the event as only partially interior, because it also contained information about the external world—though these happenings were often remote in place, if not time.[91] All this gave support for a possible telepathic attunement between two emotionally connected individuals when one was approaching death or in some other distress.

Later that month, at the November 30 meeting, four more papers touching on telepathic themes were presented by researchers active in Charcot's immediate circle.[92] The most sensational of these came from the young Pierre Janet, still a doctoral candidate, who had witnessed some strange anomalies while observing the clinical practice of a well-known physician from Le Havre, Dr. Gibert. The phenomena involved a peculiar rapport between Dr. Gibert and his longtime patient, Madame B. Quite accidentally, Dr. Gibert had discovered that Madame B. was responding to his unuttered thoughts and intentions after he had left the room, an effect that, odder still, seemed to intensify over distance. None of the hypnotic reflexes Janet described in his report would have appeared particularly noteworthy had there been contact (Madame B. was cranky, sleepy, wanted to rearrange the furniture), but that yawning bit of distance, which grew with time as the connection deepened, was a very big, disruptive deal, indeed. Such an incendiary paper from a young associate member could hardly proceed without some kind of battle-ready escort, obviously prearranged by Richet. He presented his own anecdotal support describing his experience at Beujon Hospital in 1873, when a "hypnotic rapport" gradually developed between himself and a patient he had habitually put to sleep by standard methods. Over time, she began to respond to his thoughts, also at an ever-increasing distance, like Gibert and Madame B., a fact he put to the test with the help of other interns. After Richet's paper came Héricourt with a similar story, already confided to Richet back in 1878, who had since kept it "cautiously locked away in a box for reasons easy to understand."[93] Héricourt had thought to test the growing rapport between himself and his patient, Madame D., while taking a long walk, thinking nothing would come of it so far away. Poor Madame D. fell into a lethargy so deep that her servant thought she had died. Beaunis came forward with a case of mental suggestion, willed by group of doctors into the head of a blindfolded young man. The most controversial aspect of this testimony was the where and the who: at Nancy under Liébeault's direction. Beaunis, who "never took such matters seriously," now urged the members present that "they were well worth investigating."[94]

The language of all four of these papers was deferential, and artfully designed to deflect responsibility. Everyone posed as an innocent bystander, compelled to testify about something somebody else had accidentally seen, all in order to force Charcot to witness facts he had forbidden them to study. What Richet had effectively done was arrange a quorum of evidence, publicly aired at a meeting of the society at which the "Présidence de M. Charcot" was duly noted for each paper.[95] While all this was presented as an appeal to science, the manner of presentation was also something of a threat. This particular lineup was offensive not only for its headlining account of the remote effects of hypnosis, but for its subtler affirmation of the therapeutic suggestion. Though their papers in no way forced that point, this was a tension already running through the pages of the *Bulletins*, with researchers subtly pulling at the fence containing hypnotism within hysteria. For all the opportunity associated with his clinic, Charcot's inflexible equation of trance and disease ran counter to his promise of dynamic, unfettered research. A little experimental leeway regarding some "telepathic rapport" could preserve the research momentum breaking his way and weaken the allure of Bernheim's theoretical insurgency.[96]

So Charcot let it run. Things heated up considerably in December with an extensive study of "telepathic medicines"—that is, the effects of various psychostimulating substances *without* direct patient contact—presented by Bourru and Burot, "Action des médicaments à distance"). Richet conducted a duplicate study confirming Bourru and Burot's results, thus ensuring the phenomena would have an investigative future. Richet's paper offered no explanation, only the conviction that "c'est vrai" (it's for real).[97] Bourru and Burot had just recently been making headlines of a different sort at their clinic in La Rochelle, diagnosing the first case of multiple personality disorder. The case went like a flash through the medical literature, arriving in England courtesy of Arthur Myers's report in the *British Journal of Medicine* (December 1885). The patient Louis V. became for Myers an object of fascination to which he would return again and again over the course of the next year.[98] It was all, certainly, very depressing from the standpoint of an essential human personality, and yet Myers was drawn by "the curious connections" between Louis V.'s disorder and "certain phenomena of automatic writing."[99] Was the key to some fundamental whole to be found amid these fragmentary streams of consciousness, "a self-severance which we shall hereafter trace far down into the 'the abysmal deeps of personality'?" For Myers, such phenomena were by no means limited to the asylum.[100] These alters seemed to haunt human experience everywhere, not just "inside" the head, holding us in conversation with angels or demons, or even Clelia: was this "mentation presenting itself to us as a message from without . . . a germ of externalization?"[101] How did that "externalization" fit with the deep interiority of the subconscious personality? Understanding the translocality of the other was, for Myers, a "significant precursor of deeper secrets in the fissiparous multiplication of the self."[102] What if the requirement of a truly modern psychology was not to abandon an essential human subject, but rather to radically rethink it?

In "Further Notes on the Unconscious Self" (December 1885), he offered a new way to think about the ever-changing flux of this dynamic subject in a scheme of continuity.

My contention is, *not*, as some of my critics seem to suppose, that a man (say, Socrates) has within him a conscious and an unconscious self, which lie side by side, but apart, and find expression alternately, but rather that Socrates' mind is capable of concentrating itself round more than one focus, either simultaneously or successively. I do not limit the number of the *foci* to *two*, and I do not suppose that the division of the brain into two hemispheres is the *only* neural fact corresponding to the psychical fact alleged ... Consciousness and the unconscious do not lie side by side, but are continuous capacities, separated from each other by the limiting illusion of "self" awareness.[103]

This fluctuating consciousness was a fundamentally bolder, more robust self than allowed by French psychiatry, anticipating in many ways Gerald Edelman's "dynamic core" shifting through its "functional clusters." Where Charcot's trance fractured consciousness and walled it off in a partitioned brain, Myers gave such altered states a far more protean and positive character, allowing this neural energy to range free across the mindscape. Its course was not merely plotted by neurological necessity but also reflected mysterious initiatives of a more psychological or subjective nature. Myers dared not directly disparage Charcot's rigid, three-stage progression of hysteria, but he did offer an alternative to it, characterizing hypnotic states as either "light" or "deep," with the latter signifying a maximal latitude for self-expression, freeing the mind of its limiting routines. What Charcot's *somnambulisme provoqué* had designated as an extremely degenerate state, Myers now reinterpreted as a radically liberated one.

But for all its human potential, this multifocal personality still posed a profound philosophical problem for Myers. Did this "continuous capacity" add up to some fundamental personhood, or did it merely point toward a more spontaneous, less deterministic, less fatalistic neural process? The self remained, if not in pieces, then in flux: equally unreliable and unaccountable from a moral point of view, and equally said to not truly exist. This more dynamic self was still being strung up and taken down along an ever-changing skein of neurons, parsed into multiple foci with a corresponding "personality" still conceded by Myers to be "limiting illusions." Such a self could not hope to attain a gestalt or metaphysical status in its current fluctuating, fissile state, and in the end faced the same ontological jeopardy as "colonial consciousness." In the reductive idiom of Victorian science, "ultimate essence" was to be found in the permanence of the smallest part—the stability of the thing that could not be partitioned. Unless it could prove some separate ontology of its own, consciousness remained fundamentally less real than the matter of which it was presumably made. Myers's kaleidoscopic self could not be proven to be any more real than Charcot's diseased one.

Myers conceded that his theory was far from satisfactory, though it did equally comport with the phenomena observed in Charcot's *la grande hystérie* and *le grand hypnotisme*. But all such speculation, including his own, was still premature. Psychology was necessarily awaiting something more, to be complete. Myers believed he saw the outlines of that something close on the horizon, flickering suggestively in the vast, pointillistic field of experimental facts noted thus far. It slipped between "sleep and dreams, somnambulism, trance, hysteria, automatism, alternating consciousness, epilepsy, insanity, death and dissolution," remaining ever aloof from science. As he had originally declared in "Human Personality," against the odds, "I believe there is a definite incandescent solid, but it remains beneath our line of sight."[104] As yet, he could not fix its image, he could not name it—but it seemed that now the sages of the Salpêtrière would at last help him come to know it.

The curiosity encouraged (even orchestrated) by Richet began to claim wide support from the membership of the French society, crowding the journal submissions after December. Myers made a note of this by adding an appendix to "Human Personality" before its publication in the *Proceedings*: "Since this paper was read to the S.P.R. (October 29th, 1885), very great activity has been shown in France in the direction of hypnotic research. The *Bulletins de la Société de Psychologie Physiologique* for 1885 (published in 1886) contains various cases of high importance."[105] In truth, *psychologie physiologique* was drifting inexorably into this psychical terrain, and promised even further penetration if the letter Myers had recently received from a French colleague was any indication. Janet, who had encountered some strange anomalies while conducting hypnotic research in Le Havre the previous year, was now organizing a follow-up investigation to get at the truth of these events, and invited Myers, along with Gurney and Arthur Myers, to join him in this next step round the bend of scientific investigation. Toward the end of April, Myers embarked for France—this time not as a spectator come to marvel at the Salpêtrière's astounding progress, but as one who might himself help to accelerate its advance.

TELEPATHIC HYPNOTISM

Frederic and Arthur Myers set sail for Le Havre, a town on the northern coast of France, toward the end of April 1886 to meet an ambitious, international group of investigators who awaited them there. This included Dr. Julian Ochorowicz from Poland, who had studied at the renowned Leipzig lab of Wilhelm Wundt, along with Myers's French hosts: professors Paul Janet and Pierre Janet, Dr. Jules Janet, and a local physician, Dr. Gibert. The team had gathered to follow up on the extraordinary phenomenon encountered by Gibert and Janet the previous year, almost too strange to be believed. Having stumbled into an area so far beyond the ken of established science, Janet appealed for help to the man who had himself coined the term "telepathy," hoping to submit the phenomenon to a full-scale battery of tests that might render these facts more conclusive.

Those experiments took place between April 20 and 24, and covered distances that now extended to two-thirds of a mile. Madame B. continued on as the subject, with Dr. Gibert as her operator, but now more rigorous controls were put in place to document the exact times of hypnotic transmission, along with the appearance of its symptomology. Multiple observers were stationed at either end of the study to ensure that the strange hypnotic rapport between Gibert and Madame B was entirely spontaneous and uncoordinated. Transmission times were determined randomly by those stationed on Gibert's end and were unknown to those encamped on the other side. Myers noted that while Madame B. did not always succumb to the effects of Gibert's attempted hypnosis, she nearly always seemed to sense its intruding influence: "she would try and ward him off with some annoyance by submerging her hands in cold water or verbally expressing her dismay before collapsing into hypnotic slumber." Myers was stationed at the same home where Madame B. resided for the study (belonging to Gibert's sister), and for three days observed the most curious phenomena. He was frequently awakened in the middle of the night, only to find Madame B. pursuing some strange domestic occupation inappropriate to those hours, such as sewing or watering the plants, her eyes fixed ahead with the telltale blank stare and her hands moving in a rigidly automatic manner. Just before Myers's arrival, a new twist had been added to the case: while attending to some equipment in a neighboring room, Dr. Jules Janet had severely burned his right wrist, only to elicit piercing cries from Madame B. next door. He entered her quarters and found her clasping her own wrist in the exact same place, suggesting that in addition to mental telepathy, this remote agency might also include "a curious transfer of sensation." This was reminiscent of some of the phenomena observed by the SPR's Committee on Mesmerism, which William James had ironically feared would offend this same French constituency. This particular "transfer" even introduced novel elements yet to appear in the SPR's experimental record: this was a rapport that established itself across a solid wall, between two people who had no visual contact and no hypnotic relationship.

After the research was concluded, Myers and Arthur traveled to Paris, to attend the next meeting of the Société de Psychologie Physiologique, where Myers was invited to present his own notes on the experiments, ahead of Janet's formal report to be presented at the end of May.[106] In Paris as in Le Havre, Myers's reputation had acquired a certain glow of expertise. In his field notes, he digressed long enough to address the need to organize the recent sprawl of hypnotic phenomena into a more "intelligible series." His classification system, based on known criteria, were: "hypnogeny to describe the production of hypnotic states; hysterogeny for the production of hysterical ones; dynamogeny for the production of increased nervous activity," followed by "aesthesiogens" to describe psychoactive substances, and last and most recent, "telepathic hypnotism," around which a consensus was currently building, necessitating some designation.[107] With this taxonomy, Myers provided a modern nomenclature to ease the banished contents of mesmerism back into *le grand hypnotisme*, without the appearance of any mesmeric backsliding. That Myers would take it upon himself to make such an inventory on behalf of *psychologie physiologique* reflected his growing sense of intellectual place in that community, and the necessity of overcoming its own hypnotic paralysis. Not only was he willing to follow the facts wherever they led, but he was willing to lead the facts back to science as well, in an orderly manner.

The trip was a great success, and Myers made arrangements to return to Paris that October, this time accompanied by Gurney (whose name appears on the roster of the *Bulletins de la Société de Psychologie Physiologique* for the first time that same year [1886]). The two men had traveled to the Salpêtrière together once before, back in August 1885, as the first psychical delegation to France. Among the marvelous phenomena they had witnessed was a private demonstration of "psychic transfer" given by Charles Féré, who was already a member of the SPR and shared an interest in automatic writing.[108] Féré, along with his fellow investigator, Alfred Binet, was part of Charcot's inner circle, well known for their earlier research into the "physical transfer" of symptoms with magnetized metals. But starting early in 1885, the two had begun to push metallotherapy a bit further, applying metals on purely psychological phenomena like hysterical blindness, numbness, deafness, and even autographia (the spontaneous trance writing of such keen interest to Gurney and Myers).[109] Normally the efficacy of magnets on mentally driven processes would fail to excite: an imaginary effect on a product of one's imagination. But there was a considerable twist here, in that the magnets were applied without the patient's knowledge. That the patient's symptoms were still pulled clear to the other side of their body seemed to suggest that a real magnetic adhesion was in effect. The term "psychic transfer" reflected the ambiguity of the phenomena. "Psychic" connoted an idea only, while "transfer" suggested the migration of a fixed substance, leaving one location to arrive at another (an idea, by contrast, could propagate indiscriminately). The dual aspect of the phenomenon was widely discussed, even in the Scottish review, where the writer asked, "How could the transference of psychic phenomena be possible or even conceivable if these phenomena had not a material base?"[110] As Myers himself explained in the *Proceedings*, "intention . . . possesses a material substratum and is capable of being transferred from one side of the brain to the other."[111]

But there were deeper questions to get at here as to what *caused* this hysterical malleability, letting ideas become flesh and mind become matter. In discussing psychic transfer, Myers invoked the example of the psychiatric saint, Louise Lateau, who wept tears of blood from her hospital bed. But physiologists like Charcot preferred to lean on some underlying defect of the tear ducts to mitigate the transmogrifying powers of mind. Such organicist attitudes left the field tilled by psychic transfer to lie intellectually fallow.[112] But now, in 1886, a more profound curiosity bloomed. Binet and Féré had already begun to consider "psychic transfer" (and other related studies from the French society) in the context of "animal magnetism"—not to describe the effects of *their* personal magnetism (or their magnets), but to get at something potentially beyond the ordinary psychophysical processes recognized by *psychologie physiologique*.[113] To the extent that "psychic transfer" remained a hidden process inside the patient's body, it would be difficult to coax out any more candid theoretical reflections. But now, late in 1886, at the October meeting of the French society, Joseph Babinski, Charcot's particular protégé and future successor to his neurological throne, was ready to draw such questions out into the open.[114] Babinski began with two patients, A and B, the one suffering from a severe hemiplegic motor paralysis, the other symptom free. Instead of merely physically transferring the symptom from one side of A's body to the other, Babinski made a very unusual move. Using only a magnet, he dragged the symptoms out of patient A altogether and deposited them into the body of patient B. Patient A now stood entirely relieved of her physical symptoms, while patient B manifested them in full on the receiving side of her body. What had happened here? If Binet and Féré had shown how thought could structure physical symptoms, had Babinski just shown how it could resublimate them again, as well? In other words, if mind became matter, could matter become mind? This was a kind of "telepathic transfer," a timely fusion of psychic transfer and telepathic hypnotism to create something even stranger than its parts. As Babinski pointed out, whatever this was, it was more complex than a case of "simple" telepathic suggestion: what patient A had communicated to B was more than the mere *idea* of illness, it was the illness itself. When tested, she was now fully rid of her physical symptoms, while the percipient had acquired every one of them, across a variety of sensory motor impairments.

It is important to draw attention to the obvious here: much of the experimental conditions under which these facts were asserted would not meet contemporary standards. This no doubt contributed to the extraordinary findings piling up in the world of experimental psychology in the 1880s, unreproduced in modern times. One might wonder how exactly—as in the case of hysterical or hypnotic blindness—researchers confirmed what other people's eyes did or did not see, or if, per the claims of Binet and Féré, the patient's purported lethargy cloaked the application of the magnet.[115] The Achilles heel of psychology was then, and remains now, a reliance on subjective testimony—an obstacle rendered more problematic when the patient in question was the indigent inmate of a sanitarium, given to delusion, deception, or both. The researchers at the Salpêtrière did not ignore this central difficulty of how to tell a lie from a hallucination, and indeed made it a paramount concern. The thorny issue of unreliable testimony was oddly defused by making lying part of hysteria's clinical symptomology, and making recognition of those lies a part of a clinician's medical expertise. In essence, lying about one's symptoms confirmed the underlying condition one was lying about and upheld the medical authority of the doctor presumably being lied to. What might have worked to ratchet up skepticism (the recognized pattern of patient deception) was thus deflected into a kind of circular confirmation instead.

In addition to assimilating deception into their patient profile, there was also the problem of excluding it. Madame B., the subject of the Le Havre study, was condescendingly described by Myers as an "amiable peasant woman, slow and of little knowledge who accepted no payment," a statement that clearly placed her "beneath" suspicion. Researchers, in contrast, too often placed themselves above it. A gentlemen's code not only kept researchers from asking questions of each other, but likewise bred a certain confidence in their own scrupulous objectivity: they would not deceive each other, nor deceive themselves. As to any doubts regarding the findings at Le Havre, Myers dismissed them outright: "The hypothesis of fraud on the part of operators or subject may here be set aside."[116] Why? Because "the operators were Dr. Gibert and Professor Pierre Janet," two upright professionals with advanced training to whom "the genuine character of Madame B.'s trance is readily apparent." Only the expert can judge the expert in this wonderful circle of validation.

Although Myers was rather quick to dismiss any impropriety or confirmation bias on the part of his fellow researchers, he was not devoid of skepticism, though he tended to direct it downward. In considering how the Salpêtrière had acquired "the richest collection of hysterics the world had ever seen, " he astutely observed that upon entering the asylum, the hysterical patient "learned through the atmosphere of experiment around her to adapt her own reflexes or responses to the subtly divined expectations of the operator.[117] Note that the doctors were not accused here of seeing what they wanted to see, rather it was the patient who stood accused of engineering her misperceptions. The institution itself "had been smothered in its own abundance" by this eager hysterical horde. There were other factors as well, less subtle than wanting to please their doctors, which might also motivate these "medical performances" the genteel Myers perhaps found too uncomfortable to consider.[118] Hospital beds were in short supply, and many of these patients faced homelessness, hunger, or incarceration if found insufficiently hysterical by the doctors studying them. And no doubt some more enterprising patients wanted to make the most of the opportunity, enjoying the perquisites and celebrity that an offering of consistently severe symptoms might bring, finding an easy con in the confidence of these doctors.

But all this fundamentally digresses from the main point I wish to make regarding Frederic Myers. At issue is not the merit of these experiments or the reliability of the Salpêtrière's facts, but rather how well Myers's research program fit in with that of his far more celebrated French counterparts. Myers's psychical curiosity cannot be disentwined from mainstream experimental psychology as an invading species of inquiry; rather, it arose out of a shared set of concerns initiated at the Salpêtrière itself, which were, since 1878, hypnotism, hysteria, and the strange faculties associated with these altered states of mind. As Edmund Gurney put it, "for scientific purposes, Paris is the center of the earth, for hypnotic purposes the Hospice de la Salpêtrière is the centre of Paris."[119] And by 1886, these faculties were beginning to become very strange, indeed. Telepathy, hitherto at the outskirts of French neuropsychiatry, was making its way to the heart of the *grand hypnotisme* through the rarified research of the French society, composed, according to Myers, of a "powerful group of specialists with a mass of subjects ready to their hands ... [who] are now inclining to believe that they can produce on those subjects certain telepathic phenomena." Although major historical figures like Janet, Binet, and Babinski are not particularly remembered for their psychical curiosity, those concerns did take intellectual precedence for them at the time of this conference and provided an evidential runway for Myer's own theoretical imagination to take flight. Though Charcot may have grounded this telepathic flight of fancy, he had nonetheless allowed it to take off, providing (according to my argument) a critical context for the development of Myers's first formal statement of psychical psychology, "Multiplex Personality," a paper first delivered before the SPR in late October, 1886, The research Myers used to evolve his subliminal theory of consciousness was not drawn from his own well in some psychical backwater, but from a growing pool of data produced under the sanction of the historically designated father of modern neurology. For Myers, the French foray into telepathy in 1886 signaled that a great change was at hand; he laid out this new direction at the last general meeting of the SPR for that year: "I think it possible that the facts of telepathy may compel us to extend our conceptions of physico-psychical concomitance, and to face the supposition that though forces may exist, and agencies operate, which the ordinary materialistic view altogether denies, yet these also may be correlated—though above the limit of our intelligence—with the force and matter with which our mathematical science already deals."[120]

Myers was not urging a departure from "mathematical science," but only some new modification that would include consciousness in its framework as a distinct but related order of activity. Some such notion of a new "physico-psychical concomitance" had already begun taking shape in recent years, expressed in a rash of physicalist theories that lent substance to thought (and substantiated Myers's own thinking on the topic). Myers's article referenced the variety of technical iterations that had surfaced in recent years, including James Knowles's "brain waves" (1878), Alexandre Baréty's "neurique rayonnante" (1882), Henry Maudsley's "mentiferous ether" (1883), Claude Perronet's *ondulationisme* (1884), and, closer to home, Oliver Lodge's telepathic field and syntonic brains (1884). While such a cluster of theories cannot be used to suggest that a mind-reading, matter-manipulating consciousness was by any means an orthodoxy, it did establish that such speculation was a legitimate and singularly compelling research horizon in psychology. Myers hereby positioned himself as progressive not transgressive in his thinking, and in truth, the controversies generated among the fractious schools of the Salpêtrière, Nancy, and Bordeaux concerning these ideas suggested that there was no consensus strong enough to reject the premise altogether.

One thing seems clear: psychologists of the 1880s were not so willing to "give up the ghost" as that previous generation who had indentured themselves to physiology, anxious to find room for their discipline amid the ether and atoms of Victorian science. When it came to characterizing human consciousness, science seemed more generally willing to tolerate a certain ambiguity, even latitude, with relation to physical laws so long as it did not actively contaminate its own disciplinary program. (The emergence of new disciplinary firewalls helped this tolerance along as science became further specialized in the late nineteenth century, leaving each to police their own. Yet, it also underscored the importance of psychical research's interdisciplinarity as a crucial venue to transcend these clans of knowledge and tackle the truly big questions.) For all the sense of momentum Myers expressed in his speech, in which he specifically hailed the work of Janet, Babinski, and Féré as being "of great importance for our [psychical] research" and declared his "hope that these experiments were but the first installments," the following year would already prove something of a reversal for telepathy at the Salpêtrière.[121] The topics covered by *Bulletins de la Société de Psychologie Physiologique* for 1887 (its final year in publication), turned toward new trends in evolutionary psychology regarding reflexes, heredity, and language, with only one article published on hypnosis (and that of the nontelepathic variety.) Myers's "doubt that they would drop the topic soon" failed to predict the fate of this particular hot potato.

In retrospect, it might appear that the psychical and neomesmeric dimension of French curiosity hadn't much steam to begin with and quickly petered out after its first foray. Such has been the usual story. But the case could also be made that in 1886, such psychical trends were accelerating at a pace Charcot no longer wished to encourage. The organization Myers described late in 1886 was a self-propelled French phenomenon, whose "new adhesion to telepathy" had arisen "in independent quarters and in the course of distinct and disparate lines of investigation," which was "characteristic of a true discovery."[122] It not only enrolled such French familiars as Eugène Azam, Alexandre Baréty, Henri Beaunis, Alfred Binet, Jean-Martin Charcot, Charles Féré, Pierre Janet, Théodule Ribot, Paul Richer, Charles Richet, and Hippolyte Taine, among others, but also resolutely antivitalistic German physiologists and psychophysicists, such as Rudolph Heidenhain, Hermann von Helmholtz, Ewald Pflüger, and Wilhelm Wundt. But what felt authentically like a launching point for Myers (and it was) turned out to be, as far as Charcot was concerned, the end of the road. By 1887 he had grown wary of "the forces of superstition and charlatanry to which this vast territory has been ceded for so long, bequeathing an unfortunate legacy to those who would now colonize it in the name of Science."[123] Yet even though Charcot ceased publication of the *Bulletins,* he never went so far as to retract the results.[124] (How could he? These were experiments conducted in good faith under his sanction. To have done so would have been a betrayal of his researchers and a monumental personal scandal.) So he entombed the data instead in a limited-circulation journal he aimed to keep from being read. But it remained for Myers a pristine scientific talisman, beyond the reach of Charcot's regrets, and a way out of his increasingly dead-end hysterical nosology. Despite nearly a decade of dredging the subconscious and dragging its hidden realities into the light, Charcot had preempted any substantive move away from the "ordinary materialist view" that had formed the assumptions surrounding the old "unconscious muscular cerebration." And this was the materialism concerning which Myers had warned that "it was no longer safe to assume." For all the fascination and depth of this new interior brainscape, no new theory of mind could break out of the perimeter Charcot had established with his physiological psychology. It had continuously hobbled the great leap forward primed by the research at the Salpêtrière, and now threatened to stall telepathy in the midst of its own roaring engines. But Myers was no longer a man on the receiving end of French academic authority. He had come among them as an equal. And 1886 and 1887 would find Myers stepping out as his own intellectual master, ready to lead rather than follow with his own, compelling new psychological paradigm of multiplex personality.

MULTIPLEX PERSONALITY

It is tempting to see Myers's "multiplex personality" as a grandiose digression departing on its own logic from the rest of psychology, but the concept is better understood as an attempt to start a new conversation about the facts at hand, one that could assimilate all the curious, uncomfortable data coming out of experimental psychology in the past ten years, and rework the evolutionary framework keeping its imagination captive. Since the 1860s, Darwinian narratives had been working their way out of biology and into the cultural sphere: as Edward Burnett Tylor's comparative anthropology (morality as self-interest), T. H. Huxley's sociology (survival of the fittest), Cesare Lombroso's criminology (crime as recidivism); and now, with Charcot's hypnotic exploration of hysteria, the descent of man had gone down into the subconscious mind. Myers took the full measure of this view from below in the dolorous paper, "Human Personality," (read before the SPR October 29, 1885), stripping the individual down to his or her neuroelectrical circuitry, a mere gadget to be undone by disease, a magnet, or *la pression oculaire dans l'oeil*. In "Further Notes on the Unconscious Self" (1885), he added his own multifocal spin, depicting consciousness as a roulette wheel upon which the ball of our attention spins, not unlike the whim of madness. Disintegration, at that time, was not just the study of the psych ward. Well-heeled neurasthenics and neurotics across the industrialized world reported to medical spas in Switzerland or America to get themselves shocked with electricity or dunked in ice water: all trying to repair their frayed, overstimulated, or underenergized nerves.[125] Their anorgasmic daughters might be made to straddle a trotting horse, mount a steam-powered vibrator, or take a brisk water douche in hopes of keeping hysteria at bay with a vitalizing paroxysm.[126]

Such was the modern failure of nerve, brought on by the unnaturally accelerated pace of progress, causing patients to teeter precariously between the effete enervation of Baden-Baden and now the primal insanity of the Salpêtrière. The cognitive structures of sanity seemed to be breaking apart as evolution tugged relentlessly backward on the embodied brains of Victorians in ways that the untouchable rationality of the eighteenth-century mind was totally exempt. The once "free will," it turned out, was quite superficial, a discretionary element built into the machinery of our survival, readily canceled by subconscious suggestion. Human beings appeared to be creatures of compulsions and preferences, not principles. Myers could not change the verified facts upon which the grim verdict of evolutionary psychology rested, so he would do something bolder: change its overall interpretive scheme. He would promote a science that would "teach the world that the word evolution is the very formula and symbol of hope."[127] To do so, multiplex personality would need to occupy the bull's-eye of dread in its choice of subject matter, so that no part of the modern psychiatric narrative should

remain unredeemed. The sensational case of Louis V.[128] was singularly inauspicious in the already sick, sad world of human aberrational psychology,
with the patient shattering the record low of *dédoublement* to achieve half a
dozen or more subpersonalities.[129] Myers was not alone in his preoccupation with the case. Louis's tragic tale of psychological pandemonium had by
now spread well beyond medical journals into the fevered machinery of the
commercial press, driving the public's increasingly tortured psychiatric self-
examination. But the dread and pity Louis inspired supplied Myers with the
perfect crisis to explain his new, salvific theory of human consciousness,
with the avatar of madness, "human personality," hung upon the cross.

Depending on the examining physician, the total number of Louis V.'s
subpersonalities could rise to as many as ten, with the entire cast living on
as latencies ready to emerge at any moment. But as to the question of which
of these ten was the "real" Louis V., Myers responded: all of them and none.
The person was not the personality. Here he reasserted the earlier notion of
the multifocal mind: even ordinary individuals were in a state of constant
flux, cycling from focus to focus in endless oscillations between melancholy,
happiness, dreams, rage, or drunkenness. In the unhinged hysteric "these
moods might objectify themselves each into a persona."[130] But the individual
could no more be reduced to this persona than to a passing mood. Myers
assumed, along with Reid, that there was some "true and permanent self,"
an indivisible thing that thinks and acts but is not the thought or the action,
and that is "at any rate not what [we] take him for."[131] But Reid had failed
to prove this crucial distinction between subject and object, and modern
neurology (as well as the mechanics of the multifocal mind) did not require
it. This thinker could just as well be the endogenous biological processes
that give rise to thought. And indeed, experimental psychology's scientific
mandate required a strictly physical basis, since any mental "powers" necessarily polluted deterministic physical laws. But now the facts of telepathy
appeared to finally resolve the issue in Reid's favor: if intelligent activity
could operate outside the brain, it could not be fully reduced to the brain.
Thinking was more than "thinking about"; the subject was more than "subject matter." Beyond these preliminary facts, little was as yet known.

Telepathy provided an important philosophical premise for multiplex
personality, but the idea itself remained in the background, avoiding any
metaphysical digression that might damage its relevance. Myers would have
to mine the existing narratives of delusion, *dédoublement*, and multiple personality disorder to get at some hidden psychical source. Certainly, Louis V.
seemed to eliminate all hope that there was any such "soul or spirit or

transcendental self" at heart guiding this scrambling mob toward any kind of unified purpose.[132] With a swipe of a magnet, Louis's doctors could drag him from personality to personality, each focus activating a new alter in a funhouse of mental projection and mirage. One minute, he was at his desk, calmly pulling a needle and thread, and the next he was stabbing you with the needle and garroting you with the thread. This application of magnets suggested something akin to Binet and Féré's "psychic transfer" of physical afflictions to different sites on a patient's body. Only in this case, it was the entire idea of the self that was being transferred, along with the physical modifications that were part of this repersonification. Louis's alters were not passing delusions, but had lives and memories of their own, "pick[ing] up exactly where the nervous energy had last previously rested," until set in motion again by a magnet, electric bath, or hypnotic shock, and all this was "sufficiently common, as French physicians hold, in hysterical cases to excite little surprise."[133]

Such were the facts. But now, Myers modeled another way of considering them. Autonomy and initiative were readily observable within Louis's disordered state, once properly viewed. First, each of these personalities had its own discrete rational awareness and volition. Louis's doctors might be able to choose his identity, but that identity was still able to choose. This imposition of personality was in some ways no more deterministic than heredity, which similarly cast its yoke upon identity. Free will was always thus constrained by the facts of existence; that was what gave it value as a moral exercise. Even the manner in which Louis's personalities originated and multiplied showed signs of an active will. Some external crisis would suddenly overwhelm Louis's sense of personal sufficiency, and in response, he would spontaneously become someone new, someone perhaps better equipped to meet the challenge. Unlike the more pathologically driven splits observed in *dédoublement de la personnalité*, Louis's serial personae seemed to have an almost adaptive quality. For instance, before the onset of his first episode, he had been a lonely, meek boy abandoned to a reformatory by his "turbulent" mother, but an overwhelming fright from a viper sent him into shock, precipitating a fifty-hour crisis of convulsion and ecstasy. On the other side of this, Louis was thoroughly cured of his timidity, having assumed the persona of a truculent bully. The terrorized had become the terror, with every memory of his sad childhood obliterated in the process. And so began "the series of psychical oscillations on which he has been tossed ever."[134] The next catalyzing incident redomesticated Louis as a mild-mannered tailor, and from there he became a violent criminal, an innocent boy, and an insolent drunkard expelled from the Marines (and so on and so on, depending on who was performing the inventory.

Was Louis to be judged as the thief and marauder, or the amiable, first-class tailor? Ordinary individuals acquired their morality as part of their socialization, moving from birth to maturity with progressive uniformity. But for Louis, challenged by mental illness and poverty, this maturation did not advance along a straight line; instead, it rocked violently back and forth as he staggered from personality to personality through the rough terrain of his broken mind. This may have denied Louis the appearance of moral development, but, when properly understood, Louis was indeed engaged in a battle to better himself. Somewhere within his oceanic consciousness, Louis had found the means to be patient, peaceful, and industrious, even while being constantly thrown back upon his instincts for pleasure and survival. All along his biography were signs of a will struggling to assert itself. This will was not always moral, but it was nonetheless free. (If Louis set fire to a building because some doctor placed a magnet too near his skull, it was because his inner primitive felt the need to set such a fire.) Similar tests of moral will occurred on a daily basis in modern life as instinct and ethics collided, remaining just below the threshold of people's awareness.[135] This was a normal part of the structural legacy of human cognitive evolution, but it nonetheless strained the nerves to constantly regulate these passions. Louis's peculiar condition showcased this hidden drama through the spectacle of his multiple personalities, those "priceless living document[s] nature affords to our study." The concept of multiplex personality was intended, not to make some narrow diagnosis for the psychiatric roster, but to provide a theoretical model for the whole of evolutionary consciousness:

> Let us picture the human brain as a vast manufactory, in which thousands of looms, of complex and differing patterns, are habitually at work. Now, how do I come to have my looms and driving gear arranged in this particular way? My ancestor the ascidian, in fact, inherited the business when it consisted of little more than a single spindle. Since his day, my nearer ancestors have added loom after loom. But the class of orders received has changed very rapidly during the last few hundred years. I have now to try to turn out altruistic emotions and intelligent reasoning with machinery adapted to self-preserving fierceness or manual toil. And in my efforts to readjust and reorganize I am hindered not only by the old-fashioned type of the looms, but by the inconvenient disposition of the driving gear. I cannot start one useful loom without starting a dozen others that are merely in the way.[136]

If we think about Charcot's project, it put a very negative valuation on primitivity, unpacking evolutionary history through the clues gathered in the study of brain disease. Myers wanted to tell an entirely different story, though it would rely on the exact same set of facts (thus depriving them of their power). The past he was drawing upon was not just backward but deep, mysterious, profound, and originating. Even as he acknowledged the "hindrance of these old-fashioned looms," he did not present these prerational and noncognitive states of mind as necessarily regressive. They seemed to bear some closer relation to other beneficial powers. For instance, hypnotized individuals seemed able to tap some generally unavailable source of somatic intelligence to pilot their own biological processes. This involved, Myers explained, self-anesthesia, spontaneous healing, alterations to the respiratory and cardiac systems, excessive feats of endurance, as well as vegetative states that seemed to mimic the deep sleep of death. Did such physiological modifications tap the same reservoir of power by which a newt repaired a limb, a chameleon changed color, a worm regenerated its severed half, or a fin evolved into a means of flight? It was no wonder, to Myers, that Louis had gone down to this well to effect his first monumental transformation, locking himself in the depths of consciousness for two entire days after a fright from a viper paralyzed him. When he awoke, "he was no longer paralyzed, no longer acquainted with tailoring and no longer virtuous."[137] Such an attack sent Louis down to the same levels of hypnotic sleep in which James Esdaile performed surgery, daughters shot mothers, and hysterics became delusional, to effect a profound constitutional alteration of his very being: "a change of personality which is not per se either evolutive or dissolutive but seems a more allotropic modification of the very elements of man."[138] So while Louis might have temporarily lost control of himself, he seemed to gain the chance to create himself instead. Was he "falling apart," per the degenerative nosology of hysteria, or was he putting himself back together, as partly suggested by the therapeutic framework of Nancy school? Was hypnosis the cure, the disease, or both?

For all their differences, Charcot and Liébeault each conceived of hypnosis as inhibiting brain function. A "sub-" conscious state was a lesser one in terms of its lease on the brain. But Myers saw it differently. The prime value of the hypnotic trance lay "not in what it inhibits but in what it reveals." The suppression of certain cognitive modes might "allow for emergence of outmoded sensibilities, perhaps even in the manifestation of new and centrally initiated powers."[139] That human beings had been dropping in and out of trance states long before the days of Mesmer indicated unambiguously to

Myers that hypnosis was neither the fad of the moment nor a trendy disease, but something rooted deep in humanity's cultural and evolutionary history. It was, in the words of Myers, "a process probably at least as old as Solon," prefigured in the temple sleep of ancient Egyptians and the healing dormitories of the Greek god Asklepios, as well as the Christian ecstasies of saints and monastic meditations.[140]

Evidence of this hidden power was everywhere around us in nature. What about the deep hibernation of winter, or the creative sleep of the chrysalis? Why should science accept only two conditions for consciousness, those of sleep and of wakefulness, when everywhere else nature manifested myriad variations for every imaginable utility? Why was sleep viewed as a normal brain function but sleepwalking considered a defective one, of interest mainly as "a mere curiosity"?[141] All these brain states existed in their own right for their own purposes. If it was the job of the waking self to stabilize identity across the ever-changing stream of experiences, why shouldn't an opposing mechanism (the "un-" conscious) exist to exploit some other advantage, maximizing fluidity and creativity? Somewhere between sleeping and waking, there was this hidden level in the structures of consciousness, which had been progressively walled off in recent centuries by the rise of reason. But now, it was newly accessible once more, courtesy of magnetic circles and séance rites, Mesmer's baquet and Lieabeault's suggestion, and Charcot's Chinese gong. But hypnotism, Myers argued, was potentially so much more than these rituals allowed:

> The interest lies neither in mesmerism as a curative agency as Elliotson would have told us, nor in hypnotism as an illustration of inhibitory cerebral action, as Heidenhain would tell us now. It lies in the fact that here is a psychical experiment on a larger scale than was ever possible before; that we have got hold of a handle which turns the mechanism of our being, found a mode of shifting the threshold of consciousness which is a dislocation as violent as madness, a submergence as pervasive as sleep, and yet, is waking sanity.[142]

With this, Myers had now fully wrested hypnotism from its physiological scheme, offering not a static lethargy or fractal madness but "the nascent art of self-modification, a system of pulleys by which we can disjoin and reconnect portions of our machinery which admit of no directer access."[143] In embracing this hypnotic turnstile to altered states, Myers was not advocating some mental free-for-all of psychotic disassociation. Reclaiming the shadow nature of our ancestral past did not require the rejection of the modern attributes of evolution. Rational and irrational, conscious and subconscious, waking and dreaming were complements not contradictions, coordinations not exclusions. To harness the power of the past with the discipline of the present was to multiply the possibilities of the future. Rather than dreading deviance, modernity had to embrace these strange new facts of mind: consciousness was dynamic, the self a matter of ongoing invention, and reason only one of the many productive sentient modes yet to be discovered: "I maintain the power of exaltation [and] of concentration which constitutes genius, implies a profound modifiability of the nervous system, a tendency of the stream of mentation to pour with a rush into some special channels . . . that for some uneducated subjects it has been the highest mental condition which they have ever entered; and that, when better understood and applied to subjects of higher type, it may dispose to flows of thought more undisturbed and steady than can be maintained by the waking effort of our tossed and fragmentary days."[144]

The revolutionary importance of hypnotism was that it could give deliberate access to the same "controlled subliminal uprushes" that fired the minds of Shakespeare and Newton and summoned the castles of Kublai Khan in Coleridge's sleep. The creative latitude of the unconscious mind invited the roam of genius. Robert Louis Stevenson lived his stories in the dreams, vivid, coherent theatricals created and produced by "brownies." Of the great orchestral compositions, Myers wrote, "it is not through conscious labor over the mutual relations of musical notes that the masterpieces of melody have been born. They have come as they came to Mozart—whose often-quoted words I need not cite again—in an uprush of unsummoned audition, of unpremeditated and self-revealing joy." Myers' subconscious could mediate the alchemical correspondences between the soul of beauty and the form of art: "All that exists is continuous; nor can Art symbolize any one aspect of the universe without also implicitly symbolizing aspects which lie beyond . . . the pre-existent but hidden concordance between visible and invisible things, between matter and thought, between thought and emotion, which the plastic arts, and music, and poetry, do each in their own special field discover and manifest for human wisdom and joy. . . ."[145]

This was the under-consciousness as it was first awakened by the Marquis de Puységur in his castle in the south of France, then followed by Samuel Coleridge in England "Through caverns measureless to man, / Down to a sunless sea." Here Myers found coiled within the somatic structures of perception the transcendent perceiver on its temporal detour. The philosopher Immanuel Kant offered this distinction between a transcendent subject *capable* of perceiving and the phenomenal process of perception itself as a matter of empirical philosophy; religion offered it as a matter of faith; yet none had allowed this perceiver to be perceived in the scientific sense on the strength of observational data.[146] But now Frederic Myers asserted it as an evidential fact of experimental psychology in the secular prism of a Romantic science of mind. Multiplexity healed the divide between the numinal and phenomenal self with a subject that slipped laterally from focal point to focal point in the temporal brain, but could sometimes slip altogether out of it as well, in states of telepathy and trance: "not from terrestrial evolution alone, not bounded by polar solitudes, nor measured by the sun's march through Heaven, but making for a vaster future, by inheritance from a remoter past."[147]

The uprush of psychical curiosity uncorked in 1886 giving flight to multiplexity was not so easy to put back in the bottle. Together with Myers and Ochorowicz, Richet immediately set about capturing this momentum in the form of a successor organization, the Congrès International de la Société de Psychologie Physiologique.[148] The first meeting took place in 1889, as part of the Universal Exposition in Paris, signifying a grand forward-looking endeavor. The congress was meant to be a continuing forum for elite curiosity concerning telepathy and hypnotism, but it had to operate under the colors of orthodoxy. Richet recruited an international panel of officers and sponsors, including Théodule Ribot and Hippolyte Taine from France; Wilhelm Wundt and Hermann von Helmholtz from Germany; Alexander Bain and Francis Galton from Great Britain; and Cesare Lombroso from Italy. He even managed to press Charcot into accepting a kind of titular presidency, though he did not attend any meetings and formal ties to the Salpêtrière were severed.[149] The investigative docket of the congress remained similar to that of its predecessor (hereditary mental illnesses, the mechanics of perception, diseases of the nerves, and causes of hallucination), but now with a dedicated section for its more experimental elements. The de facto withdrawal of Charcot left room for the Nancy school to become a more active presence in the community of the congress, sustaining a vital hypnotic context for Myers to evolve his theory of the subliminal self.[150]

Psychical researchers were not the only constituency of the congress trying to put the ghost back in the machine. Panpsychism, the belief that consciousness is part of the structure of the universe, lurked as a possibility even in German psychophysics, the most meticulously quantitative wing of *psychologie physiologique*. Even here, they felt the need for some kind of supplement to explain subjective experience, opening up the door to some ontic basis for consciousness. For hardliners like Wilhelm Wundt, this was strictly limited to a quality, not a force or cause. It was all right to put a germ of will *inside* matter, one that might in the aggregate become intelligent reflection, but *acting* on matter was strictly verboten. Psychical phenomena thus had little appeal for German experimental psychology, where physics ruled and "force" remained mechanical and quantitative.[151] But French clinical psychology, which faced more toward physiology than physics, had potentially more use for some such subjective power. The same psychic force that wreaked havoc on the balance of energy equations helped make sense of biological processes too complex for existing laws. We eat food and grow hair. Acorns turn sunshine into trees. How can that be? Without DNA, it couldn't have been at all clear. Thus, despite the roadblock put up by Charcot in 1887 at Salpêtrière, the broader intellectual climate was growing more open to some less deterministic thesis, even within hitherto robustly materialistic life sciences. This propelled a new phase of French psychical curiosity in the 1890s for which the SPR furnished the institutional hub, absorbing this more adventurous wing of European researchers as corresponding members.

While continental researchers like Charles Richet, Julian Ochorowicz, Camille Flammarion, Enrico Morselli, Théodore Flournoy, Pierre Janet, Albert von Schrenck-Notzing, Cesare Lombroso, and Professor Aksakoff, among others, actively collaborated with their British counterparts, distinctions remained in their general philosophical orientation. They tended to have regular university appointments and were more cautious about exceeding the perceived disciplinary mandate to find some guidance in nature. Even the entelechal panpsychism they entertained still downplayed any outright metaphysical or spiritualistic interpretations that could be placed upon psychical phenomena. On offer was a practical vitalism animating brains and bodies at the cellular level (what idealists tended to consider "back-door materialism"). By contrast, theistic or spiritual implications were generally left undisturbed in British contexts, eventually maturing to become part of Lang, Lodge, and Myers's later speculation. It makes sense that the continental turn toward séance phenomena was effectuated through physical rather than mental mediumship, specifically through the organic substance of "ectoplasm." For physiologists wishing to understand the mysterious teleology of biological development, the ability of the medium to excrete and direct her own biological *prima materia* was evidence of guidance at the mind/body boundary. The strangeness of the medium, brought on by disease or evolutionary accident, let that hidden force erupt into view, obligating investigation. This was a page ripped right out of Charcot's neurological manual, rendering virtuous all prurience and sensation in pursuit of the norm.[152]

Ectoplasm was not a topic welcomed at the congress, though reports of the phenomena were beginning to be discussed and published elsewhere by leading members as early as 1891. Having emerged from the body of a séance medium, even a purportedly biological substance could not clear the spiritual I embargo maintained by professional psychology around its institutional center. (An attempt to introduce such medium demonstrators in the Congress of 1896 caused a furor.) This taboo against spiritualism was the original and perennial obstacle faced by psychical researchers wishing to fully explore the full range of phenomena, including the mind's complete spatiotemporal transcendence or even survival of bodily death. At a general meeting of the SPR, back in December 1884, Sidgwick had addressed "the necessity now strongly felt of concentrating effort on the most difficult and obscure part of the task originally undertaken—namely, the phenomena of so-called Spiritualism."[153] All the accumulating evidence for telepathy had put tension on the bow. That was mind beyond brain, but what about mind beyond time or even beyond death itself? Sidgwick had cautioned against going too far past where others might follow. Now, it was five years later, on the other side of 1886, the year when the most accomplished investigators of physiological psychology had themselves plainly established evidence for *sommeil à distance*, community of sensation, psychic transfer, mesmeric medicines, telepathic consciousness, and so on. And was it still not time for psychology to proceed? But for Myers, the time had come to let the arrow go, to travel "that more advanced, more hazardous line of inquiry which leads without a serious break from telepathic experiments to the appraisement of phantasms of the living and of the dead."[154] In the wake of the first congress, the SPR took the initiative, establishing a committee to investigate the séance phenomena of the Boston medium, Leonora Piper. She had come recommended by William James for both her powers and her probity. The investigation would bring Frederic Myers and Oliver Lodge together for the first time as an active investigative team, in an endeavor that went far deeper than any of their previous psychical investigations, down to the depths of consciousness where the conventions of physics and psychology, time and space, self and other, God and man no longer applied, and where each would find space to build a new, reformed cosmology.

ROMANTIC MODERNITIES/SUBLIMINAL SOULS

The theory of consciousness Myers formulated in the wake of the Piper sittings took the shattered pieces of the psychiatric subject and rendered them whole as the subliminal self, idealizing not only the mind, but the world

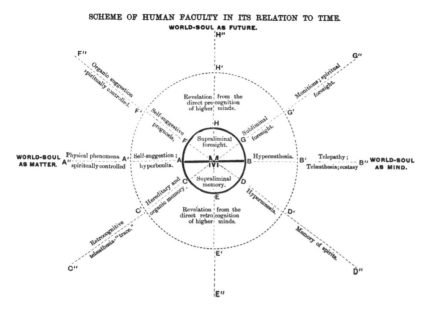

FIGURE 3 Myers's "Scheme of Human Faculty" extends the dynamic view of consciousness to the cosmos itself, pinning supernormal experience to the deepest structures of reality. From *PSPR* 11 (1895): 586.

as well in its Romantic, emanationist philosophy. This essential divinity of consciousness elevated Myers's psychology to the domain of religious meaning, dissolving its boundary with science at the edge of matter. In a series of nine thematically linked articles running in the *Proceedings of the Society for Psychical Research* between 1892 and 1895, under the general title "The Subliminal Self," Myers conducted what he called "a survey of consciousness." The series began with modest enough telepathic claims, but with each passing installment, it broached more and more forbidden aspects of psychical psychology. The chapters indexed a range of effects beginning with "Subliminal Messages," followed by "The Mechanism of Genius," "Hypermnesic Dreams," "The Mechanism of Hysteria," and "Motor Automatism," and finally breaking with all hope of naturalistic explanation with the "The Relation of Supernormal Phenomena to Time: Retrocognition" and "The Relation of Supernormal Phenomena to Time: Precognition." In his graphic depiction of subliminal consciousness, captioned as, "The Scheme of Human Faculty and Its Relation to Time," Myers created a visual grammar that used the spatial coordinates used to symbolize the physical universe, to map instead a terrain of consciousness that included both objective and subjective elements.[155] Its vertical axis ran from past to future, its horizontal axis, from matter to spirit.

At the center of the graph, the merest fragment of the whole, was a single band representing ordinary waking consciousness, bound both in time and space—that brief sliver of illumination called "now." Myers described the present as "a mere fraction of reflections of the immediate past and inferences of the immediate future," a window rather than a pinpoint in time. Cramped into this small frame was the waking self of ordinary consciousness, trapped in the limiting illusion of its own identity, unaware of the totality that lay beyond. Rational awareness was bounded by instinct from below and intuition just beyond it, expanding the human field of information to include seepage from these liminal intelligences. (This depended on the permeability at the boundary, which varied from individual to individual and between brain states.) Intuition was a kind of "idealized instinct," able to render information from immediate sensory data that were otherwise veiled. This might come in the form of a strange premonition or spontaneous insight, information that seemed to suggest some limited reach into the future or beyond the sensory horizon. This marked the spectral shift toward what Myers called the "violet" end of the psychical spectrum, where the mind began to slip the moorings of its psychobiology. Beyond intuition was "imagination," foretelling the future not through inference but through its own imaginative power—that is, the power to create future realities.

Biological instinct expanded the circle of supraliminal awareness toward the "red" end of the spectrum, honing the intelligence of the senses and directing behavior from below. This accessed knowledge that was planetary, not personal, drawing down into the body to grasp "hereditary or prenatal memories" based on the eons of interaction between life and nature. Deeper into the red, the mind began to gain access to the physiological functions of the body itself; this included the regulation of autonomic functions like heart rate, temperature, nervous reflexes, and pain response, or, in extreme cases, the circulatory system, as with the phenomena of stigmata. Red-spectrum phenomena focalized the brain within the most primitive structures of mind, marking some of the deepest states of trance consciousness where the brain was almost entirely coextensive with the body. These were among the most keenly observed phenomena of hysterics by *psychologie physiologique*, but recent, less orthodox investigations had made an even deeper penetration into the red. A mysterious excretion had been discovered oozing from the body of the medium Eusapia Palladino, whose mind seemed to reach down into the depths of her cellular marrow and shape her plasm at will. (Chapter 5 will cover this in detail.)

But most significant for Myers, in terms of this radical rescaffolding of human personality, were those phenomena that brought consciousness completely *outside* this biological matrix. Such were the phenomena he witnessed with Mrs. Piper, whose powers of clairvoyance seemed radically transcendent, moving backward and forward in time and slipping across locations at will. While "ordinary" telepathy operated with reference to the physical brain, the cognitions of spiritualism seemed to slip this anchor entirely. His "Scheme of Human Faculty and Its Relation to Time" laid out an interdimensional view of matter and mind, alongside a bidirectional scheme of time, suggesting structures of reality as fluid as consciousness itself. The atomistic experience of here and now, this not that, was fully enforceable only when fixed by the cognitive structures of the supraliminal mind. But such distinctions of past and present, mind and matter, sacred and profane lost their relevance the farther one migrated past the narrow bands that marked our own liminality. Multiplex subjectivity was not fractured, it was fluid, dynamic, transpersonal, and transcendent, the finite point from which to access the macrocosm. This, for an ordinary person, might come as a single, freak aberration: an instance of retrocognition or sudden violent premonition, before snapping back the conventional coordinates of everyday existence.[156] In such anomalous experiences, the "physically impossible" was brought within reach as the mind moved beyond the boundary of terrestrial laws. Myers labeled the totality of existence, all that ever was or would be, "the World Soul," and marked out as "personality" that infinitesimal speck of existence recognizable as "this moment" and "me." But the human being is in reality something far more than the lone lookout of this solitary lighthouse, cut off from the others on land and blind past the horizon of the darkening sea:

> I suggest, then, that the stream of consciousness in which we habitually live is not the only consciousness which exists in connection with our organism. . . . Yet it will be well to avoid the use of terms which, like the words *soul* and *spirit*, carry with them associations which cannot fairly be imported into the argument. Some word, however, we must have for that underlying psychical unity which I postulate as existing beneath all our phenomenal manifestations. Let the word *individuality* serve this purpose; and let us apply the word *personality*, as its etymology suggests, to something more external and transitory.[157]

From the tiny window of personality, the World Soul, that great ocean of consciousness and all it contained, revealed itself only partially as one's "point of view." But it could also blaze into that sliver of supraliminal consciousness, breaching the sea tower and sweeping the self into the whole, as described by mystics in their moments of *moksha, nirvana,* or ecstatic bliss.

But to be born, to be alive, to "be" was to make part of the whole uniquely our own, to contribute ourselves and our experience back to this totality, to inscribe within the eternal consciousness of the World Soul a soul of our own.

The monumental achievement of the subliminal theory established Myers as the leading intellect in the speculative psychology of the unconscious, the predecessor of Freud and Jung, who has often not been given his due.[158] In 1898, from the perch of this new scientific vista opened up by his research, Myers delivered a paper to the Synthetic Society, "A Scientific Approach to Problems Classified as Religious." (The Synthetic Society was something of a revival of the older Metaphysical Club of the 1870s, reconvened to address the urgent philosophical issues raised by Arthur Balfour's monumental *Foundations of Belief* [1896].) The "scientific approach" Myers advocated was psychical research, and the "problems classified as religious" generally specified the fate of man after he dies. This was not to be a nuanced debate about the Trinity or the appropriateness of organ music at Sunday service. Myers had little patience for traditional theism, let alone theology. Church authority, in his scathing summation, ultimately served to obstruct and not to structure man's inquiry into transcendence. Myers asked those fellow Britons who thought to defer to its authority: "The church tells you that you know all you need to know. How do you feel when they say that about Omar and his Koran?"[159] For a Protestant people who rioted to chants of "No Popery!" such accusations of obeisance would have struck home. It was not just the socially prescriptive nature of canonical religion, its endless catechisms, and moral taboos that bothered Myers. Theology structured the human self-concept at the deepest levels of the psyche, hemming in what was possible or desirable in terms of human conscious experience. Such orthodoxy posed a threat not just to science but to imagination itself.

Myers's speech identified three fundamental "streams of thought from which religions arise: 1. Christianity, 2. Maya, 3. Taoism." He was, as he stated, sympathetic to all three, but slipped in this blistering critique directed exclusively toward Christianity: "The notion that God was radically distinct from man, Heaven permanently removed from nature, and Revelation the only manner in which the two worlds could ever be linked" was, to Myers, a punishment visited by traditional religion upon the human experience. Myers had a former evangelical's loathing for the Catholic Church, which had, above all churches, usurped the proper authority of man to inquire into his own spiritual destiny.[160]

Hinduism he viewed more sparingly, allowing as it did some immanence of the divine. Maya manifested the illusion we call the world, putting us in the center of the godhead, while the atman successfully incarnated to achieve higher and higher spiritual refinement to exit that illusion and return to the source. Certainly, this fit more favorably with Myers's evolutionary scheme of spiritual self-perfection than the eschatological resurrection of the dead. But it was "Taoism," Myers asserted—somewhat controversially, considering that Christianity was among his options—"that lies closest to the truth of my philosophy."[161] Myers saw in the uncreated Tao the plenipotentiary powers of the metaethereal World Soul, both sources manifesting the duality of matter and force while ultimately transcending either particularity. The powers of telepathy, for Myers, were the powers of *wu wei*, the path of least action, achieving all through effortless yielding. At this psychical level, consciousness slipped in and out of matter's resistant framework, capable of feats that were spontaneous, prodigious, and as effortless as thought.

All that Myers offered, in his definition of religion, was the open-ended assertion that it was "the essential spirituality of the universe."[162] He offered as the "structural facts of the universe: 1. vital energies 2. our moral sense and 3. our own subjective experience?"[163] But in truth, Myers was not willing to cast off his Western allegiance so readily, accepting matter as a fiction of mind. Myers rejected what he interpreted to be the monistic pessimism of "Asiatic" antirealism (here Myers swept Hinduism, Buddhism, and Taoism together). The problem with "Asiatic Monism," as Myers called it, was its fundamental antirealism, implicit in its central proposition: the world is an illusion. This negation of the physical—and thus, the sensory—contravened an essential aspect of psychical psychology's empirical principle and its engagement with reality, not to mention its unshakable evolutionary framework for physical and moral progress. The carousel of incarnation allowed for spiritual progress as one worked one's way to escape, but the world itself was not subject to change: there would be no New Jerusalem on a hill, no victory through science, and no triumphal human evolution working its way up the violet end of the spectrum in pursuit of its own perfection. To hitch modernity's religious future to Asian monism would further, in the words of Myers, "the pessimism which dogs [modern thought] with unceasing persistency."[164] Nor could Taoism's relativistic outlook provide the same moral structure that continued to inform Myers's philosophy with Christian values.

But despite his distancing from his own religious roots, in many ways, Myers's philosophy most clearly reproduced the central message of the New Testament, an early allegiance that he may have had more difficulty in shedding than he described. Though Myers had little sympathy for orthodox Christianity, what appealed deeply to him were Christ's message (Love and

Salvation) and Christ's miracles, "proofs in the evidence of the realism of things."[165] Such demonstrations offered tangible realities that satisfied an empirical standard and interwove the sacred with the mundane, reflecting a dualistic understanding of the interpenetration—and independent reality—of spirit and matter. The subliminal self imposed not only a new psychological paradigm, but a new moral and spiritual paradigm as well: "Love is a kind of exalted, but unspecialised telepathy;—the simplest and most universal expression of that mutual gravitation or kinship of spirits which is the foundation of the telepathic law."[166]

One is immediately reminded of Myer's letter to the religiously resistant Arthur Sidgwick in 1866, pleading with him to open his heart "to such love as the gospel tells of" because "love is everything that matters," before signing off with Browning's couplet defining it as an essential element in human nature, "all with forehead bearing L O V E R , written above the earnest eyes of them."[167] This exalted love of feeling, and feeling of exalted love, that characterized the young Freddy Myers, which he had tried to shoehorn into Arthur Sidgwick and which was too great to hold within his own heart without the sustaining fires of his old evangelical piety, finally returns in a new form. It has bypassed religious creeds and personal faith to speak in the persuasive narrative of theoretical argument and is postulated as an irrefutable, unilateral universal force. Myers's Romantic cosmology repulsed the old stridencies of early evangelism and closed the emotional distance of the High Church in an all-inclusive embrace that stipulated to just one thing: Love was the key to immortality, for Love alone was fundamental. All things that were not Love were not permanent, vanishing from the reality of postmortal experience. There was no hell, and heaven was nothing more than an eternal "nearness to the ones [we] love." The essence of Love was communication, hence among its terrene manifestations was the power of telepathy. Its sacred action overcame the illusion of human separateness, allowing us "to discover how to be here and get nearer and nearer to the loveliness of the way of the Universe, constantly trying to reveal itself to us."[168] Myers, like Lodge, enunciated Christ as the highest moral and evolutionary expression of our humanity, while still denying all *supernatural* claims asserted by the dogma of his divinity. The perfection of Christ was the full realization of his supernormality. He revoked that catechism only so that he could restate it. Rather than being fully both man and God, Christ was the full expression of the divine within ourselves, holding the secret of human spiritual embodiment and transcendence within the myth of incarnation and resurrection.

Myers's subliminal self, though secular, fulfilled a religious need to affirm a transcendent source in which to instantiate one's being, something greater from which to be derived and to which to be returned. As Charles Taylor writes, "We accede to the status of human subjects through being taken up as interlocutors in an exchange that pre exists us."[169] The self, then, is a dialectic with some culturally determined source of "ultimate significance": a tribe, a place, a philosophical system, a God, a mythology, or, as in the case of the Victorian subject, one's innermost self when, through this nineteenth-century introspective turn, "finally the space of [ultimate] disclosure is considered to be inside—in the mind."[170] Ironically, just as the self assumed more and more of the burden of provisioning this "ultimate significance," the more its capacity to do so was being called into question. And without this "significance," the self did not signify (it had no import or real existence), nor could it. Psychological experience was less and less considered a faithful representation of reality. Thus, in the modern era, one's own personal subjectivity had become both the most foundational aspect of the self and the most poorly founded. The "crisis of faith" as originally conceived by Myers ("Does God exist?") had become, "Do I?" Myers proposed to restore this increasingly void "inner-most self," not theologically from the outside in, but psychologically from the inside out, drawing from the infinite well of subliminal sources. This was a Romantic response to the vacancy of "*tabula rasa*" (emptied now of the knower as well), endowing rational cognition with some deeper means of self-authentication.

We should see essentially the same undertaking in Myers's journey from intellectual pagan to "born again" Christian, to fustily reluctant spiritualist, and finally to psychical scientist. He was always in pursuit of what he called "the central question in Plato's philosophy—it must needs be a central question in all philosophies—whether there exists in man a principle independent of the material universe."[171] His magnum opus *Human Personality and Its Survival of Bodily Death* holds that question at its center. This does not mean Myers was not rigorously empirical in his methods or meticulous in the intellectual defense of his ideas—only that we need something more than disciplinary routines to comprehend his project. How else to understand the individual who creates his own discipline and puts at the center of his academic theory of consciousness nothing less than the World Soul? As Théodore Flournoy appraised the achievement of the subliminal self: "It subsumes, from its higher and more comprehensive point of view, a mass of similar but less elaborate theories, each one of which can be squared only with a certain limited domain of fact. Durand le Gros's polypsychism,

Dessoir's double self, Janet's mental dissociations, Breuer's and Freud's hypnoid states, Gyel's subconscious being, innumerable German doctrines of the subconscious and equally innumerable French theories of sleepwalking and hypnosis."[172] Myers was deeply engaged across this literature (something readily apparent in his prodigious footnotes), but he was not subordinate to it. He was not bound by its career concerns, theoretical obligations, or academic point scoring. (This amateurism did not diminish the integrity of psychical practices; it was actually meant to further ensure them.) Myers was *in* the world of academic psychology, but not *of* it. There was some more intrinsic way of knowing with which he was organically familiar from the time of his youth, drawn to the invisible world he sensed around him. He found it readily within himself, within his own consciousness, as the poet, the Platonist, and the Christian mystic. But it was a truth of which he feared he could be deprived (and the world along with him). He needed science and philosophy to furnish a language for these silent intuitions, that they might outlast their passing traditions and enter fully into modern life. Psychical research saw beyond the opposition of mystical and empirical epistemologies demanded by positivism toward some future reconciliation: one that recognized, with Myers, the dependency of the self on some metaphysical antecedent, while still upholding the secular defense of knowledge.

In the end, which came in 1901, Myers, who was once described by a friend as "almost morbidly afraid of death and extinction," refused to take the medical cure urged by his friend William James and had to hide from his wife his "half hearted clinging to earth life."[173] Instead, he embraced the slow and painful respiratory decline of Rhyme-Stokes disease because, he had at last convinced himself, "that doom meant the slow approach of sudden banishment and the shadowed portal of a blinding joy."[174] Even if Myers had failed to convince posterity of this fact, his own journey back to belief had at the last been accomplished.

CHAPTER FIVE

Knowledge in Motion

Oliver Lodge, the Imperceptible Ether, and the Physics of (Extra-)Sensory Perception

PULLING THE ETHER OUT OF A HAT

In June 1894, Oliver Lodge introduced a brand-new and curious technology to an elegant crowd gathered at the Royal Society soirée. Operating a small cable transmitter, he sent tiny pulses of electricity, struck at varying intervals and timed depressions, traveling through the air to be caught by the operator in the next room. It was the very first public demonstration of radio telegraphy, tapped out in Morse code to an astonished audience of ladies and gentleman come to ooh and aah over the latest in prestidigital science at which Lodge, in his courtship of the public, had become so adept. With a few clicks of his fingers, Lodge had become a magus of technology ushering in a new age of communication. This electromagnetic relay of information was not unlike the telepathic conveyance Lodge had investigated a decade earlier between Miss E. and Miss R. But then, the transmission had passed mysteriously between the minds of two blindfolded Liverpool shopgirls, and now, it passed no less mysteriously between a telegraph operator and a radio wave receiver. In both instances, Lodge surmised, it was the ether that provided the connecting link, supplying the physical medium through which energy waves, be they neural or electromagnetic, could be mechanically propagated.

At the time of this demonstration, Lodge's early promise as a physicist had grown into a form of academic celebrity. He was already a distinguished Royal Society Fellow and section president of the British Association for the Advancement of Science, and now his knack for explanation and demonstration was increasing his renown in lay circles as well. His career would eventually be honored by many distinctions, including the Rumford Medal of the Royal Society in 1898, knighthood in 1902, the presidency of the BAAS in 1913, the Albert Medal in 1919, and the Faraday Medal in 1932. The achievements that brought him before this distinguished audience in 1894, and which formed the basis of these future accolades, dealt mainly with the ether (that all-pervading yet ever-elusive plenum in the Victorian cosmos), as well as radiotelegraphy. But even this compelling demonstration at the soirée was of only routine interest to Lodge, just a clever manipulation of facts already at hand. His telegraph made splendid use of existing knowledge, but added nothing further to its depths. What was of abiding fascination for Lodge went far beyond such practical technology. For all his experimental gusto in the lab, he was no gadgeteer. Lodge was after something more, some deeper set of facts that would disclose what he called "the ultimate intelligibility of the universe," advocating, in his "Opening Address" delivered to Section A in 1891, for a science that "should not hesitate to press investigation and ascertain the laws of the most recondite problems of life and of mind."[1]

At the heart of solving these "recondite problems" was a fundamental reckoning with the ether, a somewhat improbable substance. It had to be rigid enough to allow for the rapid structural vibrations of light, yet supple enough that the Earth could still make its way around the sun without obstacle. It was at once everywhere and yet nowhere to be found: a tall order on which much depended. Only in 1887, after a long history of a purely speculative existence, had the ether finally been confirmed to exist at all, when Heinrich Hertz produced and detected the first electromagnetic waves in a laboratory setting. Some intervening, undulating medium (the ether) must exist to propagate this energy in that particular form.[2] Victorian science breathed a sigh of relief. George Fitzgerald, then president of the BAAS, triumphantly proclaimed, at a meeting held in Bath the following year, that "1888 will be ever memorable as the year in which this great question [was there an ether] has been experimentally decided."[3]

But for Lodge, Hertz's achievement, much like his own demonstration of telegraphy, came up short, leaving the deeper questions concerning the ether unaddressed. Lodge had himself been effectively generating these waves in his laboratory as early as 1879, even surmising that light might be some species of this same electromagnetic energy. Writing in his notebook (1879–80), he documents his moment of insight: "Light itself is oscillatory . . . The Leyden emits a spark of light. This must be the very thing [electromagnetic waves]."[4] This put Lodge's speculation ahead of even James Clerk Maxwell's, who thought of light more along the lines of an electromagnetic "effect" rather than as electromagnetism itself. The problem for Lodge was not in producing electromagnetic waves, but in proving what they were and that they were indeed there. Hertz beat him to the draw by first detecting radio waves with his ingenious radio receiver (1886), and then confirming their velocity to be equal to the speed of light, using an interferometer (1887). But Hertz succeeded where Lodge did not only because Lodge's project was considerably more ambitious. He was trying to turn Hertz's low-energy radio waves into visible light itself, ramping up the electromagnetic frequency until it attained intensities capable of illumination. This was a daring proposition, given the amounts of energy involved. Such an outcome did more than just imply some relationship through identical speed; it established a fundamental identity between light waves and radio waves, and allowed all to bear witness to this fact with the naked eye. To this end, Lodge spent years amplifying the discharge of a Leyden jar through increasingly elaborate circuits of induction (coils of wire arranged to create a cascade of interfering currents of higher and higher frequencies), only to be bypassed by history when Hertz walked his receiver across the finish line.[5] It was only recently, however, that Lodge had had the ingenious insight that Hertz's instrument could receive electromagnetic waves encrypted with Morse code, revolutionizing modern communication. But typical of Lodge, the intellectual and idealist, he never properly secured the patent for what science considers now to be his greatest contribution, ceding the credit and the profits to Marconi (as he had let the scientific glory go to Hertz).

But credit and profits were not his priorities, and as of June 1894, even with his telegraph drawing so much praise and attention, he had not yet achieved the essential thing. The elusive electromagnetic waves had been found, and even manipulated into wireless communication, but where was the promised ether? The unnerving conundrum posed by Albert Michelson and Edward Morley's negative results of 1887 would "hang like a cloud over ether physics" through the end of the century.[6] All that the experimenters had been after initially was a determination of the speed of the Earth relative to that ether. The ether itself had been assumed as part of the background knowledge. To this end, they split a beam of light and sent it along two different optical pathways, one traveling in the direction of the Earth's motion and one traveling at right angles to it. The presumption was that the speed of light should be increased in the direction of the Earth's motion. (This assumed that a certain amount of ether "adhered" to the Earth and that the light propagating through it would acquire some or all of the Earth's additional motion.) But no variation in speed was found, a piece of data relentlessly confirmed even at the most fractional level, no matter how long they extended the optical paths, no matter how increasingly precise they made their instrumentation. After six years of trying, the final negation was accepted in 1887.

That there was no detectable motion of the Earth relative to the ether seemed terrifyingly close to suggesting there might be no ether at all. This shake-up sheds light on why George Fitzgerald was so eager to proclaim Hertz's confirmation of the ether the following year, although the implications of this relatively obscure experiment in German physics were not immediately obvious to British science. Fitzgerald made sure to spell out, "for those who might not understand the problem," that this was unambiguous proof that some "intervening medium [the ether]" must exist.[7] And lest anyone present at the Bath meeting should fail to endorse that understanding, Fitzgerald went on to enumerate all that depended on their assent: ether was necessary to explaining atoms, electromagnetic force, chemical interactions, the propagation of light, mechanical motion, and kinetic force, among other essential operations. It was veritably "the thunderbolt of Jove" seized by science and gifted to modern technology, and could not be surrendered now.

But despite Fitzgerald's attempted fanfare, the mere detection of electromagnetic waves could not permanently offset the continuing failure to detect the actual ether, nor could it resolve the unnerving invariance of the speed of light. The coherence of Victorian physics was felt to be in jeopardy. Michelson and Morley's somewhat feeble defense was that the Earth "grabbed" so much of the surrounding ether with it that it became part of the Earth's frame of motion (better to move the ocean with the boat than abandon ship altogether). But this "ether grab" theory created too many problems for astronomical equations, and Lodge would not accept such intellectual alms. He rejected the theory, but would not otherwise let go of the ether, resolving instead to devise a better experiment to obtain "better" answers.[8] What Lodge proposed was essentially a simulation of Michelson-Morley on a more manageable scale, eliminating the unwieldy framework of the Earth and its movement, which was "too uncontrollable, and always gives negative results." Instead, Lodge planned to "take a lump of matter that you can deal with and see if it pulls any ether along with it."[9] This "lump of matter" involved two enormous iron discs, three-quarters of a ton each, separated by only an inch, which were spun by a warhorse of a dynamo at 4,000 revolutions per minute. The idea was to split a single beam of light, then send each ray in opposite directions around the wheel, quadrangulating the linear path with mirrors in order to keep it as close to the circumference as possible. Such was Lodge's "ether whirling machine." The beast nearly decapitated him early on, in 1891, when the wheel axle slipped its moorings, but, such was his dedication, he persevered through 1894.[10] Lodge was determined to find out whether the beam of light going in the direction of the whirling wheel traveled any faster than the beam of light traveling against it. It did not. Every time the beams completed their circuit and were superimposed on each other, a reading of the interferometer yielded not the slightest interaction between ether and matter.[11] (Any difference in speed would have put the waves out of sync with each other, a fact that could be easily determined by viewing their interference bands.)

But Lodge's endeavor was fruitful in other respects. When discussing his continuing negative results with George Fitzgerald in 1893, Lodge made the suggestion that moving objects might contract in the direction of their motion. This insight, first tossed about in Lodge's basement, became the basis of the Fitzgerald-Lorentz contraction a few months later (suggesting the creative leaps of which Lodge was capable in defense of the ether).[12] Lodge may not have gotten his affirmative results, but at least he now had a better theory than "ether grab" to hold onto: "The Fitzgerald-Lorentz contraction of all matter in motion may amount to a 3 inch shrinkage in the whole diameter of the earth in the direction of its motion; but it is enough. This slight contraction . . . I regard as the definite and interesting compensating effect."[13] Definite and interesting, yes, but still a far-from-satisfactory explanation of what the ether was or how it might operate given that no ordinary interactions could be detected. That some kind of physical or paraphysical connection between ether and matter existed, Lodge felt fairly certain, but it had to be got at some other way. The ether was far too subtle for any mechanical tests of "adhesion" and "speed" to detect, and its phenomena were too sweeping for the simulations of ordinary laboratory physics to fully comprehend. The facts of telepathy made this clear.

But there were other, even queerer phenomena than telepathy that could potentially offer an exceptionally productive venue for ether research, evidence currently under investigation in the most advanced circles of continental depth psychology. They involved the physical phenomena of the spiritual medium, Eusapia Palladino, whose body had been witnessed by trained academic observers to float, twist, compress, expand, grow heavy or light, and spark with electricity. Her body appeared to make every possible alteration to its corporeal composition at the direction of her will, as if her mind were directly "in touch" with the ether or even continuous with the forces of nature itself. Lodge himself had little knowledge of this side of mediumship; his career in physics had left little time to contribute much to psychical research beyond his telepathy experiments of 1884 and his reports on the Leonora Piper sittings of 1889. (And Mrs. Piper was a staid, middle-class lady from Boston who merely spoke to the dead.) But Lodge maintained an active correspondence with his close friend Frederic Myers, who communicated some of these intriguing developments on the continent.

Reports of Eusapia's phenomena appeared so outlandish they seemed more appropriate to a carnival barker than testimony prepared for academic consideration. But they bore out Dr. Chiaia's observations (1888), which were at first met with scorn:

> She attracts to her the articles of furniture which surround her, lifts them up, holds them suspended in the air like Mahomet's coffin, and makes them come down again with undulatory movements . . . She increases her height or lessens it according to her pleasure. She raps or taps upon the walls, the ceiling, the floor . . . something like flashes of electricity shoot forth from her body . . . [She] seems to lie upon the empty air, as on a couch, contrary to all the laws of gravity . . . She is like an India rubber doll, like an automaton of a new kind; she takes strange forms. How many legs and arms has she? We do not know.[14]

But after personal examination, Cesare Lombroso, the famed Italian evolutionary criminologist, felt compelled to announce to the scientific world that such reports could well be true. He admitted only to "the possibility of the *facts* called Spiritualistic," but otherwise assured his colleagues that he was "still opposed to the theory." Far from acting as a restraint, such demystification allowed Palladino to progress unimpeded through Milan, Rome, Paris, and Warsaw, leaping over the very disciplinary barricades put in place in part to deter any continuing professional interest in such phenomena.

This pageant of beguiling psychophysical displays proved particularly irresistible to those susceptible to a certain kind of curiosity. (Palladino's principal handlers, men like Cesare Lombroso, Charles Richet, and Julian Ochorowicz, all had ties to the original Société de Psychologie Physiologique and its later iteration as the Congrès International.) These were the same researchers who had been trying to pick the lock of the mind-body connection for the past decade at the Salpêtrière, but they had come up rather empty in the framework of *la grande hystérie* and *le grand hypnotisme*.

But Palladino's phenomena in many ways presented a far greater research opportunity than an entire hospital wing of hysterics ever could, offering a rare, external display of the most recondite processes of life hidden deep within the body. As early as 1892, some sort of protoectoplasmic materializations could be definitely added to her growing list of phenomena, including "living human hands which we saw and touched . . . that reach and touch and stroke one on the face and back, finger tips—felt firm."[15] "These "apports" were not to be confused with the cheesecloth and muslin of midcentury ghosts, but presented some kind of new, subtler phenomenon more akin to the organic structures of biology. However prone to chicanery physical mediumship was by reputation, Eusapia Palladino promised to be a far more sober object of study than any mere psychic ever could be. The communication here was not between medium and spirit, but between mind and body. Palladino's excreta were conceived as an endogenous element of her own biology, directly shaped by the most autonomic aspects of her own intelligence, inexplicably placed under her command during trance. Such a phenomenon had profound relevance for the most important questions confronting physiology and psychology: questions of organic development, hereditary memory, and the secrets of the evolutionary mechanism itself. How does an acorn become a tree? How do inheritable traits get taken up during embryogenesis to guide human development? These were the questions that allowed for so much disinhibition among these serious professionals.

And by June 1894, this all began to seem increasingly relevant to Lodge as well, when every spin of the ether-whirling machine came up zero. His close friend Frederic Myers had written him on April 16, breathless with the news that Charles Richet, who would not sign off on the Milan Report (1892) because conditions seemed too permissive, was "now thoroughly convinced of the phenomena" based on events he had witnessed in Rome earlier that year (1894).[16] There was even some discussion that Richet would host his own discrete series of séances where he could control all experimental conditions, to which Myers, and some other select associates of the SPR, would find themselves most cordially invited. The idea of a biological *prima materia* had its appeal for a physicist in search of the quintessence of the physical world. In July 1894, having just demonstrated one of the great technical advances of the nineteenth century, raised up high on the shoulders of the Royal Society, Oliver Lodge left his laboratory, his telegraph, and his ether-whirling machine to slip into a boat headed for L'Île Ribaud, a remote island off the southern coast of France as near desolation as one could get in the heart of Europe. There he hoped to discover some nearer clue to what gave rise to the landscape of the universe, turning vortices into matter and waves into energy, and carrying thoughts across its intervening sea.

AN ISLAND SÉANCE

L'Île Ribaud was deserted save for Richet's rustic cottage and a lone lighthouse on the far end. The men had gathered in this remote place, removed from the possibility of both detection and interference, to investigate the strange psychophysical phenomena alleged to occur in the presence of Eusapia Palladino. Though these were already well attested, the most stringent standards would be required to fully authenticate her phenomena, which carried with them the presumption of fraud. To this end, Richet had further tightened experimental conditions, removing the medium from the mainland to this unpeopled location. All her belongings had been carefully inventoried before she departed from the coast at Carqueiranne, and only a spare investigative team of the most trusted associates accompanied her to the island. The team consisted of Lodge, Myers, Ochorowicz, Richet, and Richet's longtime private secretary, M. Bellier. Thus, Eusapia Palladino found herself a captive of science, separated from any person or scheme that might aid deception, and outnumbered by her examiners five to one. Even for a thirty-year veteran of the spiritual trade, this would have presented an extraordinary set of challenges. She did not disappoint.

There must have been something extraordinary about Eusapia Palladino that allowed her, an Italian girl orphaned at eight who could barely speak a word outside her own dialect, to propel herself to the center of so much international attention. What exactly it was, however, is beyond recovery now, although she had been trading on this skill since she was a young girl. The scientists directly involved in examining her, though few, were significant operators in the fields of physics and psychology whose attention catapulted her celebrity—even in the midst of their own condescension. Lodge, in a letter to Andrew Lang, described her as a kind of magical Caliban, "a spoiled child of nature [and let us say, of Naples]. She hardly comes up to the ideal of respectable society."[17] And yet there she was with Richet, a future Nobel laureate, and Lodge, a future Knight of the Realm, who left off spinning his ether-whirling wheel for this promising occasion. It is hard not to imagine that for all her social, linguistic, and educational disadvantages, Palladino was more ringmaster than sideshow, leveraging largely recycled spiritual phenomena (cracking sounds, music, levitations, apports, apparitions), already combed over by the London Dialectical Society twenty years before, into a career as a financially well-compensated fin-de-siècle celebrity.[18]

But her routine was not all a matter of repackaging. On the Île Ribaud, Palladino introduced a significant new plot twist in perfect harmony with intellectual developments in science. While the Milan report had mentioned firm, fleshy materializations held in hand or briefly glimpsed, it was only on the island that this substance came into closer view and acquired its new name: ectoplasm. From the depths of her trance state, Eusapia Palladino seemed to produce a limpid, white excretion, half-jelly, half–elastic fluid which obeyed no known laws of liquid, gas, or solid. Richet, who took credit for labeling the substance, was making a deliberate reference to the protoplasmic matter found in the interior of a cell, with the prefix "ecto-" now placing that same substance outside the body.[19] (Myers wrote to Lodge regarding "ectoplasm" and pseudopods a week after the séance, so Richet's claims hold.) Whatever one might make of the phenomenon, it was tied to the most profound processes of biogenesis.[20] Lodge recalled in his autobiography Richet's exclamation as he witnessed the gelatinous extrusion: "C'est absolument absurde, mais c'est vrai."[21] This "ectoplasm" would become a more familiar part of the spiritual repertoire decades later, popularized by the sensational photos of eructating mediums spewing gauzy faces out of their mouths like so many souls.[22] But in 1894, Eusapia Palladino was breaking thrilling new ground with her phenomena. And there could be no more astute product line offered to bait the curiosity of researchers in the field of *psychologie physiologique*. Evolutionary explanations were more salient than ever, but "natural selection" was far too incomplete. It placed too much focus on the natural forces outside the body and failed to address the more profound processes within it, where much of this work was done.[23]

Lodge shared in the understanding that this was an endogenous substrate of the most basic kind, acting "as if guided by some subconscious intelligence to do outside the body the same sort of processes as are usually performed inside."[24] For Lodge, who had been fumbling after the ether for so long with only physical means at his disposal, there was an irresistible allure to these phenomena. They offered a potential access point to the ether denied by ordinary laboratory physics, but obtainable through the mind of Eusapia Palladino. It must be remembered that Lodge was not one to pick such low-hanging fruit. For the previous fifteen years he had been for the most part in his basement doing physically risky, fatiguing work with high-voltage electricity and whirling tons of iron, trying to find the ether in a wave of light. He was not successful. And so now he was willing to take on risks of a more reputational nature to satisfy that pursuit through a candid examination of Palladino, looking for empirical, and perhaps even tangible, evidence of a link between ether and matter, and, possibly, mind as well.

Lodge and his fellow investigators convened for their experiments in the late evenings, sitting around a table in the main hall, with M. Bellier seated just outside the window on the veranda, taking notes and observing that no one entered or exited the house. Eusapia Palladino sat at the same table as the investigators, her feet placed on a specially designed apparatus that would ring a bell if they broke contact with the floor. This kind of gadgetry was not uncommon. Along with the surge of scientific interest in spiritualism in the 1870s came experimental controls that were increasingly ingenious and paranoid. They had to be, in order to keep up with the ever-evolving ingenuity of the séance medium herself. Cromwell Varley used a galvanometer run with copper wires attached to contact points on the medium's hands; this way he could track the medium's energy and activity even when she slipped into her cabinet.[25] The SPR also had a protocol involving running a current through the clasped hands of séance participants to ensure they remained present at all times with hands clasped. Any failure to do so would interrupt the current and sound the alert.

Because the experimenters involved at the Île Ribaud placed themselves above suspicion (Lodge referred in his report to the notion of such cheating as a "repulsive idea"), no similar equipment was used on themselves. Instead, Lodge and his fellow investigators focused their energy on restraining the suspect, the foreign and lowborn Eusapia Palladino. Experimental laxity, whether perceived or real, could not be allowed to impugn their results. Mediums were known to slip out of adjacent experimenters' hands and join them together, or, in the dark, hold two people's hands with one of their own (pressing the backside and palm together in a single clasp), leaving one hand free to perform. Zealous precautions were made: at times there would be one man lying under a table clasping the medium's legs, another standing behind her with his arm hooked around her neck, or perhaps two experimenters flanking her with her hands pinched in their viselike grip. These efforts, especially in their extremes, are important to understanding these experimenters and the degree to which they confronted the burden of credibility. In their elaborate systems of accountability and deflection, they recognized the problematic nature of these beliefs, and contoured them sufficiently well to remain attached to the hospitals, universities, and professional associations from which their status derived. Reports on such investigations usually began with some declaration of the duty to disbelieve and some salute to official disciplinary opinion, which excluded any authority of the medium to assess her own condition.

Eusapia Palladino herself attributed her physical phenomena to the ghost of John King, a former pirate and personality of the afterlife already well known to spirit circles (along with his daughter Katie King, whom we encountered earlier in our analysis of Crookes). Such spirit guides were

common conventions of mediumship, placing a certain polite distance between the professional spiritualist and the out-and-out sorcerer. The medium was merely a host channeling other people's power, while the magus
possessed such powers for himself, perhaps courtesy of the Devil, which
was all very bad for business with a heavily Christian clientele. But such
theatrical spirits were of little use to investigators whose research agendas
dealt directly with human psychobiology, and who had no desire to deflect
these forces elsewhere. Richet, a prominent clinician at the Salpêtrière, offered the diagnosis that "John King" was a subpersonality created by Eusapia Palladino to shield her from the too-powerful psychical aspects of her
own mind. That thesis might be true enough, but for Lodge, "calling a thing
hysterical was no explanation . . . we need not shy at the idea of supernormality."[26] But Richet had to justify his interest in the phenomena in ways
Lodge was not obliged to follow. Not that it was strange for psychology to
claim interest in people who talk to ghosts, but it was acceptable only in
the context of the doctor-patient relationship. As it was, spiritualism barely
rated a mention. Pierre Janet was one of the few to do so, and then only to
classify it as an extremely aggravated form of dissociation, where the subject
split in two, externalized his or her alter ego, and then proceeded to haunt
him- or herself.[27] What he had not done was to hire a professional medium
and bring her to his clinic. Thus, there was a reason Richet, Ochorowicz,
and Myers were on a private island instead of pursuing this research as part
the International Congress they had all three done so much to create.

Richet's metapsychics (his own psychical variant) included a very broad
spectrum of phenomena, including telepathy, telekinesis, and even "crypt-
esthesia," that ultrararified cognition akin to clairvoyance. But he pulled up
sharp at an essential point. Life after death was a bridge too far. By quickly
dispatching "John King" as a hysterical defense, Richet got rid of a discar-
nate being, unwelcome in a brain-based psychology. But this prohibition
was no small point. It had immediate consequences for the self of this world
as well as the next. If there was no survival of bodily death, then the individual was by implication a contingency of matter and presumably divisible
according to its structures. So, while experience itself could encompass phenomena "numerous, varied, and mysterious" and even give the *appearance*
of pure transcendence, these were structures of the moment, not glimpses of
eternity. This kept Richet's metapsychics within the bounds of the imperative borders of his discipline: a self assembled by evolution, conditioned by
matter, under law, and, above all, secular in its politics. As long as there was
no self, there was no soul, and no need to fear claw-back from men of the

cloth. This was the flag Richet, Ochorowicz, and Schrenck-Notzing brought with them to the island, and continued to wave overhead for the next twenty years, even as mediums vomited teleplasm into their cameras and apparitions wearing fake beards appeared in their dens.

So, even while those gathered round the table at the Île Ribaud had much in common, they were yet worlds apart (a distance caught in Lodge's annoyance at Richet's unprompted diagnosis: ghost as "hysterical" defense. In the British psychical context, openness to some warranted spiritual consideration was assumed—first, because to deny that possibility was a preexisting bias; and second, because such questions were of vital personal and public interest. (This does not denigrate the empirical character of their research; it merely contextualizes their study in their many statements to that effect.) By contrast, encountering such "spirits," or even the threat of such an encounter, would have been instantly problematic for psychical research on the continent. By this, I do not mean to propose that one side was more "scientific" than the other, or even to treat them as equally biased. I mean only that "the psychology of the psychologist" matters here. To put this in the broadest strokes, then, interest in spiritualism for such continental investigators arose out of the explanatory ambitions of their disciplinary programs and had to remain confined to them. They were psychologists–cum–séance investigators. By contrast, the founders of psychical research tended to arrive at their disciplinary interest via broader philosophical questions, originally addressed through spiritualism (though not satisfactorily).[28] They were thus "séance-investigators"–cum–psychologists, a term I put in quotes because that was the venue of necessity, not choice.

The appeal of Eusapia Palladino's phenomena was the same, though, for both sides of the delegation: this was a mystery about Life, not about the afterlife. Many of the effects observed were "operations which were well within the power of the human body," but many more were, "ordinarily speaking, impossible." Little lights like glowworms flitted about, and a blue substance exuded from Eusapia Palladino's fingernail and left its mark on paper. She extracted objects from sealed boxes, untied knots on endless string, and caused cold bodies to warm and warm bodies to cool in a manner inconsistent with the "adiabatic alteration of the distribution of heat." During the experiments, a heavy table, specially designed with pointed legs to discourage sly lifting with the feet, levitated fully off the floor and began tilting. The curtains billowed forcefully into the room though the windows were shut and the air still, then mysteriously drew themselves into the "purposive arrangement" of a phantom face, while Palladino was "perfectly controlled and sufficiently visible elsewhere." A heavy escritoire slid around the room while, far from the medium's reach, an accordion positioned on a side table began to mysteriously dance and sing. On one occasion, Lodge took the music box and suspended it from the ceiling nine feet beyond the reach of the medium. At his request, he heard the music box wind up and begin to play, then it snapped off its string and floated in a spiral onto the lap of Frederic Myers. Some of these forces unleashed a great amount of energetic violence, causing the surrounding objects to shudder and quake. "The raps on the table, which were frequent, were sometimes so strong as to feel dangerous: they sounded like blows delivered with a heavy mallet . . . struck sometimes as if the table must be broken by the blows; and one could even feel nervous for the safety of our hands upon the table."[29]

There was also that other curious phenomenon, "the extrusion of ecto-plasmic material from the body [which] seems," according to Lodge, "at first a repellent object of enquiry." (He likened this to the discomfort of view-ing the body's inner organs.) The chimerical properties of this substance, however, made it somewhat analogous to the ether, materializing and van-ishing and invisibly mediating the transmission of force. Lodge witnessed "a protuberance gradually stretching out in the dim light . . . emanating from [Palladino's] side, through her clothes, a sort of supernumerary arm."[30] It approached and receded, then finally touched Myers, before vanishing entirely. At other times the "strange pseudopods" remained invisible, but their effects were strongly felt, pummeling and slapping investigators on the back, head, and neck or pinching their bodies with such force as to make them cry out. Lodge could not grab the ether with his spinning wheel, but it appeared the ether could more readily grab him.

All these inexplicable events—the movement of heavy objects without contact, the levitations and materializations—appeared to be opprobrious violations of the natural order. Or were they? Not if one accepted that these pseudopodia had a real, substantive nature rooted in ordinary matter, estab-lishing "some mechanical connexion to make matter move."[31] It was Lodge's complete conviction that "mental activity alone could never do it." In his report, Lodge speculated that Palladino unwound the etheric protoplasm bound up in her own body (or those of her sitters), and willed it to assume the form of some sort of extending rod. As preposterous as an ectoplasmic rod sounded, "perhaps," he urged his readers, "the unexpected power of protoplasmic activity was a less violent hypothesis" than the absence of such a rod (though assuming this as a stance of scientific chastity is quite color-ful). This protuberance would allow Palladino contact with remote objects, directly transmitting her own vital energy to levitate, push, and pull the object around her. This would explain the medium's frequently depleted state, as if her own energy had been "used up" to "get work done," in the parlance of physics. But she neither added to nor subtracted from the universe's existing store of energy and matter, a sustaining point for those "fundamental laws" Lodge was so anxious to leave unmolested. To secure this point of conservation, Lodge engineered a device to monitor levels of electromagnetic fluctuations in the immediate environment and periodically on the platform of a railway weighing machine to see if any mass had been acquired or lost in the process of her phenomena. Copper wires were run on and around the person of Eusapia Palladino, to see if some form of electromagnetic induction might occur in her presence, a notion suggesting that Lodge thought of her as something of an overcapacitated Leyden jar. Another one of Lodge's devices let circle participants squeeze a dynamometer in their hand before and after the séance to quantify any changes in the strength of their grasp (lest she steal away some of their own "vital power").[32]

The spiritualistic hypothesis was of little interest here. Lodge's language focused on Palladino's "vital" (that is, embodied) powers rather than on her "psychical" powers, in order to suggest a more primitive force at play, one closer to what Lodge designated as "life" rather than "mind." The idea that certain psychical phenomena were a kind of evolutionary atavism was not original to Lodge, but an established trope of depth psychology further developed by Lodge in "discussions held on the Ile Ribaud that included my friends' notions as well as my own."[33] It associated Palladino's powers with the autonomic regions of the brain, exercising the same archaic faculty accessed in major structural modifications, as when a caterpillar becomes a butterfly or a frog sprouts a new limb. Eusapia Palladino's feats were more wondrous still, allowing these earliest mental powers to harness the extraordinary build-out of the modern brain.

When Richet interpreted Palladino's phenomena, he began with neuropathology. The cause of this anomaly was a brain defect, somehow allowing endogenous energies to stream from her body, revealing hitherto unknown internal processes. Richet's discussion ended there, not digging too deeply into how this might be operationalized within the existing evolutionary scheme. Myers, by contrast, had charged into the heart of the matter from the outset, laying the groundwork for a teleological interpretation of evolution through the many personalities of Louis V. According to Myers's analysis in "Multiplex Personality" (1887), Louis V. did not subdivide his identities entirely at random, or ricochet between them in a purely reactive way. There was a will at work navigating this disease-ravaged brainscape, struggling to structure for itself some higher sanity despite the obstacles nature and nurture had put in his way. Myers accepted the details of the symptomology of multiple personality disorder, while using the disease itself to advance his own big picture. Such was the case with Palladino. Both Myers and Lodge were willing to be led by Richet's emphasis on some underlying defect of physiology. To a certain extent, that diagnostic language provided a welcome degree of insulation when it came to such hard-to-discuss phenomena, all but exonerating the researchers with the accusation pathology implied. Yet the two psychical researchers still held out the possibility that evolutionary deviance, no matter how seemingly maladaptive or backward, might also suggest nascent powers whose trajectories remain unknown. Even before they had arrived on the island, Myers had begun pressing Lodge for his most wide-reaching speculation regarding phenomena Richet had described:

> When embodied intelligences [Eusapia Palladino] act on matter by other methods than that of direct incarnation, they should be found employing that very form of action with which in our own subliminal selves we are already familiar. No wonder that they should act on the plasm/prism cell as we act on the brain cell, to shape the spirit hands as we shape the baby's or as dear baby shapes his own little self.

And also:

> Do you call it a good metaphor to say that moral evolution depends on the fact that moral combinations, as between a low and a higher sprit, tend to be endothermic, to leave the higher spirit weakened, but his lower spirit stored with a fuller portion of moral energy which is the Life of the World?[34]

The most significant fact of this identification between shaping a baby's hand and forming a phantom limb was that in one case the body was doing it and in the other, a medium did so by using the power of her own mind. This not only suggested a potential role for intelligent guidance over our own biological processes, it also opened the door for much more profound considerations: free will in evolution itself.

The phenomena Lodge witnessed on the island proved to be a highly productive framework for his thought. He emerged from that experience a changed man, forced to confront entirely new elements of reality: "However the facts are to be explained," he wrote, "I am constrained to admit there is no further room in my mind for doubt. Any person without invincible prejudice who had had the same experience would come to the same broad conclusion."[35]

And yet it is a very extravagant experimental record that would seem to overwhelm the established norms of physical science. To even consider such evidence, for many of Lodge's colleagues, would be to betray a certain professional trust. Yet Lodge maintained the validity of these events even after an investigative team from the SPR (which included both Myers and himself) caught Palladino cheating.[36] The events witnessed at the Île Ribaud were so tightly regulated and so compelling that, for Lodge, they remained immune to reassessment, forming the new bedrock upon which he would stake his physics and philosophy. He did not do so lightly. This was where his ether physics had brought him and where the empirical evidence had dug in. But there are still questions and contradictions posed by Sir Oliver Lodge's seeming surrender to this island séance. How is history to navigate such an extraordinary path as it careens along its opposing trajectories? Lodge was a celebrated experimentalist whose laboratory methods were astute enough to produce radiotelegraphy and the verification of the electromagnetic wave. He was obviously in no way delusional. Hallucinatory tendencies would be inconsistent with his record of successful scientific experimentation; he would have been left out of history, removed to his own private world. I assume with confidence that Lodge at no time was a willing participant in a hoax. His personal authenticity rings out in his private writings, in the elements of struggle in his biography, in his sense of public mission, and his abiding preoccupation with "the truth." But if he is not cheating, not fatuous, and not delusional, what then explains this extraordinary testimony?

The two obvious remaining alternatives are each unsatisfying in their own way: the phenomena were real, or Eusapia Palladino was a magnificent trickster.[37] The latter, though the most obvious, still remains problematic because of the scope of the deception. Lodge and his colleagues may have been gullible, but to what extent? Lodge claimed to "put no confidence in Eusapia's character at all but for scientific purposes treated her as being liable to deceive both voluntarily and involuntarily."[38] He and his fellow researchers understood in a more general sense that the landscape in which such supernatural claims found their home was "a jungle of deception," whether or not such facts were true. But to what extent should we be skeptical about their skepticism? I find the occasional humor in these matters reassuring. At least they could see the thing from several sides. (In a letter to Lodge, Myers joked about Mrs. Piper's refusal to hold a séance on Sunday because such spirits should be in church.)[39] Even if these experimenters were insufficiently skeptical about themselves, they had plenty of doubts about the mediums, considering all the friskings and physical restraints imposed upon the medium to offset the likelihood of cheating. To the extent that they were deceived, I would attribute it to the deceptive capacity of the medium and not, in the main, to some unconscious cooperation of their desire to be deceived. And yet, it does not all line up. There's something uncomfortable in it; it does not meet the eye. How can one reconcile who he is with what he saw?

Nonetheless, there it is. Lodge, a proven experimenter, theorist, and public man of science, came to these beliefs and eventually used them to catechize a kind of "ether theology" to a fascinated public. He was not swept up by a fad and carried over the deep end; he arrived at these conclusions along a compelling current of available evidence and reasoned analysis. And this is exactly what can help us to make sense of it: a reconstruction of his logic. To understand, we must enter more deeply into Lodge's own intellectual and methodological process, make a gestalt shift toward Lodge's point of view, a perspective both unique to its author and a portrait of the time. Because of the interest Lodge's ether theology aroused among intellectuals as well as the laity, it has a rare historical eloquence, illuminating the possibilities latent in the premise of Victorian science and providing a public record of the nation's longing for new kinds of wisdom and consolation at the turn of the century.

THE ELECTRIFYING ETHER

The love of science that sustained Oliver Lodge's cosmic quest over six decades of ardent inquiry (until his death in 1940) began in the earliest days of his youth. Born in 1851 in the county of Staffordshire, the eldest of fourteen children born to a cash-strapped clay merchant, he too, like his friend Myers, took upon himself the burden of his own identity from a very early age. His father had started off working for the railway in some clerk's capacity, but found that his growing family soon outstripped his means. With tremendous bravado, as Lodge recalled, his father, a man of energy but no particular education, threw himself into the pottery trade in a desperate bid for solvency, starting at the lowly rank of sales associate. As the eldest child, Oliver Jr. was likewise pressured to do his duty and marched straightaway into the family business. Young Lodge's passion, however, was for science, not plates and vases. Even as a boy, he spent his afternoons devising electrical experiments with wires, weathervanes, and other scraps of equipment, inspired by the drawings he saw in the *English Mechanic*. When his father fell ill, a then twelve-year-old Lodge was taken out of school and compelled to assume the full burden of his family's economic support as salesman-in-chief. (Perhaps this early market trauma explains his future reluctance to profit from science.) Lodge, a promising and passionate student, was crushed at this early demise of his dream, but his practical father saw it as a blessing, offering the consolation that there were "no prospects in a scientific career anyway." But Lodge's hopes proved persistent, sheltering in the fantastical laboratory he made out of the potter's storeroom, "where there were piles of plates arranged all over the floor," which he would picture "as the cells of a gigantic voltaic battery, and imagine all that [he] could do with the current with which might be thus provided."[40] Thus Lodge, unbeknownst to his father, refused to be cured of who he was and wanted to be.

The pottery trade brought him often to London, where he stayed with his Aunt Anne, who encouraged his abilities by bringing him to every variety of scientific lecture. Lodge became fairly well schooled in chemistry, geology, biology, and a variety of learned topics, taking advantage of the outreach that the new sciences made to educate the public. Though compelled by the business cycle to move from town to town every few months, Lodge managed to enroll in university courses along his commercial route. In this way he built the skills necessary to pass the University College London (UCL) matriculation exam in 1871, finally enrolling as a full-time student in 1874, after a decade of cobbling together his education behind his father's back and between moments supporting a large family. Lodge remembered with particular gratitude the influence of the Royal Institution of London. "It became a sort of sacred place, where pure science was enthroned to be worshipped for its own sake. Tyndall was in a manner the officiating priest and Faraday a sort of deity behind the scene."[41] Lodge went to see Tyndall at the Royal Institution on several occasions on these visits to Aunt Anne and knew him again at UCL. We can imagine the excitement such a bold champion of science held for Lodge, whose father was constantly trying to diminish its value along with Lodge's passion for it. Lodge recalled himself as an outsider looking in, with no formal education or career connections, who was welcomed to sit by the fire and enjoy the great ideas and demonstrations by way of the institution's generosity. For Lodge, "it was an opening up of deep things in the universe, which put all ordinary objects of sense into the shade."[42]

No doubt this sharing of knowledge with an intellectually mendicant clay merchant touched Lodge deeply, and influenced the man he would become. His university reforms were designed to make a career in science more professional and less elitist. He developed a missionary zeal for scientific education of the public, synthesizing each new discovery in countless addresses, pamphlets, and print columns, embedding his idea of science in a moral philosophy and rhetoric of social salvation. As a boy, he once heard Tyndall lecture at the Museum of Geology in London, internalizing at that moment the "high ideals in the attitude of a scientific man toward human life. It leads one to think that devotion to truth was the end and aim of everything."[43] There was an unmistakable language of the sacred in Lodge's description of science's truth-giving power. But for him, science did not merely bask in the sanction of natural theology, subordinate to a larger religious truth. Rather, science itself had the sacralizing power. As "the end and the aim of everything," science had the potential to confer legitimacy upon religion, and not just the other way around.

This aspirational view of science allowed it to become a platform for Lodge's ambitions as a social reformer, uniting his philosophical idealism to narratives of material progress.[44] While he set himself against scientific materialism, he did so within the framework of science and not against it, maintaining a forward-looking ethos of technological achievement and expanding human rights.[45] There was about Lodge a certain nostalgia for old, religious sentiments, a wariness of a future in which an overly granular view of individualism could no longer encompass God or fellowship. But his brand of idealism was emphatically not reactionary. He could easily have retreated to wood-paneled gentlemen's clubs with all his earned privileges, but Lodge felt for the plight of the working classes, publishing several Fabian pamphlets beginning in the 1890s and speaking out on behalf of women's education and women's suffrage. (Myers and Sidgwick were also notably involved in the cause of women's education, giving the leadership of the SPR a decisively forthright and feminist cultural politics.) In 1900, Lodge was tapped by Joseph Chamberlain to lead the new civic, coeducational University of Birmingham, allowing him to implement educational reforms in science designed to open the field to all those who sought a scientific career. Lodge was not, however, a man with political ambitions; he kept clear of the open fray in favor of a campaign from the hallowed pulpit of science. He did, once, consider running for Parliament when he felt the conservatives had too much the upper hand, commenting that "if you want a Bolshevist revolution in this country, the surest way to get it is to succeed in eliminating or discrediting the Labour Party," which does tend to show the stripes he would have worn had he formally entered the arena.[46]

This was a fighting spirit Lodge learned from his early scientific champions, who also did not politely suffer the obstruction of progress. While he may have been put off by their cantankerous naturalism as a young man, he never lost the heroic view of science he had learned from them as a young boy. In his memoirs, he named W. K. Clifford, T. H. Huxley, and John Tyndall as his most formative influences growing up, recognizing in their fact-based, truth-seeking science a program of universal progress.[47] But by the time of his graduation from University College, something of a gulf began to open up between Lodge's sense of the scientific project and that laid out by his scientific idols. Hearing Tyndall's infamous address at Belfast in 1874 forced Lodge to consider his opinion more carefully with regard to the leadership of his hardscrabble hero. Tyndall's speech celebrated the first Greek atomists for chasing away "the mob of Gods and demons" out of natural philosophy, which seemed to suggest that British science should do the same regarding the nuisance of their own clergy.[48] Whether this rhetoric was obtuse or intentional on Tyndall's part, it deeply offended the piety and sense of professionalism of many present. As Lodge observed: "I remember feeling the atmosphere grow more and more sulfurous as the materialistic utterances went on . . . and how on Sunday all the pulpits fulminated their anathemas."[49]

There was much to be angry about when confronted with so much pugilistic flair. Huxley gave the evening lecture, "wherein he promulgated Descartes' view of animal automatism and extended it to man." As a piece of oratory, it was, according to Lodge, "a brilliant tour de force . . . his manner of utterance being the admiration of all." This eloquent shocker was followed by Sir John Lubbock's compelling presentation on natural selection with beautiful illustrations by his daughter, ending with emotional praise for Tyndall's address. This one-two punch in the wake of Tyndall's knockdown Belfast address suggested something of a coordinated effort among these famous confrères of the X-Club, a dining society convened by Huxley in 1864 to defend scientific naturalism. This was 1874, after all, and their campaign against religious reactionaries, spiritual visionaries, mercenary humbugs, and professional imposters who tried to breach scientific authority was in full swing. Many scientists in their midst, including Lodge, felt that science needed defending from these bellicose naturalists as much as from anyone else. But these moderate counterparts seeking some accommodation with religion were, for the most part, unlikely brawlers. Particularly troublesome for Lodge was the fact that Tyndall, Huxley, and their ilk had a particular gift for getting their point across, an ability to mix in with the general public and stir the cultural pot. (The patrician Frederic Myers described Huxley as having "the air of a linen draper," which clearly stood him well as a popular pitchman.)[50] This gave them the upper hand when it came to shaping the public image of science. Lodge, the beneficiary of this outreach who had warmed himself at the hearth of the Royal Institution, was particularly sensitive to getting that message right. He would spend much of his mature scientific career in the spotlight, trying to oust these same ideas from the public (and professional) view of science. The tone Tyndall struck at Belfast was out of tune with the words at the Royal Institution that had first enraptured this potter's son. There was a beauty and possibility in science that this sort of mean-spirited materialism belied.

It was after this meeting that Lodge began to consciously orient himself away from this mundane matter-of-factness toward another view of science, one he would eventually come to publicly defend in the mid-1890s, as one of science's great campaigners and public personalities: "There is a naturalist type of physics which rejoices in objects that appeal to the senses such as meteorological phenomena and glaciers. Professor Tyndall was strongly of that type, a good observer of natural phenomena. My instincts seem to be more abstract, rejoicing in the hidden forces, atomic occurrences, and other things which can never be seen ... than in the material objects around us."[51] This was true of the boy tinkering in the factory wondering how electromagnetism made matter move and the man who devoted himself to spinning a giant wheel to catch at a piece of the invisible ether. Lodge's intellectual instincts moved against the grain in physics and the life sciences, as thermodynamics rushed to seal off the unknown with its reductive summary of nature's laws. Ether, the medium for pondering the imponderable, was a legitimating space where Lodge could launch an inquiry into what he called "non-sensuous reality" while maintaining the object of his curiosity to be, still, "reality."

For all his love of "things which can never be seen," Lodge faithfully modeled this space of inquiry inside laboratory science, expecting to find his ultimate satisfaction there. But when he did not, he eventually embarked for the Île Ribaud in 1894. The Piper sittings that Lodge convened in 1889 had already altered his thinking about the possible. But now he was ready to make such risks a more concrete part of his science. Up until this time, Lodge had been diligently conventional in the conduct of his career, already a class outsider. Lodge matriculated at university in the mid-1870s, entering into an academic climate approaching zero tolerance toward spiritualism. In the same year Tyndall was making his sulfurous scientific utterances, Crookes was doing him one worse by cavorting with the ghost of Katie King, a cautionary tale wafted on the penny press, sure to keep acolytes like Lodge on the straight and narrow. E. Ray Lankester made this duty clear in his letter to the *Times* in 1876: "It is incumbent upon those who consider such credulity deplorable to do all in their power to arrest its development."[52] Chastisement was reserved not just for spiritualists but for all wayward "perpetual-motionists and believers that the earth is flat" who dared interfere with settled physical law, and Lodge (at first) toed the line.[53]

The affirmation of scientific orthodoxy in the 1870s did not necessarily equate to an embrace of Tyndall's strident scientific naturalism. While conservation laws might leave no room for ghosts, there were still physicists eager to leave yet a little room for God. Lodge was not alone in wanting to disperse the whiff of atheism hanging over the Belfast meeting and, by extension, over science. In 1875, in direct repudiation of Tyndall's atomic materialism, the physicists Balfour Stewart and P. G. Tait published *The Unseen Universe: or Physical Speculation on a Future State*. This went so far as to suggest that the ether posed a link between this world and the next—not quite a staircase to heaven but at least a continuous walkway. However, they offered up their hypothesis anonymously, as a point of intelligent speculation only. This intentionally deprived it of any institutional imprimatur: as professional scientists they would have deemed it inappropriate to assert an idea for which there was no professional consensus. But as of 1882, this idealistic impulse in physics found a berth within science with the inauguration of the SPR, whose early membership, as we have seen, included many distinguished physicists, such as Lord Rayleigh, Balfour Stewart, William Crookes, William Barrett, John Couch Adams, William Ramsay, J. J. Thomson and, of course, Oliver Lodge. Psychical research offered a venue that skillfully navigated the politics of knowledge, directing questions about new forces away from canonical physics and toward the open disciplinary gateway of the mind. There was also a strict distancing from spiritualism that is often overlooked. Yes, spiritualists were admitted into this community of knowledge, but only if they did not bring their ghosts trailing after them. When mediums did make it onto the research agenda, they did so within the quarantine of a study on the telepathic phenomena associated with spiritualism, not a study of spiritualism itself. It is significant that even William Crookes felt it safe to associate himself with the SPR in 1882, despite his institutional probation.[54] Lodge, who also had to guard his professional standing because of "the negligence and ignorance of my education which will always prevent my taking the highest position," felt confident enough to enlist in 1884.[55] (However, Lodge was discouraged from accepting the presidency of the SPR until the passing of his mentor, William Thomson, Lord Kelvin, which would have put too great a strain on his professional career.) Throughout the 1880s and 1890s, the heyday of the SPR's cross-over legitimacy, there was still a perceived *odium scientificum* aimed at this interest, but psychical research did not provoke the severe criticism that was aimed at the "scientific spiritualism" of the 1860s and 1870s.[56] Even in the caution issued by the *Pall Mall Gazette* on October 21, 1882, upon the SPR's founding, psychical researchers were first praised as "the highest and most advanced European thinkers . . . risen quite above the lower childish superstitions," before being advised that "no man has hands so clean he can afford to touch pitch."[57] Given its elite membership, rejection had to remain polite, but there was also a certain mandate to resolve questions raised in psychology and psychophysics that inspired sympathy or at least neutrality from the sidelines. Even the physicist Oliver Heaviside, who publicly reviled spiritualists as "asses who talked a sort of bastard science with no powers of discrimination whatsoever," offered encouragement for psychical research in his private correspondence with Lodge: "The physical basis of abnormal phenomena of course must some day be attacked by physicists."[58]

Heaviside wrote that letter in 1895, a time when the category of "abnormal phenomena" was beginning to expand beyond the curious entries of telepathy and telekinesis and starting to encompass the nature of matter itself. Science was entering the era of radiation physics as it approached the century mark, and the solidity of the ether was beginning to fade within this fascinating and even thaumaturgical new class of phenomena. Fluorescent radiation, electron particles, x-rays, "n-rays," wireless telegraphy—not to mention the strange psychiatric tidings being reported from the edge of consciousness—all made locking down an orthodoxy with the "iron laws" of midcentury physics feel a bit premature. Even after it took this strange turn toward the Île Ribaud, psychical research continued to hold appeal for physicists looking to solve problems that had no ready theoretical platform, including pioneers of the subatomic world like J. J. Thomson and Pierre and Marie Curie, just as it had done for psychology a decade before. This kept Lodge in good company as he pursued the implications of Eusapia Palladino's phenomena, helping to stretch the framework of physics so it might appear to reasonably accommodate ectoplasmic pseudopodia and phantom faces in the drapery.

THE METAPHYSICAL MECHANIST

My main business was with the imponderables—the things that worked
secretly and have to be apprehended mentally. So it was that electricity and
magnetism became the branch of physics which most fascinated me.

OLIVER LODGE, *Past Years*

To appreciate the rigors of Lodge's later ether theology, toward which all this earlier work tended, it is necessary to understand how it was built up within the discipline of late nineteenth-century ether physics and how it was rigorously justified within the framework of Lodge's complex epistemology. In this way we can understand how Lodge eventually made the ether both an affirmative argument for a mechanical universe and the formal basis for a new metaphysics. He was interested not just in what science had to say about the ether, but also in what ether had to say about science. This was a metaphysics fully embedded in the material world, yet capable of soaring toward the spiritual and moral heights of religious revelation.

Lodge first fell for the metaphysically lustrous field of ether physics at the annual meeting of the BAAS at Bedford in 1873. It was there that he first heard James Clerk Maxwell speak, an event he recalled many years later as the most moving experience of his life. There was no need to suffer the vulgar appropriation of the ether by spiritualists, when there was this purer language by which to understand its cosmic secrets. Lodge was astonished by the beauty of Maxwell's theory and mathematics. The most exhilarating aspect of the treatise was, of course, the union of electricity and magnetism into a single force undulating through space at the speed of light, as it rapidly oscillated between its two perpendicular planes of force. Lodge did not need to await Hertz's radio receiver to know absolutely that such waves existed and that there must be an ether; his conversion was instantaneous and would guide his unwavering physics over the next six decades of his scientific career. (Loyal to the end, Lodge even tried to use the insights of relativity in the ether's defense, well after Einstein's theory had effectively obviated its existence.)[59] The year after Maxwell's speech, in January 1874, Lodge left the potteries for good and enrolled in University College London as a full-time student, where he "worked with a kind of fury to make up for lost time."[60] Lodge stayed at UCL from 1874 to 1881, working as a lecturer and demonstrator for the college. There he pursued the experimental verifications of Maxwell's equations that would prove essential to translating the abstruse theory for the wider scientific community. In a very real sense, it was Lodge's scientific labor that crowned Maxwell's achievement, helping to place the *Treatise* on its throne as the definitive culmination of nineteenth-century electromagnetic research. In 1881, he was appointed to a chair at the new University College in Liverpool, where his research continued to gain the attention of the scientific community. And by 1887, he had become a Fellow of the Royal Society.

At the start of his career, Lodge inherited a universal ether that was already somewhat sacrosanct, functionally revolutionized by the preceding generation of physicists and made holy by its own intellectual purity. The early nineteenth century was beset with a profusion of force ethers necessary to explain all the different functions of electrical, magnetic, gravitational, and optical phenomena—the legacy of Newton's *Opticks* as it was pursued in the vacuum of space embraced by Enlightenment thinkers. The failure to qualitatively connect all these effluvial ethers into a grand phenomenon gave science the appearance of what historian John Heilbron called a "permissive prodigality," making it the urgent quest of early Victorian physicists to rid themselves of this problematic plethora.[61]

Development of the wave theory of light in the early part of the century suggested the possibility of a new kind of ether, one that functioned not as an effluvial substance somehow constituting that force, but as a propagating medium transmitting it: the transverse vibrations of light. In this way the optical ether suggested a radically simplified way to ontologize force, making imponderables less imponderable. What if, instead of being each their own qualitative ether, electricity, magnetism, gravity, and light were best understood as various motions in or "of" the ether itself? However, Lodge himself identified the true "momentous beginnings" of modern ether science with Michael Faraday's notion of electrical and magnetic lines of force. These extended into the field surrounding matter, allowing the ether to be potentially understood as a vessel in which to store energy as well as a vehicle by which to transmit it.[62] Pursuing this field theory of the ether set the agenda for Victorian physicists such as William Thomson, Lord Kelvin; Peter Guthrie Tait; Balfour Stewart; Maxwell; Lodge; Henry Charles Fleeming Jenkin; and Lord Rayleigh. (Notably, these energy physicists all shared a strong idealistic bent, ranging from Thomson's deep traditional piety to Lodge's psychical quest.) Out of their efforts the ether-field theory would arise victorious, replacing the effluvial chaos of atoms in a void with the holy calm of a universal plenum. It was a philosophical victory so profound that Lodge and many of his generation were unable to give it up.

The introduction of the vortex atom rendered this universe yet more beautifully austere. In a paper he delivered to the Royal Society in 1856, Thomson, the chief architect of thermodynamics, suggested that atoms were actually vortical-elastic structures in the ether. Just as the energy conferring motion in a fountain structured water into glorious forms, so these energetic whirlpools in the fluid ether structured its plenum into particles of matter. Upon deeper analysis, Thomson determined that the rotations of this vortex atom could also give rise to stable transverse vibrations in the surrounding ether—that is, the wave structure associated with electromagnetic phenomena and the phenomena of light (as yet not fully related). Thus, this simple whirl of a vortex conferred solidity upon matter, undulations upon the ether, and even a field of force around the atom, a vision that made matter, motion, and force as continuous as the ether itself.[63] By the 1860s, the appeal of Thomson's vortex atom began to eclipse the prevailing continental idea of "atoms in a void," as the British ether propagated across the Channel with its irresistibly spare proposition: everything that existed was reducible to heterogeneities of motion in a homogeneous ether. Not just atoms in motion, atoms *of* motion. Thomson saw the dawning of a new era of scientific mastery, in which "the unparalleled train of recent discoveries is tending toward a stage of knowledge in which laws of inorganic nature will be understood in this sense—that one will be known as essentially connected with all and . . . will be recognized as a universally manifested result of creative wisdom."[64] The final steps in synthesizing all these laws into one coherent theory was made by Maxwell in 1865 when he determined that light was itself an electromagnetic phenomenon, an undulation of a particular frequency arising from the displaced energy of the vortical atom. Lodge called it an idea so beautiful it had to be true. In his first public discussion of the ether, published by *Nature* in 1883, Lodge "endeavored to introduce the simplest conception of the material universe which has yet occurred to man":

> The conception is of one, universal substance, perfectly homogenous, and continuous and simple in structure, extending to the furthest limits of space . . . which can vibrate as light, which can be sheared into positive and negative electricity, which in whirls constitutes matter, and which transmits by continuity and not by impact, every action and reaction of which matter is capable . . . This is the modern view of the ether.[65]

This put an end to the absolute distinction between matter and force, and perhaps with it, the age of atomic materialism. Matter was nothing more than a "red herring," distracting science from the deeper realities of the space surrounding it. It was Lodge who took the possibilities suggested by Thomson's atom and fully imagined the reality of such a universe.

Lodge's textbook *Modern Views of Electricity* (1889) provided a veritable tour de force of ether theory, folding Thomson's vortex atom into Maxwell's electromagnetic, undulating ether, and using it to explain the laws of thermodynamics, the properties of locomotion, the conservation of energy and matter, and even inertia itself.[66] Lodge's first textbook, *Elementary Mechanics* (1879), had lucidly explained the laws of motion but without reference to the ether, which at that date had yet to be confirmed. But now the ether flowed like a torrent onto the page, giving rise to an entire universe. The spigot had been opened, not just by the announcement of Hertz's detection of radio waves in 1888, but by Lodge's own triumphant demonstration of light waving in the ether that same year, a scientific performance held at the theater at South Kensington and done with his usual flair. With the lights doused so the room was in darkness, Lodge sent timed pulses of electromagnetic energy through a wire wrapped around the room. These pulses sent waves that interfered with and refracted the existing current, causing the wire to glow at the ventral segments of each pulse, while it remained dark at the nodes, thus demonstrating an unmistakably wavelike structure to this electrical circulation.[67] He had at last brought the undulating ether across the threshold of theory, beyond the realm of inference, and into the reality of an actual, observable effect.

In *Modern Views of Electricity* Lodge did what others had tried but failed to do: physically represent the interactions of matter, electricity, and charge, as described in the mathematical blueprints of Maxwell's *Treatise*, as a relationship between the vortex atom and the universal ether. Maxwell himself had been forced to abandon this effort in favor of a purely mathematical explanation, because, try as he might, he could not satisfy the requirements of electromagnetic behavior with pulleys and gears. But Lodge was not so readily dissuaded. To prevent, as Lodge protested, "the physics of the future from being swallowed up by a barbarous jargon of technicalities," he set about designing a model that could do just that.[68] The barbarous jargon he was referring to was mathematics. As much as he admired the mystical revelation through numbers achieved in Maxwell's *Treatise*, that abstraction had to eventually become an aspect of physical reality or it had no discernible meaning.

At the heart of Lodge's ether dynamics was something as elemental to understanding as motion itself, though of an extraordinary kind: the "perpetual rotation" that constituted the vortex atom. Thus, the Prime Mover provided the universe with its store of energy and its elemental structure using a gesture as simple as stirring a pot. The speed of this vortex rotation never diminished or increased, and was fundamentally tied to the speed of light. This fixed rate was essential to sustaining the structure of the atom (conservation of matter) and likewise explained the uniform speed of all electromagnetic phenomena, from radio waves to light. It also explained Newton's law of inertia: the energy of the vortex was not available to do anything but spin or the atom fell apart. Therefore, locomotion through space was necessarily powered by sources extrinsic to the atom, as originally observed by Galileo. That "available" energy existed in the ether surrounding the vortex atom, arising from the strain of its rotation. The "ether strain" constituted the electromagnetic field around every atom, a great storehouse of potential energy that could be directed into active forms, such as locomotion, magnetism, electrification, or combustion, depending on the particular interaction between any two "particles" of matter, out of which all the processes of the universe unfolded. This crystallized for Lodge why "it is impossible to have force without a body which is exerting that force, and also without another body on which the force is exerted."[69] There could be no simpler explanation for so many complex phenomena than the one furnished by Lodge in *Modern Views of Electricity*.

But some scientists, particularly the nonanglophone physicists, were not quite so enthralled with ether mechanics. They saw it as a mechanical *reductio ad absurdum*. In an effort to abolish action at a distance and mysterious forces, Lodge's mechanical universe resembled something of a factory, as observed by historian of science Pierre Duhem: "Here is a book intended to expound the modern theories of electricity and to expound a new theory. In it there are nothing but strings which move around pulleys, which roll around drums, which go through pearl beads; which carry weights; and tubes which pump water while others swell and contract. Toothed wheels which are geared to one another and engage hooks. We thought we were entering the tranquil and neatly ordered abode of reason, but we find ourselves in a factory."[70] He did not exaggerate (fig. 4).

For all his ridicule of Lodge's ether "factory," Duhem did not actually believe Lodge's pulleys and gears were meant to literally represent the universe as a machine. But he had reason to suspect that all this ether machinery was rather more than just a heuristic device to aid student learning. For Duhem, using a model to work out a theory was not a problem, but allowing that model to become an element of the theory was an issue. This was no longer explanation but essence. Duhem was taking issue with what he sensed was at heart a genuine metaphor: that there was some intrinsic analogy lurking amid the flywheels. As a cultural metaphor explored by historians Crosbie Smith and Norton Wise in *Energy and Empire*, this factory universe offered a cosmic sanction for industrial production and imperial exploitation. This was a thermodynamic machine engineered by God "to get work done," transforming latent energy into beneficial outcomes through applied force or otherwise allowing it to be wasted through dissipation. Such a design would seem to endorse the maximal exploitation of resources, even if those resources belonged to another nation, and to chasten any person or culture with the lack of initiative to do so. The productive forces of the universe

FIG. 39.—A provisional representation of a current surrounded by dielectric medium, either propelling or being propelled.   Section through the wire.

FIGURE 4 Lodge's Mechanical Ether, from *Modern Views of Electricity* (London: Macmillan, 1889), 186. Lodge's diagram represented matter as a series of geared wheels, mutually engaged and spinning in opposite directions so as to propel each other's circular momentum. These distinct rotational directions quite cleverly represented positive and negative charges. An electrical current was depicted as a kind of toothed bar that could abruptly move through these pinioned cogwheels, jamming the engagement of their rotating gears and directing their energy along a tangential line of force. When a medium was fully magnetized, these cogwheels were perfectly geared into each other and moved in sync, conducting all this rotational energy into the ether as consecutive rings of magnetic force. However, when a medium was acting as an electrical conductor—as in metal, for instance—the gearing was imperfect and could easily slip, allowing the energy in the ether field to escape along a linear path. Lodge imagined that in a perfect conductor these "wheels" were not geared together at all. (His illustration of this point removes the little toothed cogs from the wheels.) There is no friction, no contact, no spin; there is only pure current.

thus mirrored the forces of industrial production. (This engine in the ether was, in the words of Clifford Geertz, both a *model of* and a *model for* the political, social, and economic life brought about by the British industrial revolution, a mutually reinforcing cycle of how things are and how things ought to be.)[71]

But the heart of Lodge's mechanical reification was not so much concerned with industrial productivity as it was with intellectual productivity. In Victorian physics, contact motion (pushing and pulling) was postulated as "the completely known," something human experience intuitively grasped, which did not need to be explained further. As William Thomson put it while despairing over Maxwell's equations: "I never satisfy myself until I can make a mechanical model of a thing. If I can make a mechanical model I can understand it. As long as I cannot make a mechanical model all the way through I cannot understand . . . and that is why I cannot get the electro-magnetic theory."[72] But Lodge was able to set this right at last with his mechanization of electricity in *Modern Views of Electricity*. Mechanical push-pull modalities were simple to grasp when compared to the spooky forces of action and attraction from a distance. You push something, it accelerates; you hit something, it falls over. You twirl something, it spins. But how do you magnetize something? electrify something? illuminate something? Force without the ether was intractably mysterious; force structured through the ether was intuitively obvious, a simple affair of matter, energy, and contact force. Thus, a dynamical theory was recognized as being at once necessary and sufficient. But sufficient for what: to make complete sense to Thomson, or to be fully known by him? What was more important, the explanation or the reality?

The debate between Duhem and Lodge was not a simple restatement of the century-old conflict between French rational and British empirical strategies in scientific explanation; it was something different and deeper, evolving toward a referendum on the nature of reality and human understanding itself. But this had not previously been the case. In an article he wrote for the *Philosophical Magazine* (October 1879), Lodge explained that his aim in writing *Elementary Mechanics* (1879) was "to have all the fundamental doctrines of physics stated in ordinary language without technicalities," so long as such statements were "accurate and devoid of vagueness."[73] And yet, admittedly, "improvement of the existing language in textbooks" had to be put in "dogmatic form," with statements that were clear and concise, rather than tortuously overqualified or rendered mathematically, otherwise "they can be of little or no use." This statement of 1879 made a virtue out of utility, even if it was at the expense of accuracy. This was an attitude of expediency in science that the Lodge after *Modern Views of Electricity* would vehemently decry.

But it was not so much that Lodge had changed between 1879 and 1889, as that circumstances had altered. The mechanical models of the 1860s and 1870s had embraced a kind of common-sense vernacular at a time when the notion of "common sense" was not yet controversial. Some implied connection between understanding and reality made such models meaningful as well as expedient, reconciling the mind with reality in a way that a purely

symbolic mathematics could not, even if quantification was more technically accurate. (Continental rationalists who viewed the deepest nature of
reality as mathematical would disagree.)[74] But at the time of *Modern Views of
Electricity* (1889), the essentialism associated with representational models
in physics was beginning to come under fire. Instead of being circumspect
in his use of analogy, however, Lodge doubled down, turning the rhetoric of
Elementary Mechanics (1879) into the visual depiction of an actual machine
(*Modern Views of Electricity*, 1889). The recent confirmation of the ether
emboldened Lodge to sketch out his ether factory, giving his mechanical
imagination a definite medium and semblance of physical truth. Duhem
rightly saw this as a provocation on Lodge's part. The wheels and pulleys
were not literally true, but they were more than a mere pedagogical strategy; they offered some intrinsic analogy based on the familiar operations
of motion, substance, and contact force, asserting some fundamentally mechanical aspect to reality. The Lodge writing in the wake of *Modern Views of
Electricity* wanted to close any gap between explanation and reality that he
might have previously tolerated in less challenging times. He was no longer
after a mechanical theory that was simple and comprehensible, but sought
one "as plain as it is substantial," a point of distinction explicitly asserting
some deep connection to the actual characteristics of reality.[75] It seems that
Lodge was trying to ring an ultimate truth out of the universe that religious
devouts like his mentor William Thomson received on faith. He, who had
up until now generally viewed "settled faith as impossible and even undesir-
able," was now looking for some assurances of his own.[76]

   Modern Views of Electricity set out to be a uniquely systematic and satisfactory explanation of force physics (one celebrated by British scholastics,
if not by French philosophers), but it was never intended to be comprehensive. Its achievements signaled not the end of inquiry, but, rather, the beginning. In the words of Lodge: "The more neatly and quietly a scientist can
build his theoretical foundations, the more time he will have for building
the superstructure and the more gorgeous he may hope to make it."[77] Lodge
was already aware of other aspects of the universe that had to be left out
of his strictly mechanical curriculum because they required other forms of
explanation, which had yet to be adjudicated by science. In his discussion of
Maxwell's equations and Thomson's heat mechanics in *Modern Views of Electricity*, Lodge made no mention of "Maxwell's demon." (Though it did come
up in the context of Eusapia Palladino's psychokinetic phenomena, which
also seemed to indicate a mind directing matter at the molecular level.)
Thomson defined the demon "as an intelligent being endowed with free

will and with enough tactile perceptive organization to give him the faculty of observing and influencing individual molecules of matter."[78] It might just be that this was not "a nature fast in fate," after all. Such determinism might hold for nature's laws, but not for her tendencies, such as entropy. In Thomson's analysis of heat dissipation, it became clear that molecules moving from high- to low-energy states behaved with only statistical certainty. The actual pathway a molecule took seeking its ground state was open to possibility, even though the end state of equilibrium was as certain as death and taxes. (Thermodynamics said systems sought equilibrium, it did not specify how.) The "free will" Thomson endowed the demon with amounted to exercising this choice of pathway: the demon chose this way, not that way in the absence of a determining cause. This demon suggested there was some mysterious subjectivity acting in the interstices of more deterministic laws.

For Maxwell and Thomson, this tiny intelligence had a mostly symbolic significance, preserving a space for free will, so important to Christian faith, amid the spinning cogs of the ether. Lodge, however, confronted with the facts of telepathy as early as 1883, had to consider the demon as a potential basis for a more generalized "mental force" that could also operate at higher (human) levels of conscious activity. Lodge did not make the demon part of his mechanical theory, nor did he dare roll telepathy into the curricula of *Modern Views of Electricity*, but such considerations remained outstanding facts in his physical worldview throughout the 1880s, considered to be Lodge's most productive years in terms of his intellectual contributions to science. He was simply not ready to frame those facts as part of a formal theory or bring them to the attention of his fellow physicists. First, he needed a working model for ordinary laws, and to that end he fully dedicated his science, culminating in *Modern Views of Electricity*. But once he had established that "plain and substantial" foundation for a dynamical ether, he was ready for something more. That came in the form of the Piper séances.

Between November and December 1889, after *Modern Views of Electricity* had gone to press, Lodge held a series of twenty-two séances with Leonora Piper at his private residence in Liverpool. He had met with her at the behest of Myers, who had fallen ill and requested that Lodge oversee an investigation on his behalf.[79] There was some urgency as the lady was arriving by steamer from Boston in a few days and some trustworthy representative was needed to stand in for Myers. Until then, Lodge had remained aloof from such seemingly odious spiritualistic activity, but Mrs. Piper was a genteel medium whose table only sat the best company, and she was sailing under the banner of William James, who personally attested to her skill. Lodge was for the most part impressed by her phenomena after the arranged sitting, but his report was brief and reserved. From what he could tell, based on the preliminary facts, Mrs. Piper did seem capable of searching the minds of her sitters for private details of their lives once she had entered her trance state. Of this he was fairly sure. He had had her followed to determine if she "investigated" her sitters before meeting with them, but the watchers found no evidence of this. During the séance itself, she was placed under the usual procedural controls to prevent any unauthorized form of communication. But beyond ordinary telepathy (which was already an established fact for Lodge), nothing further could be scientifically vouchsafed, and Lodge, who had thus far remained aloof from spiritualism, wrote up a cautious report limited to a discussion of her thought reading.[80]

Nonetheless, certain aspects of her phenomena intrigued him, suggesting that something considerably more than ordinary telepathy was possibly at play.[81] Mrs. Piper might be a genuine clairvoyant. Such phenomena greated exceeded the field theory of communication Lodge asserted in 1884 to explain the telepathic phenomena of the Misses E. and R., but experimental conditions had not met Lodge's bar of proof, so his report remained silent. But there was enough there to warrant further investigation. Such a probe would require the utmost in experimental rigor, and accordingly Mrs. Piper was removed to Lodge's home. There she was subjected to a monthlong battery of sittings, each of which was written up by Lodge in a detailed report, numbering together over one hundred pages. The first test was modest. Lodge secretly ordered that a watch from his Uncle Robert be discreetly sent to him, which he then presented to the entranced Mrs. Piper. She correctly identified the watch, however, as actually belonging not to his Uncle Robert, but to Robert's long-dead brother Jerry, a fact Lodge himself did not know. If Mrs. Piper attained this information telepathically, it was not from Lodge, leaving as a source only the mind of Uncle Robert, who lived at a considerable remove from Liverpool. This was a telepathic reach somewhat greater than across the table or across the room, taking this psychical faculty to a new level of consideration. As Piper's altered state deepened, her phenomena became less easy to reconcile with even such far-flung terrestrial feats of telepathy. The information pouring forth from Mrs. Piper was increasingly filled with personal information about the boyhood of Uncle Jerry, some of which neither Lodge nor Robert had any direct knowledge of, but which was later verified:

> "Uncle Jerry" recalled episodes such as swimming the creek when they were boys together, and running some risk of getting drowned; killing a cat in Smith's field; the possession of a small rifle, and of a long peculiar skin, like a snake-skin, which he thought was now in the possession of Uncle Robert. His

[Robert's] memory, however, is decidedly failing him, and he was good enough to write to another brother, Frank, living in Cornwall, an old sea captain ... The result of this inquiry was triumphantly to vindicate the existence of Smith's field as a place near their home, where they used to play ... the killing of a cat by another brother was also recollected; while of the swimming of the creek, near a mill-race, full details were given, Frank and Jerry being the heroes of that foolhardy episode.[82]

As a result of all this corroboration, Lodge was convinced that Mrs. Piper's powers were genuine, but what exactly they entailed was unclear. Mrs. Piper herself represented that she was communicating with Uncle Jerry himself, information relayed by her spirit guide "Dr. Phinuit," whom she channeled to access "the other side." Lodge gave Phinuit little account except to note that he was afflicted with a terrible French accent, and distracted from things of more powerful interest. Piper's phenomena were framed as breathtaking leaps of consciousness into a limitless field of knowledge, spinning through clairvoyance a veritable fairytale of epistemological certainty:

> If we reject ordinary thought-transference as inadequate, it seems as if we should be driven to postulate direct clairvoyance; to suppose that in a trance a person is able to enter a region where miscellaneous information of all kinds is readily available; where, for instance, time, and space are not; so that everything that has happened, whether at a distance or close at hand, whether long ago or recently, can be seen or heard and described ... A fourth dimension of space is known to get over difficulties like this, and an omnipresent time is very like a fourth dimension. Then, again, old facts, such as the boyish acts related of my uncle, must be supposed narrated not by him nor by his agency at all ... Undoubtedly Mrs. Piper in the trance state has access to some abnormal sources of information, and is for the time cognizant of facts which happened long ago or at a distance; but the question is, how she becomes cognizant of them. Is it by going up the stream of time and witnessing those actions as they occurred; or is it through information received from the still existent actors, themselves dimly remembering and relating them; or, again, is it through the influence of contemporary and otherwise occupied minds holding stores of forgotten information in their brains and offering them unconsciously to the perception of the entranced person; or, lastly, is it by falling back for the time into a one Universal Mind of which all ordinary consciousnesses past and present are but portions? I do not know which is the least extravagant supposition.[83]

For all this wild metaphysical and psychical speculation, Lodge's report refused to acknowledge the agency of "spirits" in explaining Piper's clairvoyance, putting an almost exaggerated quarantine around the whole hypothesis. (If one is willing to propose a fourth dimension, time travel, and a universal mind, why be so squeamish when it comes to ghosts?) Even as he documented the extraordinary record of facts supplied by "Uncle Jerry," Lodge made a point of insisting that it "must be supposed narrated not by him nor by his agency at all." But Lodge performed this exclusion somewhat disingenuously, because he felt an immediate and dramatic sense of recognition when his Aunt Anne suddenly took possession of Mrs. Piper and began speaking to him in direct voice from across the table.[84] While the dialogue with Aunt Anne was included in Lodge's report on Piper's trance phenomena, he did not admit the testimony of his own inner sense, though the strange rush of feeling was the surest "evidence" he had for her continued existence. This reluctance to pursue the spiritual hypothesis was more than a lack of objective data. That clearly did not impede other trains of thought from leaving the station. But, for Lodge, spiritual considerations would have been intellectually out of turn. The séance brought his sense of scientific possibility soaring to new heights, but he was not yet ready to leave the land of the living until it had been fully explored.

It is interesting to note the passion with which Lodge made his presidential address to the Liverpool Physical Society in December 1889, a speech preceded by this packed series of séances. His topic was electricity and the ether, but his delivery was one of pure revelation. Lodge, the patient explicator of what was known in science, broke his pedagogical mold to breathlessly, almost recklessly, rejoice in what could be known and what would be known.

> The present is an epic of astounding activity in physical science. Progress is a thing of months and weeks, almost of days. The line of isolated ripples of past discovery seem to be blending into a mighty wave, on the crest of which one begins to discern some oncoming magnificent generalization. The suspense is becoming feverish, at times almost painful. One feels like a boy who has been long strumming on the silent keyboards of a deserted organ, into the chest of which an unseen power begins to blow a vivifying breath. Astonished he now finds that the touch of a finger elicits a responsible note, and he hesitates, half delighted, half affrighted, lest he be deafened by the chords which it would seem he can now summon forth almost at will.[85]

Standing on the brink of a new year, a new decade, a new century, high upon a vista provided by Maxwell, Thomson, two Liverpool shopgirls, and now, Mrs. Piper, Lodge saw this oncoming "magnificent generalization" already forming on the horizon. In his speech to the Liverpool Physical Society, he was rushing ahead to meet it, gusted by the favorable winds of ether mechanics, séances, and psychical research. But, if he had just paused to look back over his shoulder, he would have seen that others of his profession were still not quite ready to follow him toward that great epiphany in the ether where he was so eager to lead them.

KEEPING IT REAL

In a far less sanguine mood, Lodge took the podium as president of Section A to deliver the opening address for the year 1891. The predicted future had yet to arrive. Instead, what he viewed on the horizon for science was "the specter of an ossifying orthodoxy." Lodge now seized this "cathedral opportunity" to sermonize the gathering of scientists, who had become complacent, narrowly professional, and perhaps even careerist in their stewardship of physics. He warned of the dangers of consolidating discoveries into a single system of knowledge, especially, as he now suggested was the case, "when that system begins to defend itself against new knowledge in an effort to willfully blind oneself to the facts and protect the privileges of an existing creed."[86] The new knowledge to which he referred was psychical research, and Lodge, for the first time in his career, was prepared to drop it dead center of his profession, "taking the risk of introducing a rather ill-favored and disreputable looking stranger" who "was not all scamp" and whose "present condition is perhaps due to our long continued neglect." Before his captive audience, Lodge began sharing his experimental results:

> Is it possible that an idea can be transferred from one person to another by a process we know practically nothing about? I assert that I have seen it done, and am perfectly convinced of the fact. It has been established by direct experiment that a method of communication exists between mind and mind irrespective of ordinary channels of consciousness and the known organs of sense, we must urgently inquire into the process. It can hardly be through some unknown sense organ but it may be by some direct physical influence on the ether, or some still more subtle manner.[87]

It was not that this psychical subject matter itself constituted a particular scandal for a physicist. Several famous colleagues of Lodge's had been happy to append their name to such research. But here, in this hallowed hall of Section A, as a specific concern of their profession, it did mean a scandal. Lodge, an important explicator of orthodoxy, who had authored several textbooks on physics and been entrusted here with its professional oversight, was now in the process of explaining that orthodoxy in very different terms. He had never hidden his interest in psychical phenomena from his fellow physicists, but also he had never imposed it on them. Now, he did. This was not because he personally needed the affirmation of its acceptance (that is, the emotionally undisciplined view of psychical research), but because it was something he felt they urgently needed to hear.

The refusal to directly confront the accumulation of evidence was bad; but the cost to physics of continuing to do so was worse. The laurels upon which physics now rested were already decades old; particle physics and relativity had yet to break the horizon. (The Crookes tube had opened the door to a subatomic world, but had not yet unlocked any minds.) For Lodge, whose personal sense of scientific momentum had been building since 1873, beginning with Maxwell's equations and following up with the demonstration of ether waves, the confirmation of telepathy, and, more recently, the possibility of clairvoyance, the professional physics of 1891 seemed dangerously stagnant, even "treasonous to the highest claims of science."[88] It had become, to a certain extent, cynical in terms of its own possibilities. As Lodge lamented, "we hang back from whole regions of inquiry . . . with our eyes always downward and deny the possibility of everything out of our accustomed beat."[89] Lodge took a radically opposing view. No knowledge was off limits to science; even questions of the most sacred significance could be systematically explored. "Why leave it to the metaphysicians?" he asked. "They have explored it with insufficient equipment" and gained only "queer and fragmentary glimpses." Instead, he dared the program of science to mature into a modern philosophy of "Ultimate Intelligibility," renewing its original battle cry of *sapere aude*. "We must trust in consciousness," Lodge urged, "it has led us this far."[90]

"Trust in consciousness." Had the midcentury science of Thomson, with its "unparalleled train of recent discoveries tending toward some Universal wisdom," somehow lost its nerve as it approached the fin de siècle? Lodge's statement touched upon what he perceived to be the doubt creeping into the language of British science, with the tortured questioning of its own vernacular. This could make once familiar conceits seem alien: what was meant by inertia, motion, force, or mass? (This was the complaint he raised against Professor Greenhill, "who insisted on his peculiar and I may say extraordinary mode of regarding the meaning of elementary terms," in a letter to *Nature* in March 1888.)[91] This paralleled a broader trend in philosophy that was bringing this reflexivity to the most basic processes of sensory cognition. These early explorations of phenomenology were initiated by the

German philosophers Ernst Mach and Richard Avenarius, whose critique of positivism made explicit what had always been an element of empirical knowledge: sensations are a fundamentally psychological process. For a naive empiricist, these processes were essentially reliable.[92] What you see is what you get. But as early as the 1870s, psychophysicists and philosophers began to look more deeply at conscious experience, including sensory perception and its meaning for the nature of "reality." Instead of looking outward upon some (presumably) external reality, the empirical gaze was increasingly seen to fix itself upon the synthetic objects of its own mind. By the early 1890s the impact of these ideas was beginning to be felt in British science mainly through the efforts of Karl Pearson's *Grammar of Science* (1892). (In despair over the total infiltration of physics by phenomenology in the early twentieth century, Lodge protested, "Surely we should try to study the things responsible for our sensation rather than the sensations themselves?"[93] (But by that point, the philosophy of knowledge was becoming even more esoteric, occupying itself not with "sensations," rather than "things," but with the logical and linguistic representation of those sensations.)[94]

Even while Lodge was urging his fellow physicists of Section A to do more, dare more, know more about the hidden realities of the world around them, the philosophy of science itself was moving in the opposite direction, leaving behind any such notion of a "real" world. Lodge described this attitude of the early 1890s as "a general tendency to underrate certainty" that afflicted scientific knowledge with "the taint of solipsism."[95] Although the physicists themselves were not driving this course, the essential validity of their worldview was specifically under attack. Lodge seemed somewhat more attuned than his colleagues to considering the problems of knowledge as it related to "objectivity" and "reality." (In addition to his *contretemps* with Greenhill, Lodge was also at odds with one Professor Newcomb who suggested that "energy" should not be considered a real "thing.")[96] His principal challenge to his colleagues in Section A in 1891 was their institutional hostility to psychical research (and not the threat brewing in the philosophy of science), but when he cracked open the heart of their professional dilemma, he revealed a similar crisis regarding the scientific capacity to know, one as much about fear as skepticism: "We are to some extent afraid of each other but we are still more afraid of ourselves . . . We have a righteous distrust of our own powers and knowledge."[97] Psychical research was an obvious and important part of affirming this power and knowledge. It not only fearlessly claimed the "borderland between physics and psychology" as its own preserve, but also conquered for consciousness a whole new range of perceptual states.

But when Lodge asked his fellow physicists to "trust in consciousness," he was not asking his them to put their "faith" in some exotic psychical sense. He was upholding something much more elemental to science: the ordinary, observational methodology that had "gotten us thus far." Empiricism was, after all, the basis for those more exotic psychical claims, and not the other way around. He did not advocate this "trust" naively; by the time he gave this speech, Lodge had already begun to think more deeply about the mind's intuitive grasp of mechanical principles and how they might relate to the framework of time and the perception of motion. Lodge's encounter with Mrs. Piper compelled him to consider "time" as a highly contingent psychological experience, and not in the trivial sense, either, where it flew by for some while it dragged for others. He meant "time" as a causal framework, ordering "before" and "after" as a linear experience. As he discussed in an interview with the *Spectator* (1891), "time" as such was only meaningful as a structure of the material world, that aspect of our consciousness that "is of the very essence of science."[98] Trance consciousness seemed to move Mrs. Piper beyond those constraints, but it also helped draw Lodge's attention to them in a more critical way: the untrammeled perception of clairvoyance marked the contingency between ordinary matter and ordinary consciousness. Since all matter and all force were themselves varieties of motion, this was fundamentally about some deep rapport between motion and its mental perception. Before he could turn to Mrs. Piper and her encounter with some Universal Mind, he wanted to better understand the understanding of the human mind, for that was what formed the foundations of empirical science, and that was the problem that now seemed to be placed on his table. He would argue his point as a physicist, working his way from reality to perception rather than the other way around, as might a philosopher or psychologist, from perception to reality. This did not make him any less engaged, however, in the modern philosophical scrutiny of empirical knowledge.

There is a tendency among historians to intellectually dismiss Lodge, outside of his specific scientific achievement. He can become quickly tainted as the spiritualist, the literalist, the great dupe of the ether as he wanders into speculative waters. But the same individual who translated Maxwell's equations and invented the telegraph was very much present in these other endeavors. He does not cease to be an original and rigorous thinker just because some of his most prominently held ideas are now considered to be "wrong," and even absurdly so. The Lodge of the 1890s was able to take on the abstruse problems raised by phenomenology because his physics had to some extent always addressed itself to the problem of human subjectivity. That was always an intrinsic part of making ether models. Mechanistic explanations fulfilled Lodge's dictum that "before making any definitions, it is desirable and only civil to show the reasonableness of it."[99] The philosophy of knowledge he began to develop in the wake of the Piper sittings would ask much the same questions posed by Immanuel Kant in *Critique of Pure Reason* (1781): "What, then, are space and time? Are they real existences? Are they only determinations or relations of things[?]"[100] Lodge was similarly after evidence of an ultimate reality in which to anchor the perception of this one. But instead of exploring intuitions of space and time, Lodge was addressing himself to their actual structures. (Space and time were, after all, two necessary factors in determining motion and speed, and thus were part of any kinetic intuition.) This made Lodge's dynamical empiricism more grounded than Kant's intuitions, which existed only in the mind; he devised a truly scientific *realism* that did not lean on some transcendental philosophy but was part of the electromagnetic architecture of the ether. Onto this empirical foundation he would add his "gorgeous superstructure," and an additional faculty of psychical perception.

Lodge had already plainly set forth the foundations for such an argument in the chapter on dynamics in *Modern Views of Electricity* (1889), before his encounter with Piper and "Uncle Jerry." Electromagnetic motions to an extent carved out the volume of space, a fact made explicit in Lodge's three-dimensional representation of this action in the ether. The vertical axis of space arose with the elongated filaments of electricity emerging at either end of a spinning vortex atom, two streams moving in opposite directions and carrying an opposite charge. (One might imagine how a small whirlpool narrows into a conical point that trails energy into the water, but surrounded on all sides, thus trailing a filament at both ends.) The horizontal axis of space transected the vortex atom's oblate core, which sent perpendicular rings of force (magnetism) circulating outward along a perpendicular plane. The rapid oscillation between these two planes of force gave the electromagnetic wave its structure as well as its power of propagation, the ability to go from place to place, moving "up and down" and across. Thus, Lodge explained, "space is the static abstraction of motion." But time also, in turn, was an emergent property coextensive with this motion, "the kinematic factor likewise extracted from the same experience." Because one of the experiences of electromagnetic motion was that it traveled—in its transverse propagation as waves, in its linear path as a current of electricity, and in its circular expansion as magnetic rings—distance and direction were implied.[101] Distance covered by waves traveling at the given speed of light constituted the definite interval we called "time." In this manner, Lodge's empirical realism strove to connect human cognitive experience to the true nature of things. He did not do this as a naive realist might, arguing that space, time, distance, and speed were actual physical properties of the ether, or as an idealist might, presenting them as *a priori* intuitions of the mind. Instead, Lodge argued that these were psychological inferences of mind founded on dynamical properties of the ether. Thus, Lodge's science kept pace with the subjectivist trend of fin-de-siècle modernism, even while it sought to redirect that subjectivity toward something more universal. A decade before Einstein's theory of relativity, Lodge was already framing the physics of space and time within an episteme of perception, even while making it (emphatically) a foundation for realism, not relativism.

In 1893, two years after Lodge addressed Section A and gave his interview to the *Spectator*, he found himself drawn, in a highly public way, into an intense epistemological debate normally beyond the bailiwick of an ether electrician. Lodge leaped into the fray with his article "The Foundations of Dynamics," to defend the sanctity of motion, recently slandered by the physicist Dr. James Gordon MacGregor of the Royal Society Canada in his "Presidential Address regarding the Laws of Motion."[102] Lodge also called out philosophical elites, such as Karl Pearson, Ernst Mach, and W. H. Macaulay, in the same article to signify that this was no narrow skirmish among physicists. The disconcerting problem MacGregor raised for classical mechanics was this: If uniform motion was undetectable, and all awareness of motion was limited to comparison only, then motion itself had no independent existence. There could be more or less of it; it could be traveling here or there; but that said nothing about what it was. All information concerning motion was thus conveyed mathematically through speed and coordinates. These were symbols. What, then, was meant by the "reality"?

But Lodge was ready for him, extrapolating a muscular intuition of motion from the spatial and locomotive properties of the ether upon which he had already been ruminating. In "The Foundations of Dynamics" (1893) Lodge dismissed MacGregor's "whole attempt to complicate the statement of the first law of motion" as " absurd." While math was needed to precisely quantify motion, it played no role in the perception or description of it, because "motion and force are our primary objects of experience." Math was window dressing; the direct perception of motion was the window itself:

"Now I hold that all such notions as axes of reference are artificial scaffolding, necessary for the numerical specification of a velocity, but not at all necessary for the apprehension of what is meant by a uniform motion . . . The fact is that the conception of uniform motion is based upon a simple primary muscular sensation, or at any rate upon a succession of such sensations; everybody understands what it means, so far as it is possible to understand anything in this material universe."[103]

This "primary muscular sensation" of motion meant that something that could be represented dynamically was not something merely analogized, symbolized, or explained, but something fundamentally prior to representation, something *understood* at the deepest level of our physiology. But Lodge's defense of both the reality and perception of motion did not end there. In a following article, "The Interstellar Ether," published in July 1893, Lodge extended this "primary muscular sensation" of motion to encompass sensory cognition as well. Sight, smell, touch, taste, and hearing were each a specific modification of this physiological detection of motion, texturizing the cognitive experience of the material world. The ether might not be detectable, but the rotational motions of the atoms in the ether were, and thus "matter was incontrovertibly known to us" across this spectrum of sensory modalities: "Matter represents the palpable part of the ether, the only portion of it which affects our organs of sense . . . it acts on one small portion of our body [the tongue] in a totally different way, and we are said to taste it. Even from a distance it is able to fling small particles of itself sufficient to affect another delicate sense, smell. Or again, if it is vibrating with the appropriate frequency . . . the universe is discovered to be not silent but eloquent to those who have ears to hear."[104]

Though these "sensations" made only partial contact with reality, the awareness they constituted was intrinsic, tying motion, matter, and perception together as various aspects of an underlying whole, communicated through the direct perception of motion and force. (More profound than the secondary senses of smell, taste, hearing, and touch, which could only perceive the ether in its kinetic modification as the vortex atom, was the faculty of sight. The eye was "acutely, surpassingly, and most intelligently sensitive," able to directly perceive energy in the ether itself as light, rather than detecting the motions of the vortex. This made sight the least material and, in some sense, most mystical of the empirical senses.)[105] This allowed Lodge's realist epistemology to rest on a dynamical foundation, while simultaneously allowing his dynamical foundation to rest on a realist epistemology. "High up, in a range of vibration of the inconceivable high pitch of four to seven hundred millionths per second—to those waves the eye is acutely surpassingly and most intelligently sensitive."

Lodge's intensifying *contretemps* with philosophy brought him fully into public life as an advocate and interpreter of science. Physics had a new statesman. Like his early heroes Huxley and Tyndall, Lodge shouldered the burden of science's ideological defense in the cultural and professional arena. He may have rejected their brand of naturalism, "which rejoices in objects that appeal to the senses," but he continued the grand tradition of their scientism. Already known to be an outstanding orator by his scientific peers, Lodge now began speaking to women's groups, labor groups, church groups, and to every sort of lay seeker of scientific knowledge. His speeches and articles increasingly warned against the specter of antirealism and openly called for the scientific embrace of psychical research. He had ascended the public pulpit on the foundations of his "plain and substantial" mechanical science, and now, having passed the age of forty, he found that the time had come "for building the [gorgeous] superstructure."[106] It was ultimately incumbent upon scientists to take all this hard-won evidence and information, and turn it into meaning. In his centenary address before the University of London, Lodge reclaimed the science that had awed him as a boy, a science that had unfurled itself like a flag at the Royal Institution and was "the end and aim of everything":

> Now there is one doctrine that I think is wrong and that ought to be treated as a heresy. The statement that physical laws, the laws of Nature, are not so much true as convenient . . . I suppose that the idea underlying this contention is that Reality is beyond us. I do not know. It may be that the philosophers who say this are doubtful as to what is Truth. They say, "The question of true or false has no meaning in science. All that one is concerned with as in the cases of all scientific theory is its usefulness." Now that is held by not unimportant people yet I believe that this is entirely false, that it is a thing to guard against, that it is an idea to be contended against and fought as a heresy. I do not like the idea of heresies in science; but I think that this is one. We are out for truth. We may not get it, but that is our aim.[107]

Lodge was out for not just *any* truth, but for a very particular kind of truth suited to his values and his vision of the future. It had to uphold the authority of science and the reality of matter, while leaving room for some potential ascendancy of the mind. And it also had to resist what Lodge referred to as "monism" of any kind, otherwise the rest would not be possible.[108] The antirealism Lodge campaigned against took many forms, but consistent in its outlook was the collapse of any meaningful distinction between subjectivity and objectivity, matter and mind. This made phenomenology, despite its own searing rigor, a truth of entirely the wrong kind. It denied the reality of both the subject and the object in favor of some synthetic condition called "experience." This self-referential, solipsistic worldview of "those who press Monism to absurd lengths" would be forced to "take refuge in a narrow and illiterate and most unscientific variety of dogmatic skepticism or agnostic dogmatism."[109] Phenomenology was thus subject to its own doctrinalism, compelling the rejection of all certainty except for the certainty of uncertainty, the dogma of doubt. In such a scheme, science was, at best, a consensual hallucination.

But certainty in knowledge was not Lodge's philosophical requirement; only a certain kind of certainty would do. George Fitzgerald's idealism failed the test because his scientific epistemology leaned too heavily on the intervention of God. Even though Fitzgerald was an ether mechanist who believed with Lodge that "the phenomena of our consciousness are all explicable as differences of motion," his Helmholtz Lecture (1896) pressed the origins of dynamical sensation too far: "What is the inner aspect of motion?" Fitzgerald asked. "In the only place where we can hope to answer this question, in our brains . . . Is it not reasonable to hold with the great and Good bishop Berkeley that [divine] thought underlies all motion?"[110] This was an idealistic monism: here the object became unreal (reduced to pure thought), but the mind did not. While this was not solipsism by any stretch (the mind was in direct connection with God), it still made science intellectually dependent on God in a manner not unlike religious faith. This did not sit well with Lodge. It was fine for Fitzgerald, a man of unwavering faith comfortable with religious surrender, but Lodge could not and would not make that theological commitment: he wanted to put his faith in science. For that, he needed to put natural knowledge on firmer footing. Thus, matter mattered—or else reality became a kind of illusion, and science was left chasing a dream.[111] Lodge rebuked Fitzgerald's idealism in favor of an independent human subject: "There may be a sense in which all matter is evidence of and an aspect of the thought of some World-Mind but is certainly neither evidence nor aspect of my mind." God may be the totality, but humanity and its science pertained to that part apportioned to our reality: the natural world.

But naturalism alone would not suffice, because of its tendency to overvalue the object over the subject, leading to a kind of materialistic monism. As naturalism transformed the mental life of the self to pure psychobiology, so went the empirical basis upon which naturalism relied. Thus, even the cocky naturalism of midcentury science eventually eroded its own factual basis. For Lodge, both the objective and subjective aspects of the universe had to coexist and to affirm that coexistence through interaction. This dualism alone stood against the nihilistic idea that reality was "all in the mind," which first presupposed that the mind itself was "all in the brain," limiting consciousness to an evolutionary instrument pertinent to our species. Psychical psychology, however, proposed a distinct mental force alongside, and even above, matter, beggaring the limits imposed upon sensory cognition by phenomenology as those limits themselves were constantly being revised, potentially running the gamut from ordinary thought reading to clairvoyant precognition. This made psychical research the true intellectual force behind Lodge's scheme of ultimate intelligibility. The primacy of motion and force solidified the foundations of empirical knowledge in relation to an objective reality, but psychical research prevented science from its own eventual self-entombment in either materialism or psychology. So, at the same time that Lodge was affirming "trust in consciousness" through the mechanics of the ether, he was also rocking the very boat he was trying to steady with the addition of unknown psychical force:

> I have said that the things of which we are permanently conscious are motion and force, but there is a third thing which we have likewise been all our lives in contact with, and *which we know even more primarily*—life and mind.[112]

This third thing, perhaps the object of some psychical sense, was discussed briefly at the end of Lodge's lengthy preamble on mechanical motion and the five physical senses correlating it to the mind. He would only specify it as some "unknown category," which he himself connoted as "life and mind," but, admittedly, he did not "pretend to define these terms or to speculate as to whether they are essentially one and not two." Yet, for all these disclaimers, this was no elusive prospect. This "third thing" was a definite object occupying the future horizon of science, which "must also be made part of the project of physics, in order that the physics of the future may rise to higher flights and an enlarged scope."[113]

Any such extraordinary pursuit was fraught with difficulty. It required scientists to cross the threshold of what Lodge called "the transcendental or ultra mundane or super sensual region" of human experience, where "we do not know anything like the boundary conditions."[114] The existing boundary conditions were, of course, those imposed by the physical senses.

What was beyond that boundary no longer needed to remain a mystery, a fact Lodge himself could argue from experience: "the methods of science are not as limited in their scope as has been thought and can be applied much more widely to bring the psychic region under law too."[115] And so Lodge began to openly urge the scientific community to take up the gauntlet of psychical research beginning in the 1890s, based on his own, as yet limited involvement. Twelve months after the publication of "The Interstellar Ether," Lodge fulfilled this psychical mandate by embarking for the Île Ribaud and taking his seat around a table while a strange Neapolitan laundress flung her sorcery about the room. Lodge saw and physically felt the effects of this hailstorm of energy as it erupted through the ether, propelled by Eusapia Palladino's vital will. For Lodge, the investigations on the Île Ribaud fully substantiated the existence of some psychophysical vitalistic power.[116]

Lodge was the only physicist present in Richet's circle of inquiry. He was new to psychophysical phenomena and poorly acquainted with the jargon of medical hysteria. He arrived at the island by way of a grueling series of ether experiments and an increasingly urgent sense that knowledge, like the ether itself, was contracting all around him. The course of physics and philosophy had to divert from such perilous straits. The global failure to establish a physical connection between ether and matter compelled Lodge to consider a different kind of nexus between ether and matter, one mediated through the mind. In July of 1894, this threshold happened to be most promisingly sited in the person of Eusapia Palladino. From observing her ideoplastic manipulation of the ether into ectoplasm it was clear that her vital will was itself a determining power and could therefore not be explained by deterministic laws. The physical and the intellectual components of her phenomena were distinct but interpenetrating factors that allowed her to express her will upon and through matter. This seemingly psychoactive aspect of the mind/matter duality required a deep scientific reconsideration of the Cartesian divide, which psychical research was not alone in pursuing: "There is a trend," Lodge declared in a town hall address when distilling these facts for the Birmingham populace, "from the strict separation of mind and matter to interpenetration. This entire change of philosophical principles, which we find in Wundt, as we found it in Kant, Virchow, Du Bois-Reymond, Carl Ernst Boer, and others, is very interesting." This was not a return to magical thinking, but the modernization of physics and psychology, acknowledging that "mind may be incorporate or incarnate in matter, but it may also transcend it. It is neither Epiphenominal nor Parallel, neither a Subset or Non-intersecting."[117]

When first setting these ideas before the public, Lodge still hesitated to say exactly what was meant by the nature of this transcendence. He had explored the question in a limited sense before the sessions with Eusapia Palladino through his study of telepathy in 1883, in which the thoughts in Miss E.'s head had somehow found their way outside her skull and into Miss R.'s. Lodge considered this activity to be mental, not vitalistic (more mind than life), a refinement associated with the neural activity of the brain and expressing itself as "thought" rather than as "will." But such ordinary instances of telepathy, as witnessed at Liverpool, still required the presence of an agent and a percipient, indicating that some kind of neurological architecture was involved, as both transmitter and receiver. Mrs. Piper's phenomena, however, defied Lodge's field model of brain-strain in the ether, as expressed back in the 1860s. The memories she was catching out of the air did not belong to the sitters at the gathering. These were tiny specks of information: a rattlesnake belt, a dent in a watch, playing games in a field: how did she know? In what form and in what part of the universe did such memories still exist? Mrs. Piper's ability to obtain knowledge seemed to slip through the structures of space and time, in a manner that claimed that some kind of transcendence might be possible in this life—but how about in death?

Lodge's encounter with "Aunt Anne" seemed to force deep consideration of this spiritualistic possibility. He did not press this in his report, purposely diverting focus away from Jerry, Anne, and the afterlife, toward the living, and taking as his marvel the astounding cognition of Mrs. Piper. Lodge had never known Uncle Jerry, but he had known and loved Aunt Anne, and unaccountably he sensed her presence in the room. Mrs. Piper was not merely relaying Anne's information, she had somehow fetched the woman herself. This "Aunt Anne" had left her body behind long ago (she did not arrive draped in a white sheet), and she somehow persisted only as a personality. How was it that she had physically died and yet still continued from the subject view of that physical existence? Assessed in terms of biology, Aunt Anne's survival found some partial explanation. As Lodge put it, "The persistence of identity can be conceived, without the hypothesis of retention of any particles of terrestrial matter, since the identity of a person in no way depends upon identity of particles even now."[118] Cells were constantly dying and being replaced in living organisms, and yet identity suffered no interruption. Even on Earth we are not our cells. But to extend this to bodily death was to separate that psychological subject from any particle of matter whatsoever. This was no ordinary realm of consideration. This was no longer the study of a mental force or echoes of memory. This was, possibly, Anne herself. She had said as much when she presented herself at the sitting: "I am here."[119] Lodge had felt the power of her words. If this was indeed Aunt Anne, Lodge had encountered nothing less than her soul.

The forces of "life and mind" that Lodge unveiled before the public in "The Interstellar Ether" did not fully own this fact. The article only proposed that there was "yet some third thing," physically operating in the world and known to ordinary awareness, but which was not itself mechanical. "Life and mind" were presented as impersonal forces interacting with deterministic laws, belonging to no one in particular. But Aunt Anne seemed to have definitely claimed a part of "life and mind" for herself in the form of her surviving consciousness. She was not her body, but she had originally asserted herself through that embodiment. And, through her, the inchoate forces of life had become a conscious personality: "*it* is *I.*" As Lodge put it: "Experience is what we see of matter filtered through mind, but also mind is to some extent limited and conditioned by matter."[120] It was still too early to broach publicly the question of whether or to what extent a human "soul" might result from these limiting material conditions. First, some theory was needed to better explain "life" and the specific nature of its interaction with matter. Eusapia Palladino was one piece of this study, demonstrating an active will working through biological matter, shaping it, directing it, guiding its fetal development, or extruding it (in the aberrational case of Eusapia) as an ectoplasmic limb. This physiological principle extended down to the most basic level of physics itself, in the form of Maxwell's demon willing an atom move to the left or to the right.

On the other end was Aunt Anne, who was out of her body and no longer part of the material world. If her "experience" was "matter filtered through mind," what happened when that mind was completely decanted of all such terrestrial particles? There was no material element left, and yet it retained its particularity in the form of personality. Did "life and mind" become conscious by acquiring the specific awareness of a point of view? Did it have to first become a subject in a world of objects in order to be a soul upon leaving that world behind?

"The Interstellar Ether" did not list "soul" after "life and mind," as part of this new catalogue of scientific objects to be studied, but there was a visible ellipsis at the conclusion of the article, suggesting where such research might lead: "When the next century or the century thereafter lets us deeper into their secrets," it will be "no merely material prospect" that rewards the study of life and mind. Psychical research promised a "glimpse into a region of the universe which Science has never entered yet, but which has been sought from far, and perhaps blindly apprehended by painter, or poet, by philosopher or saint."[121] Lodge's physics, which had toiled patiently in the ether and molded sensory perception within that same milieu, was gearing up to push past the boundary of physical reality (even as his fellow physicists fell back from that position). Lodge's scheme of "ultimate intelligibility" had ventured past science and gone head to head with philosophy, and now it was positioning itself to assimilate the vast project of religion as well.

THE CRISIS OF FAITH IS FAITH ITSELF

The Synthetic Society was a relaunch of the Metaphysical Club of the 1870s, convening in 1896 to work out issues raised in Balfour's monumental *Foundations of Belief* published that same year. The new society was, like its predecessor, a forum in which philosophy could be applied to the most pressing problem of modern intellectual life, the proper relation of faith and reason. The point of the society, as Lodge understood it, was fostering "fundamental agreements between different schools of thought" and "minimizing unessential points of difference."[122] And yet Lodge, right from the start, took an uncompromising approach to this goal of synthesis. When Canon Gore took the podium that first year to suggest that both traditional faith and objective evidence could be complementary proofs of religious doctrine, Lodge challenged him on both points.[123] In terms of faith, Lodge suggested, "these feelings may produce conviction of the most intense character," but still, "its propositions if formulated, would usually be found seriously defective in rigorous proof."[124] As for Gore's appeal to factual support from science and history, this, too, was a meaningless formulation. For clergymen to empower science only to verify, and not to invalidate, church dogma was an exercise of theological orthodoxy and not a test of it. If Gore's "logical proofs" lacked probative value, then, by implication, religious beliefs were without basis, asserting truth as a right of faith in a way that was unilateral and ultimately ineffective. On this basis, Lodge was willing to implicate religion itself in the promotion of atheism: "An atmosphere of disbelief became prevalent and was generated by the persistence of the faithful in certain material statements which, to an age of more knowledge, had become incredible."[125]

Lodge did not get stalled in this elite circle of disputation, but actively campaigned his ideas before the nation's faithful in a manner no less direct. In "The Outstanding Controversy between Science and Faith," published in the *Hibbert Journal* in 1902, Lodge wrote with the urgency and candor of an intervention: "We must admit, even fundamental Christian doctrines such as the Incarnation or non-natural birth and the Resurrection or non-natural disappearance of the body from the tomb have, from the scientific point of view, no reasonable likelihood or possibility whatever. It may be and often has been asserted that they appear as childish fancies, appropriate to the infancy of civilization and a pre-scientific credulous age."[126] Lodge wasn't making the usual, comfortable case against fundamentalists clinging to stories of Jonah being eaten by the whale. He was bringing that same attack to the highly curated shortlist of sacred mysteries. So-called "modern" Christianity still suffered from a fatal dependency on the supernatural. Lodge's message was clear: grow up and get a belief system appropriate to the age. Even "the god of the gap" relied on mystery, the unknown origins of life. Of this last outpost of divinity, Lodge warned, "The present powerlessness of science to explain or originate life is a convenient weapon . . . but it is not perfectly secure . . . Already in Germany inorganic substances have been found to crawl about on glass slides under the action of surface tension."[127]

But even while Lodge was disparaging this popular fancy for miracles, he was also taking to task clerics who discouraged such beliefs on purely ideological grounds. This had been the strategy of High Church intellectuals since the eighteenth century, keeping God above creation and out of politics. But, as a result, God was increasingly becoming an abstraction, squeezed out of nature first by geology, then physics, then biology, and now, out of consciousness as well by psychology. In his essay "Faith and Knowledge," addressed to the Synthetic Society in April 1902, the same year he published "The Outstanding Controversy," Lodge warned that this mechanistic view of nature was tantamount to the negation of God: "The denial of free-will or self-determination to the whole Universe, amounts, it seems to me, to atheism."[128] Both these essays steered attention to epistemological questions: on what basis were religious convictions to be held? But Lodge framed the problem differently for intellectuals than he did for the general flock. In "The Outstanding Controversy," the problem was not miracles per se, but the childish grounds on which they were believed—that is, belief without evidence. In "Faith and Knowledge," Lodge pointed to a more insidious tendency, one that erased the grounds on which such evidence could even be supplied. A theology that pushed God out of the physical realm denied any means of intersecting with genuine human experience. God had to be more than an idea or a feeling for a religion to thrive. The critical juncture had come now that such traditional ideas and feelings were increasingly undermined by science and psychiatry.

While Lodge continued to frame the threat to religion as a credibility gap between faith and evidence, other members of the Synthetic Society were pushing the debate in another direction entirely. Instead of letting Lodge put faith to the test, they shifted their attention to the objectivity of science. Theologians such as R. H. Hutton, Wilfrid Ward, and Charles Bigg did not wish to deny scientific authority outright, only to restrict it, emphasizing that empiricism, too, came up against subjective limits.[129] The science based on "a multitude of specially observed phenomena" had given rise to a worldview trapped in its own methods, testing only its own predictions, and imagining only that which its laws would allow. Lodge found this skepticism about science to be intentionally obfuscating and reactionary. In response to such medieval debate tactics, Lodge demanded of his colleagues, "Do we need not a Bacon to call us from these venerable scholastic methods?" He advocated instead "for more facts and less subtlety."[130] Yet, Lodge recognized in this seemingly backward appeal to "the unknown, the dark, the incomprehensible" a tendency that was all too *au currant*, an attitude he had confronted early on in modern physics with his speech in 1891 regarding "Ultimate Intelligibility" and which was now leaching more generally into the philosophy of knowledge.[131] Taking shape here, in these Synthetic Society debates, was the future of modern religious reconciliation. Instead of fighting over which is right, science or religion, let both be equally doubtful.

Lodge's embrace of an empirical standard of truth was not the unnuanced objectivity these arguments suggested. He wanted to deepen consideration of subjective conditions, not deny them, urging the discussants to "utilize new facts from experimental psychology and other cognate regions of inquiry."[132] Nor did Lodge, as a psychical researcher, hold those conditions to be set. Lodge defended the existence of telepathy and held open the possibility that shepherds might have become "supernormally aware of spiritual surroundings" at the birth of Christ and "heard voices from the sky."[133] Part of the empirical project was to investigate the basis upon which knowledge could be held. Deepening reflexivity about the status of objectivity was built into the psychological basis of psychical research. Lodge's empiricism was a layered, open-ended subjectivism, neither a mind boxed in nor a direct observer. He upheld the reliability of sensory perception by anchoring touch, smell, taste, sight, and sound in the direct apprehension of motion. That realism was amplified by the addition of supernormal perceptual faculties, such as thought reading, and even more exotic phenomena, such as clairvoyance or precognition, suggesting a way of getting at deeper structures of reality beyond space and time. Moreover, and perhaps at the heart of Lodge's scientific epistemology of faith: truth, in order to be shared among equals, must be demonstrably obtained. The reliance on sophistry or credulity to bolster religious authority set it against the moral and political progress of which religion must rightly be a part.

While Lodge seemed to attack almost every religious convention, his own position remained unclear. Was he a theist? a deist? a "pantheistic monist?" a free thinker? Was his philosophy, at bottom, "pure Stoicism, even down to its metaphors?" (as described by E. A. Sonnenschein).[134] Lodge even incurred charges of atheism in his attempted defense of God. All this built toward a particularly heated *contretemps* in 1906 over two articles Lodge published in the *Hibbert Journal* in July and October of that year. The first edited Christian doctrine down to an empirically defensible set of core propositions. The second aimed to overcome sectarian conflict by giving primacy to this core, ascribing the remaining divisions in the national church to "superstition" rather than substance. Lodge, somewhat naively, was taken aback by the furor he stirred up when he tried to cut to the quick of these deeply held theological concerns. As he explained it, he was trying "to show how great a mass of reasonable doctrine was truly religious and yet outside the field of sectarian controversy."[135] But for many, Lodge's "statement of a common faith" looked like bargaining down the supernatural majesty of Christ to make religious doctrine more agreeable. This expedient approach to theology seemed to give weight to social considerations over sacred ones. It was that last assumption that Lodge wished to refute above all else in his *apologia*, declaring: "I am not just setting out an ethical doctrine; it is religious, not cold-blooded in spirit."[136]

One of the reasons the world had so misunderstood what Lodge called his "attempt at a fundamental religious statement from the scientific point of view" was his unconventional dualism.[137] Anglican theology tended to set the natural and supernatural realms apart, as two largely noninteracting phenomena. Mechanists tended to view all processes, even mental ones, as determined by a single set of physical laws. More recently, neovitalism had revisited the possibility that hitherto inert matter might itself contain a volitional aspect. As a scientist, Lodge recognized two distinct phenomena, which he loosely labeled "life and matter." Matter constrained life through its physical laws, but life was still "the utilizing partner" and could fully transcend material conditions. Psychical research thus supplied a far more fluid model of dualism in which mind and matter were interpenetrating and interacting, while still hierarchically distinct. There was something more here of a religious nature as well, informed by Frederic Myers's theory of a World Soul.[138] Though Lodge rarely used the term publicly, he had been germinating this emanationist outlook over the course of the 1890s as his friendship and psychical collaboration with Myers intensified.

Thus Lodge's understanding of dualism existed in the context of a greater continuity, where all such differences had their origins in a single, absolute source: "the illimitable and eternal mode of Manifestation called the Universe or God."[139] God could be a "Divine Immanence" or an "Infinite Transcendence," but what "He" could never be was *outside* the Universe because God was all. Instead, Lodge held that "genuine religion has its roots deep down in the heart of humanity and in the Reality of things."[140] What was real was God. What was false was not. Such a God was at the heart of Lodge's original scientific formulation of "Ultimate Intelligibility," which he put before the Synthetic Society with "The Nature of Proof " in 1897. Intelligibility meant that the human mind was equipped for the study of nature and that such knowledge pertained to something ultimately real and absolute. His criticism of "miracles" or "obscurity" was not a rationalist's attempt to limit divine mystery, but a seeker's call to properly engage it. When a prominent theologian like Dr. Charles Bigg argued that "from the darkness springs the very life of Religion," he was in danger of constituting mystery itself as *the* object of worship.[141] This growing strain of obscurantism in the Synthetic Society seemed to suggest that God had no basis in reality and that awe was the only reliable thing to use to make religion stick.

The position Lodge was staking out in turn-of-the-century religious reconciliation was original and complex, and difficult to root in the national conversation. He was as out of sync with the modern theologians as he was with an older generation of north British physicists, whose Scotch Calvinist tendencies clashed early on with the immanence implied by spiritualism and Crookes's psychic force.[142] But there was one constant in Lodge's decades of opposition to moderates on either side of the isle, their tendency to prioritize institutional over ideological harmony when working out the accord between science and religion. The liability of such professional elites was to think in terms of the spheres of authority in which they operated, and not in the realm of personal meaning where most people lived. Lodge, as even his critic Father Joseph McCabe acknowledged, was that "discerning observer who stands outside of both camps" and "cannot fail to see that the truce was a hollow one."[143] The overall effect of this peace-at-all-costs was a continued drifting apart, leaving the deepest questions of faith unsettled for the public.

It was this ordinary believer that Lodge most wished to reach, and it was here that he would ultimately most succeed. Lodge would become, by the end of that first decade of the twentieth century, "the archpriest of science" to whom "the religious and theological public had been accustomed to look ... for comfort and help to faith," an authority "quoted almost without limit."[144] Lodge's growing appeal was that he was willing to explain where others obscured; instead of doctrine, he proposed genuine knowledge. A truly systematic theology must begin with a mind wide open, oriented fully toward the world of appearance and rejecting all received notions of God lest they be false. "We only know of an unseen world when we glimpse it through its action on matter."[145] By implication, all other sources of metaphysical knowledge, including religion, were suspect. Psychical research alone occupied the necessary point of contact. It was here that "theology touches upon science, and science rightly claims to be heard."[146] Lodge offended many Christians and scientists alike by taking this right upon himself, especially when his religious rapprochement looked so unilateral. But those with less vested ideological interest found a genuine path out of a circular conversation, after dancing cautiously around the problem for nearly a century. Where others sought to avoid "the hampering complications of this conflict," Lodge dove straight in, bringing his formidable knowledge of physics with him.[147]

Mid-Victorian physicists, such as William Thomson and James Clerk Maxwell, pushed religious reconciliation with extreme caution, trying to protect the intellectual divide even as they endeavored to bridge it culturally. For instance, William Thomson, Lodge's mentor, only used "the demon" (that metaphoric being who pushed atoms at random to dissipate heat) to show that free will was present *in principle* as part of mechanical design. Lodge exploited this same indeterminacy to show how the mind could directly influence matter, turning a vague analogy into an authentic causal force. By 1903, however, even William Thomson, now Lord Kelvin, began to find his uncompromising dualism an obstacle to the harmony he had hoped to achieve. In a speech he gave at University College that April, he was willing to be more explicit in granting scientific support for religious faith. Thomson used "the new vital principle" in biology to insert God's "creative and directing power" into the ongoing process of evolutionary design. While Thomson had always promoted the idea that "science [was] not antagonistic but helpful to religion," he had never advocated that one could be "forced by science to the belief in God."[148] Or that God should explain what science could not. But now, the aging statesmen seemed eager to find some "proof positive" that God existed as a fact of modern science. His words were praised and repeated in the popular press, even as they were rebuked by biologists. In his own letter to the *Times*, Lodge dismissed Thomson's "creative power" as unscientific, calling it "a phrase I should not myself use, because I am unable to define it," turning the tables on his friend who publicly deemed psychical research as insufficiently rigorous.[149] But unlike other critics, Lodge made the central complaint that Thomson had failed to clarify this power in terms of existing evolutionary laws, not that he had raised the idea of divine agency in the first place. Of course, defining theistic guidance scientifically was precisely what Thomson wished to avoid. God was above, not subject to, such laws. For Thomson, the key to protecting science from religion and religion from science had always been keeping the Creator and His creation apart. But now that he wanted to stake some sacred space in evolution, Thomson had no real framework. His idea was of only "subjective interest," floating above evolution without any real anchor or argument.

At around the time Thomson was venturing his few suggestive remarks about a "creative power," Lodge was beginning to formulate his own religious scheme for evolutionary development, stating in a speech to the Nelson Street Adult School in 1902 "that evolution was the method of divine working."[150] With that phrasing, Lodge went much further than Thomson theologically, and yet remained more grounded, putting God *in* evolution, not above it. Lodge elaborated this over the next few years in papers with titles like "Sin," (1904), "Ecce Deus" (1905), and "Christianity and Science" (1906), in which he attempted to work out a "statement of creed . . . drawn up in a scientific as well as religious spirit." But Lodge was inspiring more controversy than converts.[151] In the wake of his *apologia* of October 1906, Lodge hammered his evolutionary theory into a more plainly religious statement. The work, published in 1907, was titled *Self and the Universe*, with the subtitle, "The Rudiments of Philosophy as Bearing on Christian Dogma: Being an Explanatory Catechism or Glossary for Elder Children Only." It was issued alongside another booklet directed at younger children, whose title fully summarized Lodge's new charm offensive toward traditional faith: *The Substance of Faith, Allied with Science; a Catechism for Parents and Children*. Notably, Lodge was offering his allegiance to the substance and not the doctrine of Christianity; still, the approach avoided epistemic comparisons and focused instead on correlating meaning. Several books followed shortly thereafter, rooting Lodge's "scientific religion" in the public imagination as a genuinely theological enterprise, one that might offer a beachhead for Christian faith in the very den of Darwinian iniquity.[152]

Lodge's new evolutionary theology built spiritual guidance into natural selection from the ground up, without altering its physical machinery or denying the violence and reversals that marked its development. But where Darwinians took these physical processes to be random, Lodge's catechism promised an "unfolding of latent possibilities, and a bringing out to the utmost of whatever potentiality exists in matter or life or mind."[153] But one had to peer between the gears of physical laws to spy the hidden operations tucked within: "A molecule of iron in a living body is no different than a molecule of iron in stone. There is no more energy present in a dead body than a living, the only difference is that the chemical and physical activity of the corpse lacks guidance. Death is entropy, it is the cessation of the influence which controls matter and energy."[154] This was not conventional heat mechanics. There, the entropy that overtook the body in death was presumably what faced any closed system that could no longer refuel itself. But life was not energy, "just as a painting is not in fact the paint, but the arrangement of that pigment on canvas."[155] "Life," as Lodge described it in *Self and the Universe*, animated through organization, adding the structural information that made complex living systems possible. For Lodge, this did not contradict the laws of thermodynamics, it only exploited them differently. The answer lay in "the special modification of life," which navigated entropy's stochastics intentionally rather than randomly, allowing guidance to ebb and flow into the arrangements of matter. The "cessation of this influence" resulted in a corpse, restoring the reign of entropy once again over matter.

Physicists alarmed by Lodge's *entelechal* biology need not worry about some perpetual motion heresy or other dangerous addition of energy to nonliving systems. This guidance resulted normally only in the expression of ordinary "life." Yet rare, supernormal events occurred when this intelligence extended beyond its usual biological boundary, as with telepathy or something more disruptively physical. Even in such latter cases, there was likely no violation of conservation laws. As Lodge had previously speculated regarding Palladino's ectoplasm, she was probably only manipulating her own (or other people's) biological resources, depleting herself physically much as she would during ordinary exertion. Psychical research cut the hole in the ice from which to observe, however obliquely, these hidden operations. Unlike dogma, it could only move slowly toward certainty, advancing at the pace of science and no faster. But these few facts, when leveraged correctly, added their weight to the vast human record of the fantastic, the spiritual. While science gave substance to these encounters, the mythic impressions taken of the supernormal mattered as well, "for there are no purely gratuitous inventions, every idea is based in something . . . and the ideas of humanity are embedded in the structures of the universe."[156] On that basis, *Self and the Universe* applied Christianity toward a deeper understanding of nature, and applied evolution to illuminate the figurations of Christianity. Evolutionary theology was not science, yet it was a fruitful, synergistic space in which to extend the implications of knowledge.

If we look at the psychical theology Lodge began to formulate at the Synthetic Society, we see that it holds at its core a fully modern, secular sensibility that is against all creeds and hierarchies, is centered upon individual experience and spiritual growth, values altered states and paranormal events, and conforms its theological speculation to existing scientific laws (as Lodge understood them). And yet Lodge is rarely given his due. He appears too old-fashioned, too quintessentially Victorian, too caught up in Christianity, the ether, verifiable reality, séance spiritualism (a turn he took after the death of his son, Raymond, in World War I), to be a "truly" twentieth-century thinker or relevant antecedent to the New Age. Even in his reconstructed religion, Lodge's language can appear overly conventional, especially when directed toward the layman: designating the deity as "He," or using theistic terms like "the Manager," "the Supreme Being," or "the Director" to suggest an all-powerful, masculine Other. But this language was intentionally familiar in order to ease the acceptance of otherwise radical ideas, whose commonalities with Christianity would have been far from obvious. When Lodge described the fundamental contribution of Christianity as "the perception of a human God," he did not mean to celebrate an anthropomorphic projection in the sky, or even a supernatural Jesus. This was an argument for the immanence of the divine in all humanity, symbolized in the deification of "a real and ordinary" man, whose "*un*uniqueness" was the mark of his significance.[157] (Christ was the goodness of which all persons were capable.) It also pointed to a hidden truth about the universe; it was best understood as a being, not a thing—a subject with longings yet to be realized. Lodge's statement that "evolution was the method of divine working" was meant in the most profound sense possible, "extending the ideas of development and progress even up to God himself."[158] This conferred a grander purpose on Darwinian evolution than any that could be claimed by Creationism or Design, putting it in the service of God's self-realization.

Nonetheless, even with the stakes so high, this was not a heavy-handed teleological operation but a process driven entirely from below. God was working through evolution, not above it, a deity "that could understand, that could suffer, that could sympathize" as much as any New Testament Christ.[159] And just as Lodge "revealed the Christian God as the incarnate spirit of humanity," humanity in turn acceded to aspects of the divine. These were the new realities to which modern religion must awaken: "What we have to realize in regard to our place in the Universe is that we are intelligent helpful and active parts of the cosmic scheme. We are among the agents of the creator."[160] It was not God that became man to redeem humanity, but humanity that must redeem the world on behalf of God, dependent on us for help. As Lodge explained it in "Sin": "The torture of a child or animal are a boil, an abscess on the Universe: they must be attacked and cured by human co-operators; they are hardly tractable from the other side . . . we are the white corpuscles of the cosmos and like the corpuscles we are an essential ingredient of the system."[161] While liberal reforms had made God less intrusive, less vengeful, less irrational, the idea of making God less powerful had yet to be tried. Yet, more than any other progressive creed, Lodge's evolutionary theology empowered humanity directly at the expense of God. Human beings were no longer the companions, servants, or playthings of a divine maker, but the most forward agents "in a universe struggling upward towards its own perfection."

Lodge eventually leveraged the insights of Henri Bergson's *Creative Evolution* (1911) as a way to bring absolute human agency to every aspect of the evolutionary process, including the choice to "become" at all. It was not enough for a modern faith to give humanity status and power; there must be divine justice as well, and this required a radical rethinking of free will. In the traditional Christian theodicy, death and suffering were portrayed as just punishments for original sin, and moral freedom, such as it was, was coerced by consequences. Serve God "or else" was a program with no chance for input or opportunity to opt out. Earlier Christians had swung for centuries on a violent hinge between heaven and hell, trapped in a theological apparatus meant to bring total social control. While Lodge tended to prefer Reformation theologies to Catholicism, priestly power had not been transferred to the laity so much as it had fled upward toward the omnipotence of God. If anything, the Protestant doctrines of grace and predestination lessened human agency in terms of personal salvation. *Creative Evolution* helped Lodge bring choice to the earliest germ cell of living matter. Bergson's critical contribution, as Lodge understood it, was the insight that "Matter and Mind have a common ancestry, they have arisen from a kind of cell that is neither one nor the other."[162] This shared-origin theory was consistent with Lodge's emanationist view that God was the source of all and suggested a kind of tortured dualism where mind and matter longed for reunion. The important thing here was to explain why consciousness, if it was indeed distinct and superior, would immerse itself in matter at all. The idea of a half seeking to be whole internalized this drive fully within the subject position, allowing the moral and physical perils of embodiment to be freely assumed.

Evolution was to be morally redeemed by absolute agency and by absolute necessity. This suffering, and the freedom to choose it, "were not artificial and transitory, but inherent in the process of producing free and conscious beings." At the heart of evolution was the struggle of the subject to overcome objective conditions, tunneling through matter like "water through caves."[163] An oppositional, attractive, and interactive psychophysical dualism was the key to this creative process, arising in Bergson's seminal moment of bifurcation. Lodge's evolutionary theology was an open-ended, dialectical process with agency exercised spontaneously from below, "not a being but a becoming … realized in its change in development, in progress upward and downward."[164] Yet, for all that, this was still a struggle upward: as matter was organized into more and more complex forms, life advanced its own organizing power, that is, its intelligence, moving "towards more and more self-perception." Life was bringing itself into consciousness. As such, the evolutionary arc was ultimately transcendent rather than biological, longing to be free of the body's limited subject position so as to hold more and more of the exquisite object of the universe within itself. Hence, the timeless nonlocality of "the after-life."

Lodge believed in personal immortality empirically, through his direct encounter with Aunt Anne among other extraordinary proofs furnished by Mrs. Piper in late 1889. But he also believed this to be true theoretically, because of "the conservation of Value," the idea that the universe does not surrender that which it brings forth.[165] The created value with which evolution was most concerned was not its lavish spectacle, but the exquisite structures of awareness that took it all in. The human individual, who took these sequential moments and wound them into a cat's cradle of memories, reflections, and feelings added more value still. To the extent that the self had its own character and could not be represented categorically or reconstituted through other parts, it formed a unique whole that must be conserved intact. Experience forged this subject into a soul, but matter was the anvil. Without matter, mind could not be fully realized, and without mind, matter would serve no purpose—such was the profoundly reciprocal nature of Lodge's psychophysical dualism. If the goal of evolution was to put the individual soul in touch with this source of all, then the science of Ultimate Intelligibility, in its effort to cognize the absolute, mirrored that advance from the empirical perspective. And all the while, humming in the deepest recesses of inner space, imponderable amid the vibrant, noisy variations of the phenomenal world, the World Soul chanted the motions of the ether, still too recondite for detection by either science or religion, but not, perhaps, for Oliver Lodge and psychical research.

CHAPTER SIX

Uncanny Cavemen

Andrew Lang, Psycho-Folklore, and the Romance of Ancient Man

BROMANTIC FICTION

In July 1887, Andrew Lang, a leading literary arbiter and popular wit, surveyed the state of British fiction for the *Contemporary Review*: "There has seldom been so much writing about the value and condition of contemporary literature—that is, of contemporary fiction. In English and American journals and magazines a new Battle of the Books is being fought . . . Across the Atlantic the question of Novel or Romance—of Romance or Realism—is argued by some with polished sarcasm, by others with libelous vigor."[1] And there was none more libelous than Andrew Lang himself. Henry James had originally cast this battle as one between high art and vulgar entertainment in "The Art of Fiction" (1882), introducing the status-conscious dyad that continues to this day in the curating of Barnes and Noble's "Fiction and Literature" section. James's attempt to cordon off and protect the status of the literary novel came in the wake of an explosion of popular literacy and the rapid rise of "penny packets of poison," pulp fiction's precursor, which got its foothold in the 1840s. Given the sheer volume of product tumbling haphazardly off the low-end press, the absence of any formal adjudication of literary standards rushed to the fore as a concern.[2] The "art of fiction" had no theory by which to define and defend itself, not even an academic tradition that could definitely assert if it was in fact an art form. This was all grist for the mill for a growing literature about literature, as elites struggled to guide the public taste in the face of an increasingly independent "popular culture." The power of this new consumer was asserted not from the margins, but as part of the most dynamic element of a rapidly democratizing culture. The once genteel world of letters was being disrupted from below by "the enormous appetite of readers who are prepared, like a diseased ostrich, to swallow stones, even carrion, rather than not get their fill of novelties," which threatened not only the merit of literature but, in the eyes of realists, the intellectual basis of modern, rational society.[3]

Out of this melee of moral and cultural anxiety came a new kind of priesthood: the cultural critic, a sage proclaiming from the pulpit of the art review. Andrew Lang was foremost among this new breed of chatty savants, with commentary in circulation in at least twenty-seven magazines, including *Longman's, Cornhill, Contemporary Review,* and *Fortnightly Review,* much of it in praise of adventure fiction.[4] As a conspicuous champion of "romance," Lang found himself a party to the brawl that arrayed British romancers Robert Louis Stevenson and Rider Haggard on one side, and American "realists" Henry James and William Dean Howells on the other, as factional representatives. Lang recast James's elitist polemic of "literature versus fiction" as a genre dispute between "realism and romance," emphasizing a distinction in kind rather than in quality. (James himself was actually not quite so clear about this emphasis on genre, refusing "to say definitely beforehand what sort of an affair the good novel will be."[5]) Lang saw this in clearer terms, because his was a more focused polemic, the elements of which have not been entirely understood.

Lang preserved James's distinction between popular and elite, but, for Lang, the disparagement of literary merit worked in the people's favor and not the other way around. Lang alleged himself a peacemaker, seeking equal opportunity for both sides: "Why should we not have all sorts?" he queried. "The dubitations of a Bostonian spinster may be made as interesting, by one genius, as a fight between a crocodile and a catawampus, by another genius."[6] But there was slander in this suggestion of "to each his own," in the way each was described. On the one side were those Jamesian connoisseurs who maintained that "the accurate minute description of life as it is lived, with all its most sordid forms carefully elaborated, is the essence of true *literature,*" while the romantic enthusiast (such as himself) understood that "the great heart of the people demands tales of swashing blows, of distressed maidens rescued, of 'murders grim and great,' of magicians and princesses and wanderings in fairy lands forlorn."[7]

Clearly, there was more going on here for Lang than some amicable differences in one's recommended reading. In the fevered evolutionary milieu of the 1880s, even matters of literary aesthetics could be shot through with social or civilizational consequences. Anxieties regarding heredity or regression, as well as physical and psychological degeneracy left no corner of the culture exempt from the yardarm of progress or decline, particularly those sectors in science and art that took humanity as their subject matter.[8] It was not just the content or quality of contemporary fiction that was at issue in Lang and James's transatlantic dispute, but literature's very *purpose* as a cultural form. If one were to take fiction as a safeguard of the national character (as Lang did), the catawampus had it over the dubitating spinster every time. The modern novel's overrepresentation of female subjects (marriage and divorce), female settings (boudoirs and drawing rooms), and feminine modes of thinking (fretful rumination and personal analysis) was a small and unseemly lookout onto the world, sure to dull the masculine appetite for life itself: "Were I in the mood to disparage the modern Realist ... I might add that some of them have an almost unholy knowledge of the nature of women. One would as lief explore a girl's room and tumble about her little household treasures, as examine so curiously the poor secrets of her heart and the tremors of her frame. Such analysis makes one feel uncomfortable in the reading, makes one feel intrusive and unmanly. It is like overhearing a confession by accident."[9]

For "romance and realism," one's choice in literary fare even had the power to impact one's physiology: "Homo Calvus, the bald-headed student of the future," bore the marks of realism's "unrelenting exclusion of exciting narrative from its minute portraiture of modern life," while romance was popular with big-boned, sanguine brawlers, "partly because men and boys love a good fight, not because [they] are a degraded set of people that revel in horror for its own sake."[10] (This romantic exception does, however, seem to imply that such degradation played no small part in realists' prurient appetite for doom and gloom.) Rather than soberly defend the literary standards appropriate to his station, Lang joyfully backed Rider Haggard, Rudyard Kipling, and R. L. Stevenson's "books for boys" with insouciance and gusto. In his notoriously entertaining columns, splashed across the popular press, he praised the wholesome, martial spirit of romance and returned the scorn of its hoity-toity detractors, characterizing these *fashionistas* of artistic fiction as "a mere crowd of very slimly-educated people, who have no natural taste or impulse, and who have a feverish desire to admire the newest thing."[11]

But this all made for very strange cultural politics. In this pitched battle between real and fake, *haute* and vulgar, elite and popular, Andrew Lang, an exacting snob known for his limp handshake and chilly reserve (he who took two firsts at Oxford, one in classics and the other in greats), perhaps the most credentialed critic of them all, was decidedly in favor of the people and popular taste. Navigating Lang's complex cultural agenda requires some deciphering of symbols lest he be taken at his word: a steady stream of disparaging gendered rhetoric and decided homophilia. By so loudly preferring the male adventure-romance (associated with sexism, jingoism, and racism) to the searching self-criticism of the realist novel, Lang has made himself historically conspicuous, putting himself in the crosshairs of today's feminist scholars and postcolonial critics, as well as earning himself the contempt of his fellow literary elites. Henry James, the maligned genius behind the dubitating Bostonian spinster, found Lang to be so puerile in his rhetoric that he threatened the dignity not just of literature but even of the discussion about literature: "Lang, in the *Daily News*, every morning, and I believe in a hundred other places, uses his beautiful thin facility to write everything down [Zola, James, Howell, Balzac] to the lowest level of Philistine twaddle."[12] Edmund Gosse wrote to him in sympathy: "The memory of him irritates me. His puerility, as you say, is heartbreaking." But perhaps putting the point most clearly was the poet and critic, Theodore Watts-Dunton, who stated, "I never once met a man of genius who did not loathe Andrew Lang."[13]

It was not just Lang's literary opinion that gave offense in this *querelle*, it was the factional intensity he brought to it (exacerbated, perhaps, by the preciousness of the egos that *took* such offense in the first place). Something clearly had Lang extremely exercised, but was it really dull prose or pretentious authors? The subtleties of Lang's agenda can be lost on literary critics, clouded by the chum he left in the water, as a blustering partisan. But gender is often a proxy for Lang's other concerns, recruited symbolically in ways that do not necessarily reinforce established hierarchies. Lang was not the least bit concerned about women seeking employment, riding a bike, or getting the vote. (In fact, psychical aficionados—who included Lang's protégés, Stevenson, Haggard, and Kipling—tended to be progressive in their views on gender, with founders like Myers and Sidgwick playing principal roles in educational reform.)[14] Nor was Lang active in the politics of empire (though he is heavily implicated in promoting, through his beloved "books for boys," what the critic Martin Green called "the energizing myth of imperialism").[15] Yet even such literature itself was a secondary concern of his. In the midst of the "battle," he reminded his fellow reviewers that "but a little toss of the dust that settles on neglected shelves will silence all our hubbub."[16] So, if the main object was not to oppress women, encourage conquest, or rout odious realism, what really drove Lang's battle cry? As Robert Michalski observed, there is more to understanding Andrew Lang's love of adventure-romance than merely "bad taste."[17]

Here we must return to the original case he made against Émile Zola in an article for the *Fortnightly Review* (1882), before the official fray over fiction even began. Lang was the first English critic to review Zola's work for a British audience, and he ridiculed "the joys of naturalism" rather than indulging its gravity. *Naturalisme*, as Lang presented it, was a literary style that "described passions so base, characters so detestable, scenes so unnatural in their wickedness" that "Mr. Zola has almost exhausted the dictionary in the effort to find words unpleasant enough for the unpleasantness he has to describe."[18] The issue was not primarily Zola's writing, but, more profoundly, his dislike of Zola's aesthetic theory of writing, "to the extent there is one" (as Lang maligned it). According to Zola, "the object of the art of fiction was the scientific knowledge of man."[19] Zola explained his own work as "a kind of practical sociology . . . that continues and completes physiology and substitutes for the study of man in the abstract, the study of natural man as conditioned by his environment and by physico-chemical laws."[20] There was no storytelling here, Lang complained, only factual observations offered by an author who was a "stone cold vivisectionist at heart." James may have been intrusive in his analysis of a young girl's feelings, but Zola, by comparison, excised the organs that contained them. Instead of "tumbling about a heroine's little household treasures," French naturalism confronted "the contents of a laundry maid's bus-bucket thrown to the ground for Microscopic analysis." It was worse than realism, it was realism in the wrong part of town.

Zola stated explicitly what Lang already understood to be implicit in modern literary trends: the imaginative realm of fiction was being colonized by the rational principles of science. This was being done not by the scientists themselves but by elite, literary intellectuals who imitated scientific authority, even proudly so. Zola, Henry James, and Walter Besant, who all invoked an "art of fiction," eschewed the word "creativity" and emphasized instead an author's observational and descriptive powers. Each recommended, as central to the equipment of the aspiring writer, not a fertile imagination, but a "notebook that the learner must carry always with him, into the fields, to the theatre, into the streets."[21] Lang took an ominous view of any authorship casting the writer as researcher-of-moral-disease, turning the entire realm of literary endeavor into "a second-rate science."[22] Amplifying the threat, for Lang, was that such literary elites had the cultural authority to impose their artistic hierarchy, setting it up so that the mature novel gained value by confronting reality, while the childish adventure story lost value by escaping it.[23] Lang sought to change this script, lowering James's gender status not merely by infantilizing him, but by unmanning him altogether. Calling James a girlie man writing for girlie men was the time-honored strategy of the intrapatriarchal dispute, the go-to method for contesting the terms of masculine power. Such charges of effeminacy were sure to get under the skin (especially thin skin) of one's adversaries, depriving them of their literary sanctity, and making Lang, according to Watts-Dunton's testimony, universally loathed. But while history understands why the realists hated Lang, given this record of ridicule, it is less clear why Lang hated the realists.

Understanding this requires a cross-translation of his writing within the larger field of his endeavor. By the mid-1880s, Andrew Lang, in addition to his flourishing career as a national man of letters, was beginning to gain international renown as an anthropologist, first with *Custom and Myth* (1884) and then with *Myth, Ritual, and Religion* (1887). As both a literary critic and a cultural anthropologist, Lang took storytelling and its significance as an object of deep philosophical concern, holding forth on two distinct but related battlefronts, one a defense of ancient myth, the other an attack upon the modern realist novel. For Lang, anthropological and literary narratives similarly debased the imagination and the mind. In the 1870s and 1880s, when cultural anthropology began to take off, looking to find the clues to human nature and the archetypal mind, the cutting edge of such research was located at the Salpêtrière and elaborated through *la grande hystérie* and *le grand hypnotisme*. Charcot's neuropathology was overall a gloomy glimpse into the human mind, built upon the medicalization of madness in the midcentury and inscribed into the female brain and body as neurosis, neurasthenia, anorgasmia, kleptomania, hysteria, and other mental ailments, largely built upon female patients, of whom Charcot had a seemingly endless supply. This contemporary psychiatry, with its highly gendered and degenerative diagnostic manual, also informed the naturalist aesthetics of modern literature, giving it a kind of medical authenticity. Male authors like Zola, James, and Gustave Flaubert wrote compelling "case studies," such as *Nana, Rose Arminger*, and *Madame Bovary*, respectively, from an almost psychiatric point of view, exploring their fictional women's mental and emotional states with extensive analysis. In literature as in medicine, women were the ones predominantly afflicted, but men came under similar threats to their sanity through contaminations of their virility.[24] But if human nature was the frailty, society itself was the disease, allowing male realists to confront grand sociological themes in the broken bodies and minds of their doomed heroines.

As a literary critic, Lang thought *Nana* was a lousy read, but as an anthropologist, he saw it as a deadly tale. This turned a potentially polite literary debate into the all-out donnybrook of "Realism and Romance." At stake was nothing less than the kind of grand narrative humanity would tell about itself. Were the myths and stories shared around the campfire to sound as if they had been pulled out of a medical file or clipped from the London *Times*? Lang's tirade against stories *about* women was really a tirade against the stories being told *through* them. These were stories about disease and despair, divorce and deception, and perverse rather than pure romantic sexualities. They were narrated in the claustrophobic settings of domesticity, where characters faced not heroic dangers but psychological and moral endangerment. A good adventure story, in contrast, was a salubrious outing for the mind. Instead of this medically anxious discourse of introspective fiction, repeatedly asking "what is wrong?" adventure stories moved the reader out of the sickrooms, the slums, the cloistered drawing rooms, to roam free on a continent "without a petticoat in sight."[25] For Lang, the question of what was the more authentic form, the novel or the romance, was not really about literary aesthetic criteria, but rather about the psychic experiences derived from reading such material. This was not a misogynistic discourse about controlling women's bodies, but a metaphysical discourse about freeing people's imaginations, a critique that pertained to both literature and anthropology.

While Lang submitted himself to the strict bridle in his anthropological scholarship, he let his imagination wander free when reading the anthropological romance, a genre he personally helped launch in 1885 with the promotion of Haggard's *King Solomon's Mines*, and to which his lasting infamy is particularly tied. Lang was perhaps more unpopular for his praise than for his insults. In response to Lang's glowing review of Haggard's *She* (1887), the critic Thomas Watson penned the furious "Fall of Fiction" (1888) to expressly abuse Lang's protégé. In clear reference to Lang, Watson wrote, "When we have abundance of exquisite grapes in our vineyards, is it not almost incredible that persons who pretend to some connoisseurship should be content to besot themselves with a thick, raw concoction, destitute of fragrance, destitute of sparkle, destitute of everything but the power to induce a crude inebriety of mind?" This intoxication was recklessly regressive. "Firstly," Watson complained, "there is the physically revolting circumstantial narratives of massacre, cruelty, and bloody death. Secondly, there is the element of the fantastic preternatural and generally marvelous."[26] But for Lang, the dark magic coiled up in Haggard's grotesque fantasy of Africa was not a departure from reality but a potentially deeper penetration of it. At the same time he was boosting the flourishing new genre of the anthropological romance, he found himself fighting for the soul of disciplinary anthropology itself, trying to restore some of the primal mystique to human history deprived of it by positivism.

Haggard's novels were particularly appealing to Lang "because of their air of reality," which assisted Lang's project on many levels, as both a disciplinary proxy and an alter ego. Having served in the colonial administration of Natal from 1878 to 1880, Haggard wrote not like a writer but like a witness, making readers "believe in the impossible and credit adventures that never could be achieved." He did not dream up his romances; rather, "romance tells Mr. Haggard her dreams beside the camp fire in the Transvaal among the hunters on the hills of prey and he repeats them."[27] Haggard had not only physically occupied the lands of Lang's imagination, he had even swashbuckled his way through them, organizing an impromptu cavalry to ride to the rescue of fallen soldiers in a remote outpost at Natal. In a review of *She* (1887), Lang enthused, "Mr. Haggard's practical knowledge and experience of savage life and wild lands, his sense of the mystery and charm of ruined civilizations, his appreciation of sport (especially with big game) his astonishing imagination and a certain *vraisemblance* makes the most impossible adventures appear true through imperial setting."[28] Part of the seduction effect for Lang was exactly this wealth of ethnographic detail that perfectly mimicked anthropological descriptions. And no wonder: the frontline of ethnographic accounts came from men like Haggard in the colonial service, who offered their unprocessed "eyewitness ethnography," sure to appeal to the London-based "amateur of savage life," who had never been south of France.[29] Such textured accounts were then taken by academics, squeezed of their juices, and turned into data, to be put in the service of theoretical debates expressly designed to explain away their mystery. But with Haggard, anthropological observations were dropped along a path leading to an alternative reality, where nature was not regulated by deterministic laws, but worked a terrible magic, transmogrifying humans into beasts, raising the dead, and immortalizing the living. Africa was covered over with the "grotesque upholstery of charnel-caves, anthropophagous banquets, pyramids of human skeletons and revels in which the dancers carry corpses for flambeaux."[30] Watson penned that line as an indictment, but romancers took it as praise. As Stevenson put it in his own "Gentle Remonstrance" of fastidious realist aesthetics: "Life is Monstrous, infinite, illogical, abrupt, and poignant; a work of art, in comparison, is neat, finite, self-contained, rational, flowing, and emasculate. Life imposes by brute energy like inarticulate thunder . . . a proposition in geometry does not compete with life."[31]

While *King Solomon's Mines* might not be high art, it was, by this account, a far better imitation of life.

Patrick Brantlinger dubbed this genre of grotesque supernaturalism "the imperial gothic," noted by literary critics for being primarily offensive to women rather than to racialized men.[32] In this analysis, it is gynophobia that drives the male adventure tale, with Africa as a reclining female body to be explored, dreaded, and conquered.[33] But by Lang's account, he himself thought "Mr. Haggard's savage ladies are better than his civilized fair ones . . . As for *She* herself, nobody can argue with a personal affection which I entertain for that long lived lady."[34] A statement quoted here not as evidence of feminist leanings, but to signify that this was a relationship of attraction and not dread: the anthropological romance catered to the love interest of men looking for a little strange. (The more conventional Watson was repulsed by any such *amour* with a two-thousand-year-old "commonplace coquette.")[35] Lang not only sought vicarious pleasure in the adventures of Allan Quatermain, the white explorer, but with even greater gusto, he longed for total immersion in the "savage" experience, slipping fully into the replenishing puissance and pulchritude of the tribal warrior: "The natural man within me, the survival of some blue painted Briton, [is] equally pleased with the true Zulu love story. If I were wholly civilized and cultured to the backbone that feature of the savage tale would have failed to excite."[36]

Lang's appropriation of the other for one's private fantasies, swapping out the grim realities of empire with obtuse romantic scenery, has made Lang the exemplar of political incorrectness that critics love to hate. Nonetheless, there are latent, secondary, and potential meanings of race and gender that Lang deploys in complex, even subversive ways. Lang's "savage mimesis" expresses his longing to be other than he is, a confusion of authority and desire that could not be readily accommodated in his scholarship.[37] The anthropological romance, in contrast, was a potential space in which to cultivate his heterodoxy and arrange new schemes of dominance and desire. As Homi Bhabha explains it, even unilateral power masks a certain intrapsychic complexity, stoking the prerogative of pleasure and its tendency toward fetish and fantasy. Within this "imaginary plenitude" colonial identity loses its "fixity" across "a repertoire of conflictual positions." Haggard's Africa does not so much create a fantasy as clarify it amid the polysemic confusion of this colonial discourse constructed through "tropes of fetishism-metaphor and metonymy."[38]

Hitching a ride in Haggard's fantasy let Lang travel to a world populated with "noble savages" and suffused with supernaturalism, without abandoning the imperious throne of the Oxford armchair anthropologist. Ali Behdad writes, "Such a desire makes the orientalist subject surrender his or her power of representation and pursuit of knowledge by becoming a hedonistic participant in the immediate reality of the oriental culture."[39] Lang tries to have it both ways: he is at once the anthropologist creating and controlling knowledge about the imperial object in affirmation of Western identity, and the romantic subject embracing his shadow "desire for/of the orient" that questions that identity, exploiting the ambiguities within orientalist discourses to satisfy the contradictions within himself. Clearly, the most important valences of his text should in no way be understood as politically engaged on behalf of "the racial other" (any more than his literary criticism was an attack on women's rights). The status being protected was Lang's: the metaphysical, aesthetic, "romantic" Lang whose cultural authority was pressured by the rising eminence of reason and calculation in elite epistemology. Lang was writing about an ideological crisis internal to the existing male regime of power, while fantasizing about "the racial other" in order to force a "productive split" within white masculine subjectivity that could question the *telos* of scientific modernity rather than affirm it.[40]

A strong element of "native magic" was not just part of the orientalist *vraisemblance* of Rider Haggard's imperial fiction, but an actual element of his experience. Such accounts favorably contrasted the potency of arcane, irrational powers to the mundane efficacies of modern science. Haggard recalled one such incident from his colonial days in 1879 when an old washerwoman, "in a great state of perturbation," ran to tell him that "terrible things had happened in Zululand; that the 'rooibatjes,' that is, redcoats, lay upon the plain 'like leaves under the trees in winter,' killed by Cetewayo."[41] Haggard in this case was merely the colonial bystander, but Lang himself had directly experienced some such brush with the supernatural. In 1886, at the same time he was composing *Myth, Ritual, and Religion*, Lang submitted a personal anecdote to *Phantasms of the Living* describing an event from his college days at St. Andrews. According to Lang's entry, he had been out strolling one evening, when he clearly saw Professor Connington, leaning against a lamppost on Oriel Lane. As Lang walked by him, passing quite close, he called out a greeting, and oddly got no response. He later found out that Connington had died around the time of this sighting. Lang's aesthetic affinity for the power of myth ran deep, intersecting with a life not untouched by mythic powers and with a deep respect for wonder. This kind of firsthand experience also had deep significance for Lang the anthropological scholar. It naturally raised for him the possibility that similar testimony, which wrote across every verse of ethnographic history, might also be subject to a veridical interpretation, a supposition of revolutionary import. Lang had already begun to aggregate and analyze such spiritual encounters in "A Comparative Study of Ghost Stories," which appeared in *Nineteenth Century* in 1885, the same year his occult imagination took flight with Haggard's *King Solomon's Mines*. Of course, he did not assert any psychical or spiritual hypothesis. Academic anthropology, with its ties to evolutionary theory and neuropsychiatry, strictly framed ghostly encounters as delusions known to afflict primitive or otherwise disturbed minds. To embrace psychical phenomena would have thus "lowered" Lang to the level of the peoples he studied, so he stayed close to the script laid down by Sir Edward Burnett Tylor, T. H. Huxley, and Sir James George Frazer, which tamped down on all such speculative flights.

Nonetheless, this *frisson* of psychical and even spiritual possibility must be seen as part of the context in which Lang thought and wrote in the 1880s, embracing the anthropological romance and its magical alterity. "Books for boys," according to Lang, were "legends not novels," reinforcing his identification of romance with Tylor's classification of superstitions as ancient "survivals." But in the mid-1880s, such fantasies were not necessarily fiction, as psychical speculation became more and more part of the regimen of experimental psychology. Lang, like most anthropologists, paid attention to these developments, but took particular interest in Frederic Myers's "Multiplex Personality" (1886), a revolutionary attempt within the field to theoretically incorporate the outstanding data of psychical research and the Société de Psychologie Physiologique (the Salpêtrière's most forward sector). Ancient storytellers imbued their world with such spectral phenomena as second sight, telepathy, and spirit contact. Did these mythmakers merely imagine other realities and other beings, or did they actually commune with them?

Both realism and anthropology actively cut consciousness off from this mythic creativity, adding to the intensity of Lang's battle of the books. Lang's critique of modern literature, not unlike his critique of Tylorean anthropology, was an objection to the extension of scientific values into the irrational realm of imagination. Lang's goal was to reverse this flow of influence and to extend imagination into the realm of science—something he would eventually do through the platform of psycho-folklore. Lang's psychical anthropology did not seek tolerance in the margins of disciplinary investigation, but hoped to seize its platform altogether, confronting the modern dilemma (realism or romance, myth or science, mind or soul) in a way that orthodox anthropology, literary art, and cultural criticism alone could not. Yet these were all vital aspects of Lang's thought, from which he forged the higher synthesis of psycho-folklore. This was an act of philosophical will that can only be known through the unfulfilled aesthetic and ideological drives that make up its history. To understand this, it will be helpful to turn to Lang's subject formation as an anthropologist and *littérateur*, germinated as it was in the hothouse of the Oxford aesthetic movement and his own melancholy Highland heart.[42]

THE AESTHETICS OF ANTHROPOLOGY

Lang was born in 1844, in Selkirk, Scotland, a "borderer" who grew up near to "the half-world of legend, of kelpies in the loch and elves in the green-wood, and of miraculous happenings beyond mortal compass."[43] Even in adulthood, Lang's imagination was shadowed by the fantastic; he confessed himself an *"incredibilium cupitor,* an amateur of savage life, fond of haunt-ing, in fancy, the mysterious homes of ruined races, a believer too in the moral of the legend."[44] He had once danced around an Oxford bell tower intoning from the *Malleus maleficarum* (the "Hammer of Witches"), trying to raise demons and the dead. (His hopes had been raised by a course in medieval classics.) Lang differed from many other SPR members in that his interest in the supernatural did not have roots in a wounded Christianity, but here, in the magical worldview of Highland life. Unexplained phenom-ena were something of a tradition of "the place of the Gaels," particularly in the form of "second sight" and "scrying," which remained widespread folk practices and beliefs in Lang's youth. There was an almost offhand, norma-tive quality about Lang's relationship to the supernatural that differentiated his attitude from that of Lodge and Myers, who were more driven by the need to prove and justify. Magic in Lang's upbringing had been a kind of cultural backdrop, with Christianity and science going about the business of progress while keeping the past always in view.

The romantic influences of his Highland youth informed Lang's deepest nature, that part of him expressed in his poetry, aesthetics, and childhood longing.[45] But Lang's public persona was also acerbic and droll, a critical wit who loved to milk a sacred cow. Despite Lang's conspicuous association with romantic themes (folklore, fairytales, romantic fiction, supernaturalism, and unquenchable longing), he did not outwardly possess the associated charac-ter attribute of the earnest spiritual seeker. But it is there in his wistful poetry pining after the lost wonder of childhood, seeking to regain "the enchanting glimpse of eternity in paradise." Such transcendent yearnings defined the essence of a Romantic disposition, impelling him, as with Myers and Lodge, along a psychical trajectory toward the beautiful and the unseen. Of his own character, Lang remarked, "I have a gay mind, but a melancholy heart."[46] But "Merry Andrew," as he was called in the press for his perpetually jesting manner, often rubbed others the wrong way. To some extent, the umbrage of men like Watts-Dunton and James had as much to do with themselves as with Lang. They not only missed the joke (Lang was, in fact, being funny), they also missed the more serious point, buried as it was in the anthropological scholarship they might, perhaps, have neglected to read.

At Oxford, Lang was something of a wunderkind, setting the madly prolific pace that would mark his publishing career. He could translate hundreds of French ballads, Greek myths, and European folktales for publication, while writing his own original poetry, essays, and stories. In the light of Lang's extraordinary gifts of letters, Joseph Weintraub, a modern scholar of English literature, saw him as something of a failure: "Lang, with his background in anthropology and folklore and his consequent perception of the relationship of romance to myth, legend and epic, had the tools, perhaps as well developed as any critic of his time, to create an aesthetic of romance."[47] Yet, he did not do so. Instead, he conducted this debate in a droll patter that fostered ill will, but no serious scholarly engagement. But Weintraub takes Lang's primary concern to be the world of creative fiction, not the world that fiction creates. That would make Lang's folkloric anthropology his essential sphere, and in that context, Lang was a significant and serious contributor to the debate. Without this hermeneutic, the genius of Lang—such as it was, not as it should have been—is invisible to an academic community that, more than a century later, has long since lost sight of him.

Lang made a study of classical and medieval literature as a graduate student at Balliol College, Oxford, in the late 1860s, after taking his degree at the University of St. Andrews in Scotland. He eventually became a Fellow at Merton College, Oxford, before leaving for London in 1874 to marry Leonora Alleyne and begin the life of a Fleet Street wit. His years in Oxford coincided with the height of the aesthetic movement, while his interest in classical antiquity and the original medieval "romance literature" brought him directly into the inner circle of the art critic Walter Pater, which included John Ruskin, Algernon Swinburne, Dante Gabriel Rossetti, Edmund Gosse, and William Morris, among other neo-Romantic artists, critics, and aesthetes of the period. Lang would have his anthropological awakening within this highly perfumed intellectual milieu, where pleasure was the highest calling of language, and poetry and myth its most potent forms.

Under the spell of Pater and these other British aesthetes, Lang became "A la-di-da Oxford kind of Scot," who went about with an eyeglass and a wan air of decay, collecting blue china and Renaissance furniture.[48] But the flow of influence went the other way as well. Through this friendship, Pater acquired an anthropological dimension to his romanticism (what the literary historian Robert Crawford calls Pater's "anthropological romanticism"), while Lang's critical interest in myth and folklore set the course for a decidedly romantic anthropology.[49]

Crawford argues that Pater's friendship with Lang proves the aesthetic movement was not as willfully "cut off" from scientific sources as has been thought. However, it is questionable how much of a "scientific source" one could consider Lang at this point, and perhaps this relationship can be just as fruitfully read as an aesthetic source for science. Certainly, Pater's idea of consciousness as a whirlpool of perception that attained its highest subject formation through aesthetic ecstasy offset the gloomy influence of medical psychiatry and evolutionary psychology on Lang's early anthropological tutelage by Tylor. This took the creative mania that modern science had mistaken for madness and gave it an elevated role far above reason in self-determination. Pater's fundamentally libidinous and aesthetic understanding of the act of subject formation imposed its own evolutionary "selective criteria" in the assertion of its point of view, which could then be projected onto the world as art. Within the code of this friendship, Pater and Lang encountered a virile, vital past as a potent source of this will-to-beauty, celebrating the intense and even violent emotions through which the self was brought into being. Certain episodes of artistic achievement, for Pater—specifically, the classical impulse in the Renaissance—expressed not just high cultural creativity but the power to create culture itself, enacted through these dominant, imaginative subjectivities imposing change upon the world.[50] However, Lang, through his growing interest in anthropology, expanded the framework of Pater's aesthetic theory to more fully incorporate distinctly non-Western, preclassical forms, extending the search for this Hellenic wellspring of civilization to include mythologies throughout time and around the world. Lang had what he called his first "savage epiphany" in 1869, while reading an essay by the anthropologist J. F. McLennan on primitive totemism: "It led me to examine the other fragments of antiquity. The poetry and the significance of them are apt to be hidden by the enormous crowd of details. Only lately we find the true meanings of what seems like a mass of fantastic, savage eccentricities."[51] Tribal myths, like Norse sagas, Sumerian epics, and medieval ballads, along with Sir Walter

Scott's *Ivanhoe* and James Fenimore Cooper's *The Last of the Mohicans*, were an archetypal celebration of potent, primal virility, expressing a universal troubadour tradition born on the battlefield amid blood and danger. Heroes were men. Romanticism was phallic, expressing its "joy of living" in its love of regalia, excitement, self-exhibition, and boasting. In this lyrical take on violent masculinities, *l'homme sauvage* could be celebrated alike for his physical "savagery" and for the boldness of his mythic creativity.

In contrast to this aesthetic view, the dominant cultural tendency elaborated in the wake of evolutionary theory was to render the past a fundamentally tainted source of male self-imagination. To be "savage" was to be regressed, and in a unilinear model of evolution, what could be worse? If the best is yet to come, then the worst is still behind us, and it is a threat that must constantly be outrun. Scientists imposed atavistic interpretations on criminality, idiocy, and hysteria, placing new hereditary stigmas on mental illness and mental disability. Newspapers gave this evolutionary angst a more sensational twist with tales of fur-bearing human bodies and boys raised by wolves, blurring the line between man and beast. "Savage exhibitions" put tribal peoples on display in living dioramas complete with educational notes describing a keener sense of smell, a quadrupedal tilt to the pelvis, protruding jaw, night vision, smaller brain casing, and larger teeth.[52] Explorers sponsored by academic, government, and scientific organizations collected their ethnographic data with calipers, tape measure, and plaster of Paris, ignoring culture as part of the ecology in the fevered pursuit of racial taxonomy. As Johannes Fabian points out, this quantitative "objectivity" foreclosed the possibility of any kind of mutual encounter between two "peoples," reducing them to observer and observed, instrument and anatomy.[53]

While Tylor and J. F. Maclennan somewhat turned the tide on this with the beginnings of a distinctly "cultural" anthropology, they could not escape a fundamental pessimism about the past, captured as they were by evolution's developmental framework.[54] But Lang occupied his own ideological fortress inside the Oxford classics department, and it was here that his interest in anthropology first took shape, with its own sense of intellectual purpose and approach to methodology. Where evolutionary anthropology muted its subject, Lang's folkloric approach took expression as its principal concern, focusing on the stories people shared rather than the taxonomies that divided them. Myth had for Lang a sacred quality as the tabernacle of a people's indigenous voice. To hear and understand this voice opened up the possibility of an imaginative cathexis between oneself and the other. But this was also the search for archetypes, a deep-down identification that went beyond "mutual understanding," to erase distinction altogether. Part of the appeal of this study for Lang was to extend the ambition of his literary criticism. Anthropology was not just about reading texts but about writing the universal story of man. (That was the hope; the reality of this scheme tended to be more autobiographical, in the form of "You are me, but a lesser version of me.") In the developmental model of anthropology, ethnographic histories were all collapsed into one narrative of progress, charting a single path for social evolution en route to scientific modernity. This meant that the entire history of the modern West was still extant in the various contemporary cultures arrayed along this spectrum. Lang and Pater did not have to content themselves with torn pages and broken vases to reconstruct the "masculine pulchritude of the past"; it had been conveniently preserved for them in the warriors of Africa or the surviving Samurai of Japan. This gave *vraisemblance* even to theories touching upon those otherwise physically untraceable roots of humanity's most ancient history: "We are not obliged to fall back upon some fanciful and unsupported theory of what primitive man did, and said and thought," Lang enthused.[55] Thanks to the indigenous populations of empire, prehistoric man was only a ship ride away. But the past also lies within us. "Nature," according to Lang, "leaves us all savages under our white skins."[56] This savage did not pose the usual threat of evolutionary regression, but rather offered an anchor point to prevent the further degeneration predicted by Havelock Ellis with the inevitable feminization of the modern male: "The large headed, delicate-faced, small-boned man of urban civilization" was now "much nearer to the typical woman than he is to the savage."[57] The racial Adonis of Lang's anthropological speculation and literary lust remained untainted by the marks of effeminacy. In the stories the Zulu told about themselves (and the stories Haggard told about the Zulu,) modern man could partake in the "joy of adventurous living," a zest belonging to a warrior's virile biology rather than a colonial administrator's sedentary ways.[58]

Both Pater and Lang have been linked to the writer Charles Kingsley's "muscular Christianity," but this is no guide to Paterian poetics or "romance and realism." Kingsley's no-nonsense men of action had "romantic feelings" of a strictly matrimonial and appropriately goal-setting kind.[59] These were rather aggressively heterosexual, can-do pragmatic heroes (and fathers) who brought the foreign and the strange under their own firm control, extracting the prosaic from the sublime.[60] They did not scramble into the jungle to hide from civilization, they went in order to make a clearing for it. This outward engagement of the here and now is not to be compared to a literature of aesthetic longings directed inward, upward, or to the past, which took as its object the masculine rather than the feminine ideal. The Oxford aesthetic movement pursued its refinement of intellect, beauty, and pleasure in contexts that were decidedly homosocial and tacitly homoerotic, and this tendency found its fullest bloom in the corner claimed by the Oxford aesthetic movement.[61] To what extent it was actively homosexual remains unclear, but at the very least, emphasis on the physical pulchritude and athleticism of the (male) Hellenic body was built into the basic curriculum of the classics department. Aesthetes in Pater's circle turned this academic sensibility into a kind of personal and social mystique, and eventually, through their work, elevated it into a cultural movement. Pater is known to have had a relationship with another Oxford undergraduate, but matters are less clear for Lang. Lang refused to allow any sort of biography to be written about him, and ordered all his letters destroyed. Thus, much of his personal life remains behind a carefully cast screen. Lang's eventual biographer, Roger Lancelyn Green, hints at some sort of issue with sexual orientation.[62] While this insight could be important to understanding his subversive attraction to the anthropological romance, and add to the stakes in this intrapatriarchal dispute, it is something of a dead end. He both covered his tracks and distanced himself from that heady social atmosphere after leaving for London and putting himself in the public eye.[63] Regardless of what may or may not have passed between Lang and Pater, Lang's early incubation in Pater's aesthetic enclave helped push him further outside the mainstream and anchored his anthropology within this unique perspective. The synergies of their relationship were highly productive for both Lang and Pater, helping to formulate a decidedly subversive masculine aesthetics that flowed through their writing into late-Victorian culture.

Lang was formally drawn into the scholarly debates of anthropology as a participant when he attended a lecture given by E. B. Tylor at Oxford in 1872. Lang was astounded by what he called Tylor's "masterpiece," *Primitive Culture*, and immediately began to recognize the relevance of Tylor's work to his own study of folklore.[64] Tylor had amassed in a single text an unprecedented amount of ethnographic data drawn from around the world, including European folktales. Through systematic comparison, Tylor was able to transcend the merely ethnic, tribal, or national inferences, to offer instead a "science of man." In so doing, Tylor initiated a methodology for a thrilling new anthropology, one built upon the analysis of culture that aimed to reconstruct the evolution of society. *Primitive Culture* inspired a powerful coterie of leading anthropologists, including Henry Maine, John

Lubbock, J. F. Maclennan, Robertson Smith, T. H. Huxley, and Herbert Spencer, to pursue Tylor's new disciplinary project within the methodological framework he provided.

Lang wanted in; he was on fire with Tylor's theory of "survivals," appealing as it did to his passionate appetite for atavism. Through some Celtic wolf tale or Zulu prayer, he could enter into his own enchantments, nursery rhymes, ghost stories, and jump-rope jingles still tucked away in the cultural interstices of modernity. The elemental structures of human thought d were conserved in storytelling's narrative remains, , the living imprint that evolution had left on language. These stories still formed the sacred bedrock for the tribal cultures that ringed the empire, making modern-day anthropology a way for Lang, a folklorist, to directly grasp an element of the past in the study of the present. Whether it was African folktales, pre-agrarian superstitions, Greek mythology, stories of the saints, or high Vedic Hinduism, myths expressed the cultural logic appropriate to their phase of social development, stages that were then neatly arrayed along a unilinear progression toward the *telos* of European modernity. Culture was thereby given a Darwinian evolutionary framework, and implicitly grounded in the uniform determinants of human biology and psychology. In this developmental model, today's "savage" was yesterday's Briton, and this ethnographic mirror let Lang gaze upon the face of his own prehistory.[65]

Another part of the appeal of Tylorean anthropology for Lang was its stark rejection of Max Müller's then dominant theory of Aryan diffusionism. Müller's famous thesis posited that the Vedas were an exalted source for the diffusion of legends, myths, and folktales along the branches of the Indo-European language tree. The dross they left behind was nothing more than the garbled remains of a once noble Vedic storyline drifting along the currents of migration. The hold of diffusionism was strong. It explained both the irrationality of myths (their meaning had been lost in translation) and their similarities (they all arose from a single source), and had a complex appeal. It offered a reconstruction of pre-antiquity in the absence of an Abrahamic timeline, and allowed textual analysis (a strong suit of biblical scholarship) to come to bear on the problems of their discipline, suggesting a way past the limitations of hominid bones and shards of pottery. This all made diffusionism a kind of Casaubonian key to world mythology, an ambition pervasive enough among contemporary scholars of antiquity that George Eliot saw fit to ridicule it in her novel *Middlemarch* (1871–72), depicting a romantically feckless husband lost in this same vortex of self-importance. Lang had never been impressed: "After taking my degree in 1868, I had leisure to read a good deal of mythology in the legends of all races and found my distrust [formed earlier in his undergraduate days] of Max Müller's reasoning increase upon me . . . I very well remember the moment when it occurred to me, that the usual ideas about some of these matters were the reverse of the truth, that the common theory had to be inverted."[66]

That moment came as an epiphany during Tylor's lecture. With Tylor's comparative method, Lang recognized a powerful tool to bring down the Aryan hypothesis and liberate the authentic voices buried beneath this suffocating conceit. The deepest nature of man could be discovered in the stories he told. These primal tales of lost peoples were words as old as the bones themselves and promised to give Lang a way into a past once thought irrecoverable: "The student of this lore can look back and see the long trodden way behind him, the winding tracks through marsh and forest and over burning sands. He sees the caves, the camps, the villages, the towns where he has tarried, for shorter times or longer, strange places many of them and strangely haunted desolate dwelling."[67]

Lang was more than one such "student of this lore," he was arguably its master, with a knowledge of mythology and insight into the structures of language perhaps unique in his generation. Finally, Müller's theory had met its match in a prodigy of comparative literature whose philological prowess matched Müller's own—someone who could read across ethnic traditions and historical episodes to find meaningful patterns that others, including Müller himself, did not see. In 1873, Lang launched the opening salvo of his attack on Müller in the *Fortnightly Review* with "Myths and Fairytales," beginning an argument that was to rage over the next three decades. That Lang—a nobody in the world of Sanskrit philology, coming from a second-rate discipline of folklore—should dare to attack the towering machinery of German philology was in itself an act of mythic valor.[68] He was at first easily ignored as an interloper and kept on the margins of official debate, but Tylor was impressed with Lang's prodigious scholarship and with the strength of his arguments, which pinpointed the nonsensical aspects of Müller's argument that others were too impressed to see. The cultural contact required by diffusion was out of the question for many of the cases Müller cited, and those correlations too obviously off the pathways of Aryan migration were ignored. Beyond these methodological objections, Lang was philosophically offended by Aryanism. Müller attributed folklore to "the disease of language," an infection spread throughout the peasant tale by misunderstanding and mispronunciation, resulting in scurrilous nonsense. In "Myths and Fairytales," the young Lang set out to rescue the *vox populi* of folklore from the disdain of Müller's elitism, establishing the defining purpose of his anthropological career. He redefined folklore for anthropology as the self-originating voice of a culture and a people: the foundation upon which higher forms arose and not the debased stratum to which they sank. In Lang's universe, to aim high—to get at "the stuff of all our poetry, law, ritual"—one must first search low, looking to the voice of the people:

> The natural people, the folk, has supplied us, in its unconscious way, with the stuff of all our poetry, law, ritual; and genius has selected from the mass, has turned customs into codes, nursery tales into romance, myth into science, ballad into epic, magic mummery into gorgeous ritual. The world has been educated, but not as man would have trained and taught it . . . We have a foreboding of a purpose which we know not, a sense as of a will, working, as we would not have worked, to a hidden end.[69]

Lang's notion of a will that works "to a hidden end" was not one that could readily be incorporated into British evolutionary anthropology, even in its cultural variety. But German literary critics like Novalis and Hegel had recognized such a creative spirit working its way through literature, with the fairytale and other such *Märchen*-forms as a kind of perfected vessel.[70] Such ideas were familiar fare for a well-read folklorist like Lang, providing a way of explaining similarities between myths without yielding to Müller's diffusion thesis. Tylor gave Müller's thesis a more practical modification with the "borrowing theory," which allowed myths to migrate between cultures as long as they were at compatible stages of development. Lang, in contrast, refused to surrender an inch of ethnic authenticity to the idea of foreign contact, sticking fast to his own thesis of "independent invention," sometimes beyond the point of all probability. Here Lang's reason was clearly mystical, not practical, in his anxious defense of the indigenous tale. Later, Lang had to explain away this somewhat irrational stance in order to "get right" with his discipline. He did so in the preface of *Custom and Myth* (1884), his first serious work of anthropology: "In 1872, I was probably more under the influence of Hegel than at present, and I may have somehow been inclined to a mystic theory of *Märchen*-forms, everywhere present in the human intellect . . . but I have since become more inclined to the borrowing theory." (Putting aside love poetry and fine china was not the only backpedaling Lang had to do from his unconventional past.) While Lang eventually did let go of mystical *Märchen*-forms, he would not surrender the mythic status of folklore, holding that unwavering position even under Tylor's mentorship.

Tylor, like Müller, regarded Lang's beloved European folklore as occupying the basement of cultural production, but in *Primitive Culture* he approached things from the standpoint of development rather than diffusion and decay. In other words, where Müller's hypothesis had denigrated European folklore as somebody else's garbage, Tylor treated it as Europe's very own "leftovers." It was rudimentary and backward, but still an authentic expression of the earliest (and the most primitive) thought of a people. As such, these "survivals" (Tylor's term for folktales, myths, and legends) showed the cast of consciousness of a long-ago age, bearing the lingering impression of "the mental condition of earliest man," and offered a glimpse into the past. Unlike Lang and Pater, Tylorean anthropology was not looking for culture's engendering myth; rather, it sought universal stages of social development, trying to lock in the sequence of a scientific formula. As Tylor took the lead in the field, English anthropology became more uniform in its outlook on religious development. Müller's sentimental thesis that religion was born of "man's sense of the infinite" was discarded alongside Vedic diffusion. To be fair, Müller was not claiming any "divine revelation." However, he did commit the intellectual sin of arguing that man was innately driven to search for a higher divine truth (which Auguste Comte had already made clear was a waste of everybody's time). Instead of this sighing, contemplative creature offered up by Romantic German philology, the British saw the primitive mind at work with a kind of practical industry. Religion was an effort to manage the threat of unknown natural forces with knowledge seemingly useful to human survival. Thus, as broadly thematized from Tylor to Frazer, religion was a kind of bad science, just as magic was an inferior technology. The "moral considerations" that Müller attached to qualifying religious thought had no place in this evolutionary framework. In its place was a kind of moral deception, where individuals promoting their own self-interest encouraged others to sacrifice themselves by calling it "good."

In *Custom and Myth* (1884), Lang was able to cautiously challenge some of these grim, utilitarian conceits without breaking with the evolutionary storyline driven by practical survival. The book's emphasis on customary behaviors rather than religious belief tilted the discussion toward mythic function, getting a jump on Émile Durkheim's sociological analysis. As early as 1884, Lang recognized religion as a social force, galvanizing the formation of community around its matrix of values. These ideals were inculcated through storytelling and myth, helping, in the words of Lang, "the Tribe, the Family, Rank, and Priesthoods to grow up, and to form the backbone of social Existence."[71] This made the fundamental utility of religion its supply of moral vision, not rational explanation, thus moving it out of the frontal lobe and giving it a deeper seat in human nature. This was about meaning, not understanding.

Lang followed *Custom and Myth* (1884), with his most remembered work, *Myth, Ritual, and Religion* (1887), which brought more original thinking to the venerable conceits of developmental anthropology. Even as Lang recognized in his preface that "mythology was a thing of gradual development corresponding in some degree to the various changes in the general progress of society," he did not agree on how that progress first proceeded. In a direct challenge to Tylor, Lang denied the animistic thesis. Religion did not arise through the observation of the animating forces of nature all around early man, but was a response to a unique cognitive experience arising naturally from *within* him. (Thus, Lang anticipated both Durkheim's functionalism and the *homo religiosus* of Mircea Eliade.) That "epiphany" was not accompanied by an awakening of some moral or mystical sense, a thesis already scorned by the utilitarians Lang was trying to court. It derived instead (at least according to the Lang of 1887) from "the idea of Power," a "recognition of God's divine magnitude," producing a religious belief system Lang characterized as "selfish, not disinterested."[72] But still, recognizing the power of God and wanting to use it are as different as Heaven and Earth. Lang's theory, unlike Tylor's, was arguing for a distinct divine principle, God as something greater than ourselves and of an entirely different (divine) order. (Tylor's God aggregated "God" out of lesser deities, who were projections of ourselves, making "him" only us writ large.) Lang's theory also advised some humility on the part of science when approaching questions of God: "No man can watch the idea of God in the making or in the beginning . . . since the actual truth cannot be determined by observation and experiment, the question as to the first germs of the divine conception must here be left unanswered."[73] Lang was not just questioning the propriety of inquiring into the "origins of religion" because there were no "eyewitnesses." He was also implying that such a mystery was beyond the powers of anthropology to observe.

There is also the provocative challenge, appended to the end of *Myth, Ritual, and Religion* (1887), that Tylor's grasp of the anthropological record was incomplete. In Tylor's "ghost theory" of the human soul, a component of animistic religion, the belief in an indwelling human spirit arose from "the reasoning of early men on the phenomena of dreams, fainting, shadows, vision caused by narcotics, death and other facts which suggest the hypothesis of a separable life apart from the bodily organism."[74] The belief in spirit deities was for the most part based on the empirical observation of natural events, but the belief that we ourselves possess a spirit is a matter of delusion and confusion. While Lang did not deny the merits of this as a partial theory, it was not complete, he acknowledged, because "it fails to add the kind of facts investigated by the psychical society":

> Such facts as the appearance of men at the moment of death in places remote from the scene of their decease, with such *real or delusive* experiences as the noises and vision in haunted houses, are familiar to savages. Without discussing these obscure matters, it may be said they influence the thoughts even of some scientifically trained and civilized men. It is natural, therefore that they should strongly sway the credulous imagination of backward races in which they originate or confirm the belief that life can exist and manifest itself after the death of body.[75]

Given that Lang himself had had one such encounter, he is very cryptic in defining the nature of these psychical events. In some sense he is merely emphasizing how altered-state experiences would overpower the primitive mind, which does not depart from Tylor's point. When Tylor set out to determine the "mental state of early man," he had to factor in all of man's conscious (and unconscious) conditions as well, which included an unsavory class of phenomena thoroughly vetted by the extensive medical studies and thus a legitimate form of causal speculation regarding the origins of religion. Mechanical paradigms, such as "unconscious muscular cerebration," neuroelectrical seizure, or psychiatric accounts of "brain disease" validated the phenomena as real, without encumbering them with any metaphysical baggage. But Lang throws in the phrase "real or delusive," reopening a door slammed firmly shut by Tylor: that spirits had a factual existence. Such a conceit pertained to primitive culture only and was fully incompatible with the modern scientific mind. It was, therefore, an impossible thesis for the discipline of anthropology to even remotely entertain. But Lang was not so sure, although currently there was no such platform.

VIRILE MAGIC

In 1894, Lang released *Cock Lane and Common-Sense*, his manifesto and field guide for a new kind of cultural anthropology. This was a hybrid discipline integrating the philological analysis of folklore studies, the psychological and evidentiary frameworks of psychical research, and the comparative methods of Tylorean anthropology. He dubbed it "psycho-folklore" to mark its revolutionary intersection of subject matter and methodology. Upon publication, Lang lamented in a letter to a friend, "I have scandalized the folklore society etc." But it was Lang's original thinking that had rescued British folklore from German diffusionism a decade earlier, and here he merely applied the same interdisciplinary facility and panoramic grasp of the data—sweeping from high to low, from culture to culture, from past to present, and jumping between literary forms and ritual customs—to make connections others could not see. This allowed him to confidently assert that "such evidence as we can give for the actuality of the modern experiences will, so far as it goes, raise a presumption that the savage beliefs, however erroneous, however darkened by fraud and fancy, repose on a basis of real observation of actual phenomena."[76]

Such a possibility was unlikely to alter the entrenched skepticism of Tylor, who referred to spiritualism as "a monstrous farrago," but Lang was no longer committed to adapting himself to the framework of *Primitive Culture*. What had felt revolutionary over two decades ago was now becoming its own irrational superstition. Lang tried to put it to his reluctant colleagues as plainly as possible: "It was stretching probability almost beyond what it will bear, to allege that all the phenomena, in the arctic circle as in Australia, in ancient Alexandria as in modern London are always the result of an imposture modeled on savage ideas of the supernatural."[77] Psycho-folklore surveyed the entire field of anthropological evidence, without filter or exclusion, without foreclosing obvious possibilities:

> Such things as "clairvoyance," "levitation," "veridical apparitions," movements of objects without "physical contact," "wrappings," and "hauntings" persist as matters of belief in full modern civilization and are attested to by many otherwise sane, credible and even scientifically trained modern witnesses. To what extent are some educated modern observers under the same illusion as red men, kaffirs, Eskimo, Samoyeds, Australians and Maoris? . . . Even if we could, at most, establish that people like Iamblichus, Mr. Crookes, Lord Crawford, Jesuits in Canada, professional conjurers in Zululand, Spaniards in early Peru, Australian blacks, Maoris, Eskimo, cardinals and ambassadors are similarly hallucinated, as they declare, in the presence of priests, diviners, D. D. Homes, Zulu magicians, biraarks, jossakeeds, angakut, tohyngas and saints and Mr. Stainton Moses—still the identity of the false impression is a topic for psychological study. Or if we disbelieve this cloud of witnesses, if they voluntarily fable, we ask why do they all fable in exactly the same fashion?[78]

Lang referenced the SPR when he mentioned "otherwise sane, credible and even scientifically trained modern witnesses," using the society's growing stature to support the plausibility of less reliable testimony. The scientific character of the SPR was itself a direct challenge to Tylor's developmental orthodoxy; elites capable of conducting such a disciplined program of rational inquiry should be immune to such spiritual curiosity. To dismiss all such "savage beliefs" as a result of irrationality, credulity, or deception necessarily touched upon the reputations of these highly respected academics as well—maneuvering Tylorean anthropologists into an orientalist dilemma par excellence. In his preface to the book, Lang asserted for the first time his total independence from these positivistic orthodoxies, confronting rather than working around the biases of Tylorean anthropology after two decades working in the field. *Myth, Ritual, and Religion* had brought him to the center of the discipline, and he was ready to fully exploit the opportunities his assiduity had created, clearing a new, unencumbered space from this vantage point, properly conducted. Lang even called upon his wit to write down his disciplinary rivals, the scientists of "common sense," a tone he usually reserved for fiction and Fleet Street: "Of all the wanderers in Cock Lane, none is more beguiled than sturdy Common-sense, if an explanation is to be provided. When once we ask for more than 'all stuff and nonsense' we speedily receive a very mixed theory in which rats, indigestion, dreams, and of late, hypnotism, are mingled much at random', for Common-sense shows more valour than discretion when she pronounces on matters (or spirits) which she has never studied."[79]

This science, "the science of common sense," interestingly, was a "she." It was the science of the safe and obvious, and yet totally blind in all it saw. Like the realists of literature, the rationalists of science found themselves unmanned by their lack of imaginative vigor and their dearth of intellectual courage, attributes that Lang's aesthetics cultivated as fundamentally masculine energies. Common sense, instead of boarding the ship bound for the new world, rowed a toy boat around a tub, circling its past discoveries. Such a science no longer served to enlarge upon knowledge, but rather worked to keep it small. This protest paralleled Lodge's address to the BAAS (1891) and Frederic Myers's "Science and a Future Life" (1891), which likewise urged psychical research as a heroic alternative to continued confinement in an increasingly impotent creed. But, Lang urged, "if there be but one spark of real fire to all this smoke then the purely materialistic theories of life and of the world must be reconsidered."[80] Tylor's assertion that "the first shamans were lunatics and sufferers from convulsive disorders or hypnosis, who communicated revelations during the fits of their disease," plainly observed the correspondences between ancient shamanic and modern medical phenomena, but failed to calculate the deeper possible significance.[81] Tylor had the "common sense" to acknowledge these "barbaric parallels to our modern problems of this kind . . . but he does not ask, 'Are the phenomena real?'"[82] The scorn of Tylor's attitude had cowed anthropology into a state of intellectual timidity, not just denying the obvious but refusing even to see it. Lang set out to ensure that it was no longer possible to do so.

Lang "proposed to do by way only of *ébauche*" what neither anthropology nor psychical research nor psychology could do independently: to put the savage and modern phenomena side by side. He documented how the "magnetic sleep" described by the Dene Hareskin (the aboriginal peoples of northwest Canada) appeared to be much the same phenomena practiced by the Marquis de Puységur in early nineteenth-century France. Similar parallels could be noted between the ancient shamans who could heal the sick, talk to the dead, and see into distant times and places, and the modern medium who claimed much the same. What exactly was happening with the hypnotic cures at the Nancy school, if not the workings of this medicine man? Indeed, Lang speculated, certain tribal rituals involved entire villages coming together in a communal trance, so was this the same "group hypnosis" active at the séance table or in the mesmeric trance of those gathered around Mesmer's "animally magnetized" tree? And might the hypnotic anesthesia induced by the Scottish surgeon Eisdale in his Indian patients likewise explain the power of natives around the world to perform the firewalking ceremony?

For Lang, these parallels demanded deeper consideration than as part of the confused fiction of "spirit philosophy." These were not just ideas and delusion but effects as well. The ability to move objects, discharge sounds, radiate light, levitate or elongate one's body, retard fire, manipulate a divining rod, effect physical cures, and so on formed much of the ceremonial basis of ancient religious beliefs around the world. These had to be appreciated independently as practices, not just as elements of a creed, and as such, events appeared to establish a pattern of behavior around the world associated with extraordinary brain states and psychokinetic faculties. Lang compared an account by Delacourt, an eighteenth-century French missionary who witnessed the exorcism of a Chinese boy "who was transported in the twinkling of an eye to the ceiling," to the testimony of Lord Adare and Lord Lindsay, who solemnly vowed to have seen D. D. Home float out of one room through an open window and insert himself, elongated, into another.[83] The seventeenth-century Jesuit, Père Lejeune, who traveled among the North American Indians, wrote about medicine lodges shuddering and

flailing beneath the might of a shaman's incantations, just as Lodge and Myers attested to a séance table rocked by the gale force of Eusapia Palladino's
psychical powers. A Kutuchtu Lama "who mounted a bench and rode it to a
tent of stolen goods" appeared to call upon a capacity similar to the Scottish
diviner following his rod to water or "the willer" compelling the discovery
of an object in the willing game.

All these physical phenomena linked the powers of mind and the structures of matter in a hidden sympathy little understood (and thus actively
denied) by modern science. This intersection was what Lang called "the X-
Region of anthropology," marking the spot where the uncanny aspect of
the human psyche spliced into a larger, unknown field of force. Borrowing from the Melanesians, Lang used the concepts of *mana* and *nunuai* to
express the idea of a "magical rapport" between *mana* and man. *Mana* was
"the magical ether." (In a letter to Lodge, Lang wrote, "I would like to see
the Ether proved," desiring to frame his own usage of the ether-as-*mana* in
a manner consistent with Lodge's definition as a physicist.) *Nunuai* were the
sense impressions this energy field left on the mind, enabling the mystical
exchanges that formed the core of man's most sacred beliefs and ceremonial rights. Holy men, seers, healers, and perhaps the mad—these were the
people especially marked by the gift of *nunuai*, a psychic sensitivity that
gave them special awareness of *mana*'s presence in the world and even the
ability to possess some of its power:[84]

Mana is the uncanny, is X, is the unknown. A revived impression of sense
is *nunuai*, as when a tired fisher, half asleep at night, feels the draw of a
salmon, and automatically strikes. The common ghost is a bag of *nunuai*, as
living man, in the opinion of some philosophers, is a bag of sensations. To
myself it rather looks as if all impressions had their *nunuai*, real, bodiless,
persistent, afterimages; that the soul is the complex of all of these *nunuai*;
that there is in the universe a kind of magical ether, called *mana*, possessed,
in different proportions, by different men.[85]

Cock Lane and Common-Sense urged belief in the factual basis of this X-
region, but emphatically did not subscribe to any specific religious interpretation surrounding it. Lang was no more a believer in ancient animism than
he was in modern spiritualism. But even though Lang denied that these
phenomena were supernatural (connected to another world not our own),
the events of the X-region were indeed *supernormal* (strange and marvelous).
Man's intuition of this spectral materiality connected him to a fundamental
wonder, the productive source of all mythology. Only recently had a civilizational myth sought to exclude such phenomena from society's major belief

system, and that myth, for Lang, was modern science. As science assumed more and more the dominant role in cosmological explanation, its philosophical outlook would come to shape modern culture and society.

As Lang gloomily observed in *Magic and Religion* (1901), "in the long run magic and religion are to die out perhaps and science is to have the whole field to herself."[86] If so, that field must include the X-region, or imagination was lost. The dilemma for Lang was not that science was to become the new mythology replacing Christianity—every society must imagine itself anew. Science could provide as heroic a narrative as any Homeric epic—but not if its current doctrinal materialism prevailed. This kind of biologism, in both science and storytelling, sapped the imagination, a danger amply demonstrated by the wretched spirit of Émile Zola's "science of literature." The fantastic, the unfamiliar, the fearsome, the ecstasy of the unknown: these were the crucial meditations for the generative mind. Without the X-region, an essential part of the human psyche was left fallow, a danger uniquely posed to society by positivism in its systematic suppression of the idea of the supernatural.

Lang's defense of the "supernatural in fiction" can be seen to parallel his defense of the supernormal in science (the X-region): "Perhaps it may die out in a positive age—this power of learning to shudder. To us it descends from very long ago, from the far off forefathers who dreaded the dark and who half starved and all untaught, saw spirits everywhere and scared discerned waking experience from dreams. When we are all perfect positivist philosophers, when a thousand generations of nurses that never heard of ghosts have educated the thousand and first generation of children, then the supernatural may fade out of fiction."[87]

By denying the existence of *nunuai*, the X-region, *Kalou*, *wakan*, *fée*, *tamate*, telepathy, Lang feared that science cut man off not just from the facts but from the philosophies that flowed from such *mana*. This "she" of common sense represented the science of realism, increasingly alienated from the very forces it should wish to probe. But there was a science of romance. The masculine and imaginative spirit of psychical research remained mythologically "intact," venturing into the demonic X-region where science could remain replenished by mystery. Masculine daring was required to tackle the unknown, to break free of the tidy taxonomies of effeminate science, and to push the mind beyond armchair analysis and into the jungles of a dangerous and unpredictable cosmos. Certainly, *Cock Lane and Common-Sense* embraced such daring, being profoundly ambitious in its challenge to entrenched positions and setting a new course for an emboldened anthropology. And yet, it was really just a prelude to a far more comprehensive challenge Lang was readying against the entire apparatus of Victorian faith and its most cherished notions of progress.

ANDREW LANG AND UPRIGHT RELIGION

In 1898, drawing upon the data of *Cock Lane and Common-Sense*, Lang published *The Making of Religion*, mounting a comprehensive, dedicated attack on the animistic origins of religion and radically revising the previous positions he had taken regarding the nature of faith. For one who titled so many volumes with the word "religion," Lang had little interest in the Scots Calvinist faith of his fathers. Religion for Lang was an object of study, not a subjective creed, and he had little struggle pulling up anchor and setting course for new kinds of metaphysical speculation. While the six-year-old Myers was busy documenting every little moral deviation in his "God diary," Lang was daydreaming through his religious instruction: "I got no harm from The Shorter Catechism, of which I remember little, and was neither then nor am now able to understand a single sentence . . . For most children, one trusts, Calvinism ran like water off a duck's back; unlucky were they who first absorbed, and later were compelled to get rid of The Shorter Catechism."[88] As such, Lang was not prone to the kind of angst suffered by Myers, whose loss of faith triggered a "crisis," nor did he have the evangelical drive of Oliver Lodge, who, though less religious than Myers by upbringing, was driven to find new saving theological formulas. For Lang, the "crisis of faith" was not how best to recover from its loss but the danger of ever believing such punishing creeds in the first place.

But if there was no childhood attachment to the promise of religion, Lang was deeply bound to another hope: "Memory holds a picture," he wrote in *Adventures among Books*, "of a small boy reading the *Midsummer Night's Dream* by firelight, in a room where candles were lit, and someone touched the piano, and a young man and a girl were playing chess. The fairies seemed to come out of Shakespeare's dream into the music and the firelight. At that moment I think I was happy; it seemed an enchanted glimpse of eternity in Paradise; nothing resembling it remains with me, out of all the years."[89] This might not be a longing for biblical salvation, but it was nonetheless a yearning for another "glimpse of eternity," one populated rather by fairies than by saints. Lang did not so much miss religion in the world as mourn its lost magic. This sense of longing and loss is a defining aspect of Lang's Romantic metaphysics, which by its very nature lacks specificity since its object is beyond this world's ability to articulate or fulfill. As mature thinkers, Lang, Lodge, and Myers set themselves against the claims of orthodox religion and its attempt to control and represent this region of sacred longing. And yet, each, in his own way, needed to believe in the reality of a transcendent world and man's immortal spirit, set loose from its traditional framework.

Lang's psychical anthropology focused on *why* people believed rather than on *what* they believed, shifting his focus from religion's objective criteria to its deeper, experiential underpinnings. In this way, Lang hoped to understand the hallowed aspect of human nature that could first imagine a world beyond itself. This search for the epiphany that would become religion put Lang on the trail of the most ancient and elusive strata of an oral tradition many millennia gone. Lang's project had a kind of reverence for humanity's deep antiquity, as opposed to theorists like T. H. Huxley, E. B. Tylor, J. Lubbock, and H. S. Maine, who saw only a morass of confused polytheistic beliefs and wanton rituals in these remains. The bias of this developmental model correlated the most barbaric religious practices (infanticides, cannibalism, priestly tyranny, blood sacrifice) with the earliest states of divine belief, allowing for the development of moral regulation over time. In this long unquestioned model of social evolution, religion began in violent tribal squalor and peaked with Protestant Christianity, which was itself to be phased out by a secular humanistic creed.

But Lang, using the very same textual sources, was able to turn out a very different account through the careful application of his rigorous philological training. He demonstrated how anthropologists' biased reasoning led them to ignore crucial evidence that could challenge the developmental theory and interpret other evidence too favorably in that light. A. W. Howitt, Mary Kingsley, Bishop Callaway, William Ellis, and countless missionaries and traders gave continuing testimony that even among the most remote tribal people, there often existed the notion of a moral and omniscient God. Whether He was called "Darumulun, Bunjil, Baiame, Cagn, Unkulunkulu, Yama or Hetsche," around the world, a Supreme Being yet haunted the pantheons of tribal mythology as moral and transcendent as any Calvinist Yahweh. Lang's controversial assertion was that this Righteous Creator was not a recent religious arrival, but a *survival* of the most primitive stratum of supernatural beliefs. Lang theorized that corruptions arose only in later phases of the anthropological record, when "the usefulness" of a deity became the primary religious focus. Lang offered for an example the self-interested character of the religion of the Zulu, who were described in *The Making of Religion* as "a comparatively developed and practical military race":

A deity at all abstract was not to [the Zulu's] liking. Serviceable family spirits, who continually provided an excuse for a dinner of roast beef, were to their liking. To the gods of the Andamanese, Bushmen, Australians [more "primitive" tribes], no sacrifice is offered. To the Supreme Being of most African peoples no sacrifice is offered. They are not to be "got at" by gifts or sacrifices. The amantongo [lesser Zulu deities] are to be "got at," are bribable, supply an excuse for a good dinner, and the practical amantongo are honoured while in the present generation of Zulus, Unkulunkulu [the original supreme Maker-God] is a joke.[90]

Unkulunkulu, along with the Supreme Beings of other tribes, demanded no sacrifice, promised no favors, and remained transcendently aloof from the affairs of man. In essence, God's relationship to early man was purely one of grace. Lang's first "savage" operated much like a proto–Scotch Calvinist, a chaste disciple of a remote and lofty God to whom he offered worship and thanks, but from whom he sought no advantage. It was only in the later stages of African social development that this gorgeous austerity was crowded out by vulgar supplicants pressing God for favors. Thus, native religion paralleled the decline of early Apostolic Christianity which "sank" into Catholic priestcraft, Popery, and intercessory sainthood, with its own similar descent into witchcraft, divine kings, and spiritism. That this transcendent Maker God sounded suspiciously like the God of New Testament–toting Evangelical missionaries led Tylor and others to believe Unkulunkulu and the rest to be the product of missionary contact. But many of these tribes were completely isolated, which Lang felt preempted the possibility of the "borrowing theory," be it from Christian missionaries, Muslims, Catholic Portuguese, or even other tribal sources, to explain the inspiration of this Maker God: "I note that the lowest savages are not yet guilty of the very worst practices—'sacrifice of other human beings to a blood-loving god, and ordeals by poisons and fire' to which Mr. Darwin alludes . . . There are as regard to these points in morals, degeneracy from savagery as society advanced and I believe there was also degeneration in religion. To say this is not to hint at a theory of supernatural revelation to earliest men—a theory which I must *in liminae* disclaim."[91]

But if there was no revelation, no missionary contact, no worldly agenda, then where did belief come from? In animism, the spiritual theory was derived from humanity's own animate, purposive nature projected onto the world. Grown over time, these spirits of nature ascended to the status of an omnipotent God (quantitatively but not qualitatively different from man). But Lang disclaimed this notion of God as well. Religion, for Lang, began

not with man impressing himself upon the universe, but with the vast-
ness of the universe impressing itself upon man. Lang explained "origins"
through the words of St. Paul: "God that made the world and all things
therein . . . that which may be known of God is manifest in them, for God
hath showed it unto them . . . being understood by the things which are
made." Like Lodge, Lang, too, saw that true religion had "its root deep down
in the heart of humanity and in the Reality of things."[92] Lang was careful to
add a clarification lest he be misunderstood: "St. Paul's theory of the origin
of religion is not that of an 'innate idea,' nor of direct revelation. People, he
says, reached the belief in God from the Argument for Design."[93] Lang's natives reverse-engineered God, working backward from Creation to Creator,
using observations of the natural world to derive their theory of a spiritual
one. (This made them good scientists, as opposed to Tylor's bad ones.) The
God whom Lang recovered from the primal layer of savage religion was "a
being who created or constructed the world, who is eternal, who makes for
righteousness and who loves mankind. To him no altars smoke and for him
no blood is shed."[94] The Maker God was apprehended when man reflected
on creation. In that moment, the mind of man transcended his circumstance
to understand a Being greater than that which had been created. The religion of early man thus reflected the magnitude of a God so recognized. It
was only later that man, annoyed with this aloof, unresponsive Almighty,
began to prefer lesser gods, gods with needs, whose palms could be greased
with the right propitiation, more user-friendly gods with a compatible human interface, such as tasty animal sacrifices and festival dances.

Despite the apparent theological perfection of this Maker God, Lang was
emphatic that the idea arose purely from our natural cognitive processes; no
burning bush, no star in the East, no "sense of the infinite" was necessary.
But it was not an epiphany based on the recognition of God's naked power.
Lang also pointed out that the theory was argued through purely *anthropological* methods, inferring its conclusion strictly from the textual record.
What Lang saw in this primal epiphany of God was a monumental phase-
shift in the mental development of our hominid ancestor, his humanizing
moment of "Eureka!" when he stepped out of nature and into complex consciousness. That Lang should so vigorously exclude divine revelation in religious origins seems odd in a theory that included psychical data, but this
framed things empirically and focused control in human cognition, both
in terms of direct knowledge of God and academic discussions about God.
Lang would trust no *deus ex machina* of revelation to bail out his theory of
religion. God was thus excluded from his own origin myth, to be discovered

by man through the process of his moral reasoning and without the counsel of priestcraft and theology.

Lang's anthropological and psycho-folkloric science preserved the two essential elements of the Victorian religious *weltanschauung*: an immortal soul and a moral God.[95] Each of the two parts of *The Making of Religion* upheld one of these two pillars of faith. The Maker God was a primal epiphany of an intellectual nature, disclosed by Lang in the philological analysis of textual sources. But the birth of the soul was a different matter, originating not through the rational contemplation of God, but out of the depths of the intrapsychic human experience that brought primitive man into direct connection with his own, indwelling, unseen power. This power was not the transcendent intelligence of the divine mind, but the panpsychic life force of the creative will. Savage lore paid witness to a demiurgic power alive in nature, suffusing the world with an electrifying *mana* in which the consciousness of early man was immersed. Psychical phenomena erupted at points of contact between the mind and *mana*, through channels of perspicuous emotion and psychical intuition that could momentarily connect with or even briefly contain and control the chthonian will of virile nature within the continuum of an individual consciousness. This was the psychokinetic thunder summoned by shamans and warlocks, mystics and saints, and now by physical mediums like Eusapia Palladino, who were conduits for the vital power tunneling through nature as the seminal substance of nature's primal virility.

"The evidence of all this," Lang wrote, "deals with matters often trivial like the electric sparks rubbed from the deer's hide, which yet are cognate with an imitable, essential potency of the universe."[96] These "electric sparks" tapped into "the essential potency of the universe" and were the key to something profound, and at the same time all around us and connected to us. It was not that maternal aspect of nature, studied by biologists who patiently observed its biochemical processes grinding away according to insuperable laws. It was the creative force that worked through those laws as an expression of will, imposing the shock of destruction and the horror of creation at work in the heart of Haggard's African netherworld, a sublime, intoxicating, and vitalizing terror with the power to awaken those same awesome energies within ourselves. Father nature is creative power, "the essentially penetrating" potency of the universe, witnessed in the "savage lore" of books for boys. It is the masculine counterpoint to mother nature—that material, mechanical process by which creation is achieved, set forth in literary realism, which likewise creates through the observation of natural laws.

Lang's theory made subtle distinctions that were often lost upon his critics, who could not quite escape the framework of Tylorean animism: religion as bungled science and man as primitive reasoner. *Nunuai* were not to be confused with observational data asserted by witnesses of supernatural events. This reduced Lang's revolutionary thesis to merely a "spirited" variation of Tylor's own ghost theory of religion—ancestral man trying to deduce the cause of the animating effects—only this time such events included the interpretive possibility of being real.[97] A key element of Lang's argument for the psychical origins of the theory of the soul was the subjective nature of this spiritual connection. While the grounds of these psychical intuitions were given objective verification through contemporary psychical research, *nunuai* were mainly intrapsychic phenomena that might manifest external effects, unfolding within human consciousness in the interpenetration of *mana* and mind. As an advocate of psychical research, Lang had no objection to the empirical engagement of such phenomena: observing them, analyzing them, and intellectually rendering them into theory. That was the procedure appropriate to undertaking a modern scientific study; it did not belong as a guide to understanding the religious consciousness of early man. This was, in a sense, a magical psychology whose most important aspect pertained to alterations in human states of mind, not physical states of the natural world. For Lang, the essential problem with positivistic anthropology was that it saw Tylorean "savages" as evolving within the brutal imperatives of Darwinian natural selection. They were thus the practical, ruthless people they had to be, with no deep piety or mysticism at all, as that was an unconstructive use of mental energy. By its very nature, a world suffused with *mana* would have its own evolutionary logic, distinct from the purely biophysical laws of survival. For Lang, a psychical sense meant, "that the *will* and *judgment* are less closely and exclusively attached to physical organisms than modern science has believed."[98] This opened up a space in the mundane world for more ideal considerations to take root and even suggested some continuity between the self and this Unseen Power.

Lang interpreted these many distinctive aspects of religious experience with reference to Myers's subliminal psychology, arraying the various psychophysical and perceptual phenomena along his continuum from red to violet. (In fact, Lang asked Myers to proofread his manuscript to assist him with this framework.) Within this magical tangle of *nunuai* and sensory phenomena scattered through the anthropological record, Lang attempted to tease out a primal narrative for the transcendent soul. Only the most metaphysical species of psychical events—those at the violet end of the spectrum—furnished evidence that "human personality possibly survives bodily death." This was particularly true for clairvoyant episodes that saturated custom and myth with soothsayers, oracles, sibyls, and astrologers. If such episodes were veridical—and evidence of psychical research seemed to testify that they were—the mind appeared to exit material and spatiotemporal frameworks of ordinary cognition to merge with the mind of God. (It thus became continuous with what Myers called the World Soul and Lodge called the Universal Mind. Both Myers and Lodge were intellectual tributaries for Lang's thought.) This at once demonstrated the independence of the subject from the body and its cognitive frameworks, and seemed to directly participate in aspects of the divine. Perhaps immortality was one of them. At the red end of Myers's spectrum were the psychophysical phenomena associated not with the Apollonian mind of God, but with the primal magic of his demiurgic will. These two polarities of mind and *mana* were at once distinct and related phenomena, arrayed along the continuum of Myer's subliminal spectrum in which mind and matter were interactive, immersive, and transcendent. This overturned the dualistic Christian (and scientific) understanding that mind and spirit were separate phenomena, liberating the circumstances of humanity's first religious awakening from these oppressive constraints. (And this likewise opened up a prospect for some resurgent spirituality made possible by the psychical retooling of the mind/matter relationship within science, aligning *The Making of Religion* with the observable patterns in psychical theologies.)

The incorporation of psychical considerations transformed the nature of the statements Lang was willing to make concerning religion. If we look at how Lang defined "religion," in "Mythology," an entry for *Encyclopaedia Britannica*, in 1890, it still falls within the evolutionary conventions set by Tylor. His definition in *The Making of Religion* (1898), however, reframes the entire discussion, departing toward a new anthropological paradigm on the powerful conceptual leverage gained by adding psychical evidence:

> Religion may be defined as the conception of divine or at least supernatural powers, entertained by men in moments of gratitude or of need and distress, in hours of weakness, when, as Homer says, all folk yearn after the Gods.[99] ("Mythology," 1890)

> By religion we mean, for the purpose of this argument, the belief in the existence of an Intelligence, or Intelligences not human, not dependent on a material mechanism of brain and nerves, which may or may not powerfully control men's fortunes and the nature of things. We also mean the additional belief

that there is, in man, an element so far kindred to these Intelligences that it can transcend the knowledge obtained through the known bodily senses, and may possibly survive the death of the body.[100] (*The Making of Religion*, 1898)

In the first definition, belief in God was rooted in human psychological needs and shaped by punishing physical realities. It was a form of wish fulfillment directed at this world, and thus, conceptually, the God of its creation was also of this world. But the God of *The Making of Religion* was "an Intelligence, not dependent on material mechanism of brain and nerves . . . at home among the stars." This religious instinct arose from a noble impulse of mind that recognized the magnitude of God's creation and flourished in those liminal spaces of consciousness where the psychical flashes of *nunuai* lit up the darkness that lay beyond the waking mind.

The Making of Religion was not just a study *of* religion but also an argument *for* religion. Men like Tylor, Frazer, and Huxley saw no place for religion as a cultural form in the modern world; viewed without sentiment, it was false knowledge taking up room. Lang, in contrast, held open a perennial space for religion in culture's most untouchable sanctum, not because it was true but because it was mythic. He was not merely defending the mythos of religion on the functionalist grounds he had used in his earlier anthropological contribution to religious debate. Nor was he highlighting the imperative aesthetic and psychological needs served by such fantastical beliefs, made clear in his defense of romance. His anthropological analysis of religion found its deepest authentication in man's testimonial encounter with the divine spirit in nature. Such contact was collected in the accounts of psycho-folklore and psychical research: "Not being able to explain away these facts, or, in their place, to offer what would necessarily be a premature theory of them, we regard them, though they seem shadowy, as grounds for hope, or, at least, as tokens that men need not as yet despair."[101] This "despair" was the inevitable scourge of an unredeemed Darwinian anthropology and the failure of science to find any truly foundational beliefs. (Lang, too, became an elected member of the Synthetic Society the year he published *The Making of Religion*, putting him with Myers and Lodge at the hub of this eclectic twentieth-century push for a new idealism.)[102] As Nietzsche saw it from across the way, the humanism furnished by the "old cold boring frogs" of British academia, who were "forever hauling the *partie honteuse* [buttocks] of our inner world into the foreground," focused with a "certain disillusioning tenacity on the low beginnings of our race."[103] By the late 1890s, there was a general feeling among intellectuals that humanity needed out of that biological swamp.

Despite the controversial nature of his arguments, Lang tried to model his intellectual dispute as one *within*, not against, anthropology's developmental orthodoxy, because he wanted to carry the discipline along with him. Similarly, he had no wish to abandon evolution, but rather desired to seize its platform and claim its legitimacy for his own theory. Lang proposed four stages in the development of religious thought, but reversed the directional vector proposed by the dominant school. Lang's developmental process began with "(1) The Australian un propitiated Moral Being, (2) the African neglected Being, still somewhat moral, (3) the relatively Supreme Being involved in human sacrifice, as in Polynesia, quite immoral and (4) [the eternal return to God's fundamental nature]: the Moral Being reinstated by Christianity," and thereafter humanist philosophy.[104] Here the positivistic value of progress intersected with the Romantic appreciation of resurgence to describe an upward trajectory that spiraled its advance in the repetition and perfection of its essential form. The divine *anthropologos* of true religion was never lost, but it was inevitably corrupted, compelling the destruction, rebirth, and renewal of its divine trope. Societies at the furthest remove from their origins suffered the greatest amount of moral alienation in proportion to their degree of modernity. Lang intended this as an indictment not of modern Christianity, still the popular myth of the people, but of decadent European high culture, rotting in an "atmosphere like that of the boudoir of a luxurious woman, faint and delicate, suggesting the essence of white rose."[105] They had traded in their stories of gods and heroes in favor of "emasculated specimens of an overwrought age, with culture on their lips and emptiness in their hearts."[106] Hence, the reverse of Lang's earlier homily (going low to aim high): to really troll the bottom of one's culture, one must go straight to the top.[107]

Lang's cure for "progress" was to tap the redemptive power of the past through his strategic championship of magic, romance, folktale, and myth: "The people, the folk, is the unconscious self, as it were, of the educated and literary classes who, in the twilight of creeds are wont to listen to its promptings and return to the old ancestral superstitions long forgotten."[108] For Lang, resurgence was the new resurrection. But despite his attraction to deep antiquity and the voluptuous atavism of the anthropological romance, Lang was in no way a reactionary. Resurgence was a theme intrinsically embedded in alternative notions of progress and future, as-yet-unimagined forms. Psycho-folklore was essential to realizing these other evolutions, using the same empirical methods currently accused of stifling modern culture, to liberate its future trajectory. Lang did not formally join the SPR until 1907, eventually serving as its president in 1911, a year before his death. This does not make psychical research the final phase in an otherwise eclectic career: it was rather the gestalt form in which he could finally assert the fundamental relatedness of all the different aspects of his thought—the poetical, analytical, anthropological, and metaphysical. Biographies of Lang tend to compartmentalize his various interests as the divergent pursuits of a talented polymath. A psychical biography joins all these aspects of Lang's praxis and his passions to render a more holographic history. Lang also makes possible a more holistic understanding of the psychical project, adding the dimension of anthropology to physics and psychology within its dynamic perimeter, as psychical research sought to incubate not just a new paradigm for science but a new modern worldview, one weighted toward the Romantic wing of the Enlightenment.

Through Lang's friendship with Pater, the past became an enchanted grotto of masculine fabulism, one he would disclose through anthropology and relive through romance. It was a myth he would build out for the people in more than four decades of anthropological scholarship, from "Mythology and Fairytales" (1873) to *The Secret of the Totem* (1905). Lang's history puts him in his university days inside the Oxford aesthetic movement, the epicenter of neo-Romanticism, but at the same time he was the protégé of Tylor, a towering figure of positivistic anthropology. He was always something of a conflicted figure straddling these two worlds looking for synthesis. He found it in psycho-folklore. Psycho-folklore, like Myers's subliminal self and Lodge's forces of "life and mind," put human consciousness in touch with the noetic aspect of itself, radicalizing the implications of both human knowing and human being. This revolutionary anthropology allowed Lang to hold and defend "that element which gives a sudden sense of the strangeness and beauty of life; that power which has the gift of dreams," as part of the natural world and of human nature.[109]

With *Cock Lane and Common-Sense* (1894), Lang pivoted directly toward the philosophical object hitherto approached only tangentially, to address himself and his work, "in an age of fatigued skepticism and rigid science, to the imaginative longings of men who will fall back on savage or peasant necromancy."[110] To supply these longings, Lang cast his line into a netherworld of *mana* and imagination that ran from the occult margins of the oriental imperium all the way to the primal forests of druid religion, bringing up this virile magic from the past to enchant the modern world.

CHAPTER SEVEN

Psychical Modernism

Science, Subjectivity, and the Unsalvageable Self

In a letter to Frederic Myers dated 1872, a decade before the psychical proposal had even come to mind, Henry Sidgwick wrote: "I sometimes feel, with somewhat of a profound hope and enthusiasm, that the function of the English mind, with its uncompromising matter-of-factness, will be to put the final question to the Universe with a solid, passionate determination to be answered which must come to something."[1] Yet, despite 130 years of effort aiming to do just that, it seems the work of the SPR has come to nothing, so far as mainstream science is concerned. If anything, the early momentum of the society, which tripled its membership in the first ten years around a nucleus of intellectual and social luminaries, slowed after the 1920s, as it loosened its ties to mainstream psychology. Even the stripped-down, largely quantitative field of parapsychology, the straight arrow meant to pierce the armor of academic resistance, has failed to root itself in university culture. There are only a few degree-bearing programs offered through academic departments at major universities, mainly in the United States and England, though these opportunities appear to be expanding in the past few years. In the United Kingdom, universities in Cardiff, Bristol, Hertfordshire, Lancaster, London, Manchester, Northampton, and York have established programs in or related to psychology. In the United States, there are course offerings and degrees available in paranormal subject matter through the Rhine Education Center; the Institute of Transpersonal Psychology (now Sophia University) in Palo Alto, California; the Theoretical and Applied Neuro-Causality Laboratory (TANC Lab) at the University of California, Santa Barbara; the University of Virginia's Division of Perceptual Studies;

and the University of Arizona's Sophia Project, part of the Laboratory for Advances in Consciousness and Health.[2]

Much of the funding that sustains this research comes from private sources outside of academia, chasing the hope of some certified spiritual discovery. Among the most interesting of these legacy grants was made by the novelist Arthur Koestler upon his death in 1983. A member of the SPR since 1952, he echoed in his bequest the yearnings of the original program: orthodox science must take up psychical research lest, in the words of Koestler, "the limitations of our biological equipment condemn us to the role of Peeping Toms at the keyhole of eternity."[3] That same plea to science, made over a hundred years earlier, remains, unanswered. But clearly, Koestler, ringside in Europe for the 1930s, the "low dishonest decade" when disenchantment with democracy gave rise to nakedly ruthless strategies of power, understood parapsychology to be of existential significance to the project of modernity. And this was where he chose to take his last stand in a lifetime spent on the front lines defending human decency.[4] To this end, he left his entire estate toward the establishment of a fully academic doctoral program, what he called his "Trojan horse," to gain access to the citadel of knowledge.[5] But like the original empirical impulse of the SPR, this was not an assault on science, but rather an obeisance to its exclusive claim on intellectual relevance. The endowment eventually went to the University of Edinburgh, after demurral among other leading universities about "the scientific propriety of endorsing research in fields such as clairvoyance, telepathy, and levitation."[6]

While Koestler described himself as having a "mystical temperament," like Lodge, Lang, and Myers, he directed this metaphysical impulse into a reforming secular vision (though his energies were more political than scientific).[7] As a Communist activist and sometime party diplomat between 1931 and 1938, he traveled as an envoy to Stalinist Russia. Instead of a social order where "the truth had set them free," he found a regime with the power to warp all sense of reality. He had witnessed the same in fascist Italy, Germany, and Spain: first the conquest by fantasy, then by force. Koestler came to believe that, even absent religion, a new "metaphysics" would slip into its place. In the case of Stalin's state-run Communism, chillingly reproduced in Koestler's *Darkness at Noon* (1941), this unreality ruled in the form of an immaculate political theory and its apotheosis, "Number One," the dictator-cum-messiah.[8] And even as the secular tyrannies of the twentieth century made gods of their leaders, such governments proceeded to annihilate their citizens in both body and soul. Koestler portrayed the failure to ground the real and sacred existence of the individual as the most insidious aspect of Stalin's regime—that which enabled his slaughter at the "statistical" level. The nine million people that Stalin is estimated to have killed (let alone the millions enslaved, starved, or deported) began with the death of "the one" that mattered: zero times nine million is zero.[9] In the secular eschatology of *Darkness at Noon*, the revolution was the object of salvation, not the citizen and not the soul. Koestler was trying to make sense of the Moscow show trials, where loyalists confessed to crimes they had not committed and died at the convenience of the master they served. This was what it meant to be, in the refrain of the book, "a grammatical fiction," written in and out of party history as needed. All that remained of Rubashov's conscience was a "silent passenger" that spoke only in the haunting form of his somatic pains.

What is relevant here is the way Koestler explicitly advances the earlier psychical defense of the self, seen in the writings of Lang, Lodge, and Myers. There was a hollow left at the center of humanism, the intended space of private conscience that citizens could not fill. Koestler witnessed how fascism, nationalism, Stalinism, consumerism, and, finally, war rushed into the spiritual void liberalism had left empty by design. My subjects' fears of a godless future seem less overwrought considering that they were directed toward the moral community: a concern about who would succeed the Christian subject and the kinds of societies they would convene. In the tormented political anthropology of the 1930s, identity was variously derived from economic activity (Marx's *homo faber*), from the fascist state ("everything for the state, nothing outside the state"), from one's racial *geist*, and, problematically for meritocratic societies, from one's bank account ("the rich are different from you and me . . . very different").[10] By the 1980s, when Koestler was making his bequest, poststructuralism was already reconstituting the self as a "grammatical fiction," in the very humanist bastions that were supposed to be its defense. The western subject had survived with its civic boundaries intact, but it was vanishing nonetheless, this time in an untethered cloud of language. That Koestler's final political and poetic act was to endow a parapsychology program goes to the heart of his deepest concern: the frailty endemic to all formulations of the modern self that made no compelling case for its real existence.[11]

THE GRIN WITHOUT A CAT

In "Charles Darwin and Agnosticism" (1888), Myers cautioned against the idea of an evolutionary self "summoned out of nothingness into illusion and evolved but to aspire and to decay."[12] At the same time, Myers considered natural selection a "majestic conception that set science on the track of truths so great and new that they seemed to fill the whole horizon and transfigured life with their glow." As evolutionary biology raced ahead, advancing other disciplines, it gained the world, but at the soul's expense. In that same article, Myers wrote, "Unless some insight is gained into the psychical side of things . . . some light thrown upon a more than corporeal descent and destiny of man, it would seem that the shells to be picked up on the shore of the ocean of truth will become ever scantier and the agnostics of the future will gaze forth ever more hopelessly on that gloomy and unvoyageable sea."[13] According to Myers, this was the blighted future of science under the current rule of biology. While enough was still missing from Darwinian explanation to provide a gap for God and a role for religion, the materialistic implications were plainly there for those who looked hard and long and took the theory down to the bone. Instead of trying to overturn evolutionary theory, psychical researchers co-opted it, pushing on its narratives from the inside. By the 1890s, the accumulation of psychical data allowed Myers, Lang, and Lodge to advance speculation beyond the basic proposition "Was any of it real?" to "What did it all mean?" In summarizing his life's work, Myers explicitly framed psychical research as a potential successor to religious inquiry: "an incipient method of getting at Divine knowledge, with the same certainty, the same calm assurance with which we make our steady progress in the knowledge of terrene things."[14] While Myers was clearly transferring his allegiance from creationism to evolution, he did not surrender some eventual knowledge of the divine.

This idea that science and not religion would furnish the myths of the coming age was discouraged by moderates looking to preserve what they could of the status quo. Yet this was the path down which psychical research was moving, making it just as culturally subversive as the so-called militancy of Victorian scientific naturalism, or perhaps even more so. There, religion was mostly being asked to get off the lawn. Psychical research, on the other hand, used so-called religious phenomena to fertilize its own speculation, substituting their science for hallowed explanation. Even though Myers and Lodge, and Lang, to a lesser extent, had sympathy with traditional faith, they still saw it as their task to relieve religion of duties for which it was no longer qualified. Story time was over; it was time for religion to get real. In the words of Myers, "What use is there in fondling hallowed traditions or in juggling with metaphysical terminology? Unless the human race can find more facts, it may give up the problem of the Moral Universe alto-gether."[15] That insistence on an empirical basis for moral values put Lodge, Lang, and Myers in the most forward position of Comte's historical pro-gression, promoting science as a cultural system, not unlike positivists like E. B. Tylor, James Frazer, or Sigmund Freud; the difference was that their scientific worldview retained a halo of supernatural possibility. The psy-chical researchers of the 1890s were tilting psychology toward an ultimate truth, while their colleagues were mapping the conditions of a permanent uncertainty. Neuropsychiatry and psychophysics, the two great initiatives within experimental psychology, had built consciousness into a tight cor-ner, where evolutionary theory crossed energy conservation. A brain shaped by natural selection was not geared for knowledge but for survival. Sensory perception was the economic reduction of stimulus required to get that job done. Useless information was a tax on finite mental energy. A science an-chored within such a mind could never strive to be more than useful. The turn toward phenomenology had begun with this more emphatic distinc-tion between perception and reality.

It was not that prior to Victorian psychophysics, thinkers had failed to consider the psychological basis of empiricism. The role of observation shadowed "objectivity" from the earliest beginnings of Greek natural phi-losophy. A sustained critique of human cognition turned a few new force laws into a scientific revolution. Yet, for all the radical doubt expressed by early modern skeptics, they found a way to be sure of themselves. René Descartes and John Locke both invoked divine guarantees for human cog-nition (a deity that would not deceive and God-given reason). When Locke got rid of Descartes's innate ideas, he had something just as good on hand: Newton's laws. Scottish common-sense philosophy reinforced Locke's pre-sumptive realism with more well-reasoned arguments, preserving objec-tivity for the Victorians. And August Comte only set realism aside because there were more practical ways to affirm certainty. Scientific results were proof of its methods; thinking about scientific thinking was not. His posi-tive age began "once the mind set aside absolute notions . . . and the causes of phenomena and applies itself to the study of their laws."[16] There is no mood of resignation in Comte's words, no yearning for some unattainable metaphysics. Comte's "positive spin" set empirical limits on human knowl-edge like a fence keeping bad information out. Looking back on his youth, Myers assessed the boogeyman of positivism with more mature insight, see-ing it now as a salubrious impulse to "compromise between old beliefs and new." In his essay written in 1889, Myers wrote, "Positivism is a religion consisting simply in the resolute maintenance of the traditional optimistic view, when the supposed facts that made for optimism had all been abandoned."[17] Comte was, in Myers' assessment, attempting to replace an untenable religion of salvation with a modern cult of progress. But it had not worn well. Its earlier confidence lingered ghoulishly "like a grin without a cat ... persuading us that all in this sad world is well."[18]

By the time Myers wrote those words, psychology was already sending positivism in a more subjective direction. Ernst Mach, a physicist interested in perception since the 1860s, published his landmark *Contributions to the Analysis of Sensation* in 1886, having realized the full epistemological implications of his research in psychophysics. Instead of approaching perception as having two parallel aspects, one subjective and the other physical, Mach recognized a single, synthetic phenomenon. There was no outer dimension of sensation. Knowledge of that external input was lost in the translative act of perception, a symbol of reality the mind presented to itself like a program turning source code into images. The real world was a mysterious cause knowable only through these experiential effects. Mach's insight would begin to shift the epistemic locus of science back behind its own perceptual boundary. What was once a slippage between perception and reality was now on track to become an unbreachable void. Instead of going after fundamental truths, the project of science was to document the rules of this shared hallucination that was the only reality we could ever know.

For Mach, sensory determinism eliminated all questions of metaphysics equally. If there was no such thing as "matter," then the dyad of matter and the spiritual-thing-that-is-not-matter became "a superfluous and therefore misleading addition towards our concepts."[19] Ironically, even materialism, once the virtuous rejection of metaphysics, had become a superstition in its own right. Mach's antirealist, anti-idealist stance might appear to deepen the futility of the psychical project, but the subliminal self was uniquely positioned to amplify phenomenological knowledge, building upon rather than altering its synthetic framework. While ordinary consciousness was limited to a sensory mirage of physical inputs, the mind itself was a separate structural predicate. Under extraordinary conditions, human subjectivity could slip deeper into matter or even fully transcend it, putting it in direct touch with larger frames of reality, including its own nonphysical basis of existence. Phenomenology, by contrast, was left watching shadows on the wall. Such a science signified not the conquest of the unknown, but rather conquest by it.

There is a tendency among scholars to take a rosy view of "separate spheres for science and religion," emphasizing what is moderate, cooperative, and conciliatory. But that obscures what was jarring and radical about this emerging conceptual landscape. The older trajectory trying to reconcile science and religion becomes something else entirely when underpinned by psychology and limited to deconfliction. Still, this subjectivism offered certain advantages. Skepticism lost its sting, and faith could be deemed a necessity rather than an indulgence with such limited options. Theologians no longer had to battle it out with science over the nature of reality, competing only with other supernatural beliefs in its category. Here, Christianity seemed to have the clear advantage as an incumbent power. But for how long? The continuing projection of Anglican authority in the first half of the twentieth century may have looked like a successful balancing act, but that disguises just how compromised the category of the supernatural had already become. For most of its history, Christianity had held this "higher" ground as the kingdom of heaven, regulating who might opine and who might enter. But what began as strict theological enclosure ringed by heretics on holy fire was fast becoming, by the end of the nineteenth century, the most promiscuous region of modern intellectual life. It included not just canonical religions and their sectarian heresies but also the *Hermetica*, folklore, and Vedic mysteries opened to eighteenth-century curiosity and tilled by Victorians into ever more heterogeneous forms. Christianity's coercive power was in retreat in governance and education, while the supernatural, by law if not custom, was becoming an unrestricted free-for-all. Science, meanwhile, was just getting serious about locking down its own discourses with a growing hydra of strict disciplinary certifications.

PSYCHICAL MODERNISM AND THE UNSALVAGEABLE SELF

Buried in the footnotes of Mach's *Analysis of Sensation* was a far more devastating observation, implicit in the constructed nature of reality. It offered a glimpse into postmodernity so terrifying that even Mach felt the need to shield his readers from what was otherwise hiding in plain sight: "The Ego is unsavable. It is partly the knowledge of this fact, partly the fear of it that has given rise to the many extravagances of pessimism and optimism and to numerous religious and philosophical absurdities. In the long run we shall not be able to close our eyes to this simple truth which is the immediate outcome of psychological analysis."[20]

While Mach stressed the utility function of the ego in various passages throughout *The Analysis of Sensations*, calling it a "practical unity for purposes of a provisional survey" and "a necessary postulation," only in the footnotes did he spell out that absent this function, the self should not be considered to exist. Like the sensations that gave rise to it, the self was a useful fiction and nothing more. As such, Mach went on to say (in that same footnote and with an optimism not likely to be shared), "we should no longer place so high a value upon the ego" and "be willing to renounce individual immortality ... In this way we shall arrive at a freer and more enlightened view of life which will preclude the disregard of other egos and the over estimation of our own."[21]

But this happy forecast of 1886, especially the "freer and more enlightened" part, was not what came to characterize the 1890s. Mach, like Hume, seemed constitutionally impervious to the despair he caused in others, although he was correct in predicting that "we shall not be able to close our eyes to this simple truth." Literary scholar Judith Ryan identifies a mood of intellectual panic that seeped outward from this text as the implications of "the unsalvageable self" worked their way through fiction, philosophy, and psychology (including psychical research).[22] This phenomenological impulse spread through Europe, Britain, and America, taken up by philosophers like Karl Pearson, Théodule-Armand Ribot, and William James, who profoundly influenced the understanding of "understanding," as it pertained to science, psychology, and education, as well as by art critics like Hermann Bahr, who interpreted Mach for a literary audience. Though "modernism" is primarily viewed as an aesthetic cultural response in art and literature, its obvious preoccupation with subjectivity and cognitivity has recently directed more attention to its links to cognitive science.[23] Hermann Bahr's widely read article, "The New Psychology" (1890), helped to define the new "self-conscious" genre of modernism by drawing the aesthetic gaze inward to consider its own origination in the new neural framework of perception built out by modern psychophysics and psychiatry. Such crossover considerations gave psychology a formative role in modernist aesthetic values, which above all tried to fold the private activity of representation into its representative object.[24] This "new psychology" spread what Lang called "the green sickness of pessimism so popular among literary authors today," through which, Myers proclaimed, referring to the works of J. K. Huysmans, "we [were] led into a world of joyless vice from the sheer decay of the conception of virtue."[25]

In *Beyond Physics, or the Idealisation of Mechanism* (1930), Lodge, too, related the perceived collapse of moral values to the new relativistic philoso-phy he associated with "Mach and Karl Pearson and Poincaré," declaring that "the distinction between right and wrong measurements is a step in the same direction as the philosophy of Nietzsche who attempted to go 'beyond good and evil.'"[26] But Lodge resolutely maintained, in dogged opposition to the new philosophy: "The aim of science is absolute truth and though it is difficult of attainment, nothing less should be admitted."[27] Despite his urgings, Lodge and his fellow psychical researchers were more and more menaced by a new framework that found human beings increasingly rela-tivized in relation to reality and to each other, and defocalized in terms of the self. The historian Dorothy Ross sums up the scholarship on modernism as "the recognition that no foundation for knowledge or value exists outside the meanings that human being construct for their own purposes."[28] But psychical modernism would prove an exception, not by enlarging knowl-edge beyond human understanding, but by making that understanding itself limitless.

Mach's empirical psychology drew heavily upon Gustav Fechner's math-ematical analyses of sensory perception in the 1860s, which attempted to quantify the degree of physical stimulus as it related to the extent of sensory experience.[29] This brought consciousness into the domain of physics, plac-ing matter and mind under the purview of mechanical law, even as Darwin-ian theory effected the transfer of psychology into the life sciences. In the 1870s medical psychiatry added the mysterious terrain of the unconscious to this biophysical brain, giving consciousness an on-site evolutionary his-tory. Clinical hypnosis brought forth atavistic and inchoate states of mind that crossed regression, aberration, and disease in a troubling Darwinian storyline.[30] In the words of Frederic Myers issued from the front lines: "We seemed to have tracked mental life to its inmost recesses and have found it everywhere enwound with an organism [the brain] which tells us much of our bestial origin, nothing of our spiritual future."[31]

This idea that modern man was the product of ancient and open-ended biological processes required some sort of saving cultural and intellectual response to sort through the implications. This was in many ways the project of modernism, an unprecedented tour de force in science, art, and philoso-phy, to explain and creatively express what evolution meant for humanity. The cross-pollination between art and psychology between 1870 and 1930 is complex and defies any simple model of origination versus appropria-tion, but this exchange took place in a milieu saturated with evolutionary anxieties about the nature of mind and the moral status of humanity. Henry

James, Virginia Woolf, Marcel Proust, Rainer Rilke, Guillaume Apollinaire, J. K. Huysmans, Thomas Mann, W. B. Yeats, and others set about capturing a human condition that now included the unconscious mind, multiple personalities, automatism, identity-as-memory, sexual pathology, neurosis, neurasthenia, hysteria, atavism, conditioned reflexes, stream of consciousness, and more, as explicated by theorists and researchers such as Richard von Krafft-Ebing, Max Nordau, Josef Breur, Sigmund Freud, Jean Charcot, Frederic Myers, Pierre Janet, William McDougall, Charles Richet, and Théodule Ribot. This was also the context in which philosophers, such as Franz Brentano, Richard Avenarius, Ernst Mach, Karl Pearson, William James, and Edmund Husserl, "psychologized" empiricism, drawing not on cellular biology but on energy physics to explain the mechanical cause and effect of consciousness.[32]

These two streams of psychological analysis converged to create a modern self caught in a downward cascade of fragmentation. While psychophysicists atomized perception into bits of sensory data (providing the basis for William James's "streams of consciousness"), the evolutionary psychologists disassembled personality into its hereditary pieces of hardware. Hysteria, automatism, atavism, and *dédoublement* all underscored the fragile unity upon which identity rested and the will was exercised. Thus, the modernist crisis of subjectivity was catalyzed in many ways by evolutionary psychology's first anxious encounters with a single individual's "subjectivities," all unfolding within the medical setting of the Salpêtrière in the 1870s, inside hysterical patients' altered states of mind. This was cultural and intellectual space soon to be occupied by psychical research as well.

The attraction of unconscious trance states was that they seemed to suggest a hidden brain archaeology capable of its own, distinct intelligent activity. By exploring these structural recesses, aberrational psychology promised to open the black box of evolution and reveal the secret phylogenic history of consciousness itself, as it coiled through the human brain's many millennia of development. As doctors lowered patients deeper and deeper into these trance states through the skilled application of hypnosis, making even greater inroads with the progress of mental disease, they seemingly brought the past directly into view. For Myers, the anomalies encountered in these subconscious states offered "our first intimation of the true extra terrene character of our evolution."[33] But this view was not shared by the majority of medical researchers involved with hysteria and other altered states, cued as they were by Charcot's continued focus on neural anatomy and its pathologies. For neurophysiologists like Charcot, the mental energies powering consciousness operated within conservation laws and were guided strictly by the logic of evolutionary theory, making the hard problem of consciousness all that much harder. Even Freud's personality-driven psychiatry was based on an essentially thermodynamic account of instinctual energies charging and discharging through axons and neurons.[34] Thus, even the mysterious, irrational, and unmapped unconscious of dynamic psychology could not escape the "iron clad laws of physics," laid down in the mid-century. Where Myers hoped to find glimpses "of our spiritual future," colleagues saw merely our bestial origins, dispersing the brain into a billion parts at the cellular level and replacing the unified rational will with the promiscuity of instincts and the mindless reflexes of "unconscious muscular cerebration."

More pantheistic reflections upon biology, driven by German thinkers such as Arthur Schopenhauer, Friedrich Nietzsche, and Ernst Haeckel, put the will of nature yet more firmly in charge: the subjective intention, traditionally seen as "intellectual" and sited in the mind, now migrated into atomic matter, where it asserted a creative but brutal terrestrial force. While the British anthropology of the 1870s had preserved the primacy of reason by linking rational self-interest to fitness for survival, such period views soon became swept up in the dark, vitalistic will of modernism. The striving to become and to overcome was the new "purpose" in nature, replacing both the divine telos of salvation and positivism's "faith" in human moral and intellectual progress with the visceral aspirations of power and survival. Personal motivation derived not from the higher self (whether identified with religion or natural reason), but now oozed out of what Myers called one's "protoplasmic substructure," impersonal, amoral, and bubbling up from below. These were not the biomechanical energies of British naturalism, but something more troubling. This was what Myers called "the bitter philosophy of Schopenhauer," in which the intellectual principle of free will had become *die welt als wille*, rooted in "the witchery of nature."[35] These new philosophical reflections replaced the spirit/matter duality envisioned by Christian theism with a panpsychic monism giving matter, not man, a soul.

This assertion of an overmastering irrational will, though morally problematic for many late Victorians, proved crucial to modernist aesthetics and to the human sciences. The modern mind had fully broken free from the Enlightenment's utilitarian moral and psychological conventions in favor of a more dynamic view of the self. But because this modern self could no longer be schematized within a progressive evolution working its way in one direction from instinct to intellect, it became caught in the reverse tide of biological degeneration and civilizational decline, supplementary evolutionary storylines that came to characterize the fin de siècle. Medical and psychiatric discourses entwined with social and cultural criticism to invoke a future marked by physical impotence and moral depravity, iconically rendered by Max Nordau in *Degeneration* (1893) but initiated earlier by new, hereditary constructions of madness. Fey intellectuals, like Andrew Lang's slope-shouldered "Homo Calvus," were to be trampled underfoot by the animal vigor of anarchists, atheists, criminals, egotist, and hedonists, as well as "l'homme sensuel moyen," whom Myers described in "The Disenchantment of France" (1888) as "at the mercy of an instinct which they can neither comprehend nor disobey."[36]

This pessimistic strain in literary and cognitive modernism roped across ethics, art, psychology, and medicine, tracing its way back to a single source: the self, "summoned out of nothingness into illusion and evolved but to aspire and to decay."[37] This self was not just the creature of Enlightenment materialism, striving to be free of religion, nor can it be ascribed to nineteenth-century evolution and the grinding governance of mechanical laws. This unsalvageable self was intrinsic to the earliest formulations of skeptical philosophy in the 1600s. Two hundred years later the competence of scientific explanation finalized the loss of the subject's transcendent referent with evolutionary psychology, a self that was in truth, beneath the face of things, already gone. The essential elements in Mach's argument (*das Ich ist unrettbar*) he identified as already present in *Cogito, ergo sum*, which tried to reverse-engineer the "self" from the act of "thinking." Despite the weakness of Descartes's first axiom, all other truths, even belief in God, were to proceed from it. Empiricism did away even with the weak metaphysics of innatism, pushing the question of an ultimate self beyond philosophical view where it would never be seen again. Locke was willing to affirm the existence of God, but he remained evasive about the immateriality of the soul. So, while *tabula rasa* proved extremely fruitful as a secular epistemology, it willfully set out to fail as the basis of an ontology. Its emphasis on sensation turned "thinking" into an increasingly psychobiological process of passive physical perception and procedural, mechanistic reflection. It did not, however, provide "the thinker" with any existential basis independent of that process. Could the self survive the inevitable "self-" investigation put in motion by such an epistemological revolution?

Thus, it is no wonder that less than two hundred years after Locke's *Essay concerning Human Understanding*, the self and its contingent knowledge (not to mention Descartes's derivative God) were buckling under the pressures built into the episteme of cognitive psychology. The conclusions that Mach saw as "the immediate outcome of psychological analysis" in 1886 were already obvious to Hume in 1739, when he observed: "When I enter most intimately into what I call myself, I always stumble on some particular perception or other, of heat or cold, light or shade, love or hatred, pain or pleasure. I never can catch myself at anytime without a perception, and never can observe anything but the perception . . . could I neither think, nor feel, nor see, nor love, nor hate after the dissolution of my body, I should be entirely annihilated."[38]

While Hume is frequently identified as the philosophical ground zero for unsalvageable selfhood, the problem was, as Mach observed, endemic to the psychological emphasis of empiricism itself. Its outward, empirical gaze always kept one eye inward, inevitably extending skepticism about knowledge to the knower himself. Kant tried to entrain this empirical introspection toward the demonstrable proof of *a priori* intuitions, but this was mere child's play for Mach's modernist mind. Mach was fifteen when he read the *Prolegomena to Any Future Physics*, grasping the watershed concept that the mind structured perception, but a mere two years later Mach had a postmodern epiphany: "The superfluous role played by the 'thing in itself' abruptly dawned on me. On a bright summer day, under the open heaven, the world and my ego appeared to me as one coherent mass of sensation," to be understood exclusively through "physics and in the physiology of the senses and by historico-physical investigations."[39] Metaphysics, like God, was dead, and with it, the self.

When SPR founder and first president Henry Sidgwick went on the attack against Mach and what he called the "pulverizing scepticism of Hume," he did not even bother to turn to Kant or metaphysics for ammunition. His spirit, like that of the society he helped to found, was resolutely empirical. But by 1895, when Sidgwick delivered a lecture to the Glasgow Philosophical Society entitled "The Philosophy of Common Sense," Sidgwick, the once philosophical powerhouse, appeared quaintly Victorian and irrelevant.[40] There was no turning back the tide to the days of John Hutchinson and Joseph Priestly. The philosophy of empiricism now required empirical evidence. Even Oliver Lodge's charge against "solipsists like [Ernst] Mach and [Karl] Pearson" put the problem somewhat optimistically: given Mach's deconstruction of the ego, the self was in no position to confirm even its own existence, let alone a physical reality beyond itself.[41]

To the extent that the various "self-" conscious crises of modern history (the Victorian crisis of faith, the modernist crises of subjectivity and language, the postwar existential crisis, and the postmodern crisis of meaning) are rooted in the intrinsic limitations of empirical epistemology, they must also be seen not just as a cognitive issue, explaining the failure of reason and the unreliability of the senses, but as an ontological issue as well: the collapse of the self as guarantor of the empirical subject. Psychical research sought to reform this epistemological deficit at its source, at the level of the self. This was unlike Mach, Pearson, and their neopositivist heirs, who dealt with the problem of knowledge by bypassing the fatally flawed subject and focusing instead on the primacy of logical analysis. This merely compensated for the self; it did not try to save it. Psychical research, in contrast, reinforced this problematic subject, supplementing synthetic sensory cognition with a noetic supersensory mind. This subject could posit not only its own objective existence—that is, it could "have" a body rather than be just an emergent property of that body—but also the possession of the body's metaphysical corollary: it could "have" a soul.

If we look at the research program of the SPR—its emphasis on the irrational, the visceral, the unconscious, and the unexplained—it contends with the same regressive and relativistic subjectivity that bedeviled continental psychology. And indeed, the SPR was intimately part of that disciplinary conversation. But while it was fully immersed in the problems of what Dorothy Ross calls "cognitive modernism," its response to the problems posed by psychology was wholly unique in the human sciences. The key to psychical modernism was its resolutely optimistic view of evolution as progressive. Psychical theory augmented evolution—not with a Nietzschean panpsychic will—but with a psychical mind, one that could fully transcend evolution's material scheme and thus escape its dark biological dicta. But yet, the psychical self still fully participated in the uncertain evolutionary drama of the struggle to become. It was a supplement to, not the negation of, the dangerous and dynamic naturalism of the nineteenth century. As Lodge explained, "The disintegrations of personality, the painful defects of will, the lapses of memory, the losses of sensation such as are manifested by the hysteric patients of the Salpêtrière and other hospitals, the lesson to be learnt from those pathological cases is not one of despair at the weaknesses and ghastly imperfections possible to humanity rather on this view it is one of hope and inspiration . . . some specimens of our race have already transcended [evolution] and have shown that genius almost superhuman is possible to man and have thereby foreshadowed the existence of a larger personality for us all." Even this atavistic underbelly of the unconscious was a resource for further creative evolution, a forgotten part of our psychical inheritance by which "we have got hold of a handle which turns the mechanism of our being." Psychical research looked beyond the fragmentation of the self, endemic to modernism, to acknowledge instead the "underlying psychical unity existing beneath all our phenomenal manifestations" that made us ultimately whole and ultimately transcendent.[42]

By rescuing progressive evolution, psychical modernism saved the unsalvageable self, and in so doing, reaffirmed heroic science and the epistemology on which such a science rested. The principal means and end of this salvific strategy was a spiritual (as opposed to merely cognitive) defense of the self, although these theological and epistemological crises are so deeply related as to define a single problem. This is not to impugn the SPR's empirical practice as a science, but to clarify the original project of psychical research in intellectual history as the scientific reinstantiation of the metaphysical self. Psychical research began (and continues) as something far more imperative than the stated research object of confirming what Myers termed (in *Phantasms of the Living*) a "supersensuous" faculty of perception. More than a question about the nature of subjectivity, it was an inquiry into the ontology of the subject itself; an attempt to satisfy existential concerns regarding this life and the next; an affirmation of our participation in something greater than our "selves" as described by biological existence and a utile psychological awareness.

Beginning in the late 1890s, Lang, Lodge, and Myers amplified psychical evidence into a theoretical defense of the transcendental subject, the vulnerability of which they linked to the deepening pessimism poisoning culture, philosophy, and the modern self—the flaws of empiricism coming home to roost. In *Reason and Belief* (1910), Lodge unequivocally affirmed both the spirituality and the individuality of the self with his theological assurance that "we are chips or fragments of a great mass of mind, and of the great reservoir of life and . . . each fragment of spirit is supposed to become a separate individual through incarnation."[43]

Lang, too, offered hope. In *The Making of Religion* (1898), Lang "ventured to suggest that 'we are not merely brain'; that man has his part, we know not how, in we know not what—has faculties and vision scarcely conditioned by the limits of his normal purview," which could be regarded, "though they seem shadowy, as grounds of hope or at least as tokens that men need not yet despair." And in his autobiography, *Fragments of Inner Life*, Myers likewise expressed his assurance that "science is now succeeding in penetrating certain cosmical facts which she has not reached till now. The first of course is the fact of man's survival of death."[44]

Thus, psychical research promised to forestall the negative implications of modern psychology both for the self and, thus, for knowledge conditioned on the self. Myers, Lodge, and Lang not only prosecuted this program but also to a large extent developed it: Myers through the core hypothesis of the subliminal self, Lodge with his dynamical epistemology and entelechal evolution, and Lang through the use of global mythology to replenish the wonders lost to man and mind in Tylorean anthropology. These theories converted that which remained in excess of satisfactory explanation, such as hypnotic trance, spirit communication, the unconscious mind, and tribal magic, among other liminal truths that could not be let go, into a consolidated challenge to both materialism and relativism. The category of "psychical phenomena" was an explanatory rubric that could assimilate rogue information troubling to other disciplines, using it as a basis to restore the transcendent subject to a still fundamentally modernist discourse of mind.

THE SLOUGH OF DESPOND

While visiting the SPR archives in London, I decided to attend the twenty-sixth annual International Congress of the Society for Psychical Research. I was struck by something of a cultural divide in the midst of the gathering. The parapsychologists, mostly graduate students, communicated their finding in terms of z-scores and mean chance expectations. They had a scientifically chaste, unsentimental air. The psychical researchers, in contrast, were more ardent, holding fast to spirit photographs, poltergeists, and the survival hypothesis. Occasionally, a donnish research veteran would speak to larger moral and philosophical themes in a manner worthy of Myers himself. It was a tale of two cultures.

All these contradictions had briefly synergized in 1882 in the context of still fluid disciplinary boundaries, philosophical naturalists, and great expectations for "Science." But things did not hold together long. Professionalism and credentialing were increasingly the grounds upon which one spoke, narrowing the sweeping vista required by the project of psychical research. The grand cosmological authority wielded by Oliver Lodge was increasingly viewed as extradisciplinary amateurism, as the Stanford psychologist, Frank Angell, observed in 1917: "It must be said with the utmost frankness that the mantle of Sir Oliver Lodge's great reputation as a physicist cannot be stretched to cover his work in Psychical Research . . . what is necessary for the advance of Psychical Research in the eyes of the scientific world is precisely what all other kinds of scientific work demand; that is, the undivided time and attention of investigators possessing a special training for their work."[45]

In Angell's view, it was not just the lack of professionalization in, but also the spiritualistic encumbrances upon, psychical research that caused him to deem it "unduly hampered in its investigations by religious complications." When deliberating with his colleagues in the Stanford psychology department as to whether or not they should accept an endowment to pursue psychical research offered by Thomas Wellton Stanford in January of 1912, "the question arose of whether, in view of Professor Sidgwick's authoritative utterance that Psychical Research had made no discernible progress in the last twenty years, the field was not a slough of despond through which no scientific progress was possible."[46] John Coover and Frank Angell's solution was to create a new "statistical method of experiment in psychical research," with practices and questions so narrowly defined that all such extrascientific considerations would be structurally excluded from the research going forward. This entry into university life signaled the beginning of what was to become the professional discipline of parapsychology, named and formalized by J. B. Rhine in 1930 as part of the Duke University psychology department, then under the leadership of former SPR president, William McDougall.

But despite the early momentum of parapsychology, propelled by Rhine's promising laboratory protocols and affirmative results, academic receptivity to parapsychology soured by the late 1970s. Scientists, pushing back against parapsychology's rising star, began to aggressively question its methods and results, calling public attention to some troubling discrepancies. While hundreds of articles referencing psychical research appeared in *Science*, the official journal of the American Association for the Advancement of Science (AAAS), between the 1880s and the 1920s, slowly tapering off through the end of the 1970s, such references became exceedingly rare after 1980, notably in the wake of two major scandals involving high-profile researchers.[47] But parapsychology's response to every scientific rebuff and disciplinary failing has been to incorporate the criticism and to refine its own methods. It has become more quantitative, more computerized, and less subjective, tamping down on any experimental interference from "wish fulfillment" or other "religious complications." Zener cards have given way to ganzfeld experiments, while ganzfeld studies have become increasingly automated.

The research expanded beyond its traditional domain of academic psychology when the Stanford Research Institute began conducting remote-viewing experiments on behalf of the US Central Intelligence Agency in the 1970s. In the 1980s, mathematicians and physicists at Princeton's PEAR Labs incorporated random-number generators to test for "psi" as a subatomic effect that could interact with both living and nonliving information systems, giving it a quantum context. In 1999, Richard Wiseman used this technology to conduct the first mass-participation experiment with his Mind-Machine, an electronic "coin toss" that tested the predictive powers of some 27,500 subjects in shopping malls across the United Kingdom. The results, however, published in the British Journal of Psychology, were only consistent with chance. Undeterred, parapsychologists designed subtler, more ingenious tests, going after the existence of a "gut sense" or "heart sense" by measuring precognition with EEGs (electrogastrograms) and EKGs.[48] (Presumably, "emotion at a distance" would be a less developed, and thus more pervasive and detectable, effect than "thought at a distance.") More recently, in 2011, the well-known Cornell psychologist, Daryl Bem, put this idea of a psychic physiology to the test, asking subjects to determine which computerized curtain concealed an erotic image.[49] The publication of the resoundingly affirmative results in the Journal of Personality and Social Psychology dismayed some psychologists concerned about professional standards, but most conceded that Bem's work appeared sound on the surface. The editorial staff defended their position to publish on the grounds that Bem met the experimental criteria of their referees. In response to the seeming legitimacy and scale of Bem's nine-part study, NPR science blogger Robert Krulwich advised: "If serious scientists can repeat his results, this story is going to be big." Initially, the failure to replicate Bem's findings placed the controversy at the boundary of psychology: should "Feeling the Future" ever have been allowed to breach the hold of parapsychology and enter the academic mainstream? But the real crisis ensued when "replication" itself became the issue under scrutiny. As it turned out, more than half of the published studies, including even canonical ones, had not met this replication standard now being used to discredit Bem. In contrast, Bem, with a team of three other scientists, authored a meta-analysis of ninety precognitive studies in 2015 (further revised in 2017), which roughly upheld his own p values for "Feeling the Future."[50] Still, the status of his results remains in parapsychological limbo because the terms that might be used to certify them were put into disrepute in the process of appearing to do so. In other words, if Bem can be right by professional standards, the standards themselves must be wrong.[51]

It seems ever thus, for parapsychology. However near it comes, the quarry always withdraws.[52] Either the object itself vanishes under scientific scrutiny, or the scrutiny itself comes under fire. Bem's survey discussed the possibility of an "experimenter effect"—the idea that some individuals are psi-conducive and others psi-inhibitory, appealing to the notion of "quantum consciousness" for support. This points to the deeper theoretical divide that ensures these methodological debates remain unresolved. There is no medically sanctioned model of "quantum consciousness," only a physico-chemical one, mooting both the experimental phenomena and the experimenter effect. Skeptics view such quantum schemes as so much New Age "woo" (the preferred term of colloquial disparagement), and the idea of an experimenter effect as a sad measure of parapsychology's depth of denial, with its endless call for more research and refusal to take "null" for an answer. A former graduate student, freshly decamped from his Ph.D. program in parapsychology, described it as "a house of cards and one whose very foundations are extremely shaky and yet people continue to build on top."[53] Yet, despite his tone of emancipation, his choice of title for his blog, "everythingispointless.com," seems to ratify at least one part of earlier psychical speculation: that the failure to affirm some higher plane of existence would erode meaning on this one.

It is important to note, however, that those professionally trained scientists most intimately involved with this research (and most familiar with the data) maintain that a small but consistent effect is in evidence, and that support for psi is growing more robust, not less, as it adopts more stringent tests in response to criticism. Instead of questioning the legitimacy of quantum considerations, they turn the tables and ask, Why are twenty-first-century models of the mind still so stuck in Victorian times? Especially given the inadequacy of its mechanical physics to the task? Dean Radin devised a novel "double slit experiment" meant to definitely tie the "observer effect" to human consciousness, incorporating meditators in the collapse of the wave function. If his results hold up, it will no longer be possible to perpetuate ambiguity around the notion of observation—that is, is this merely some aspect of mechanical detection or a genuinely mental event? Like Bem, Radin has prepared a replication packet to facilitate broader engagement, with promising results among the few "straight" physicists willing to affirm its data.

Given its ideologically loaded content, parapsychology's commitment to a purely quantitative idiom makes tactical sense. But still, this remains a science that is trying to get at what is fundamentally a personal force in an otherwise impersonal universe. As such, its statistical methodology is strangely dehumanized and alienated from the heart of its own propositions. In his autobiography, Frederic Myers raised a note of caution regarding the SPR's future: "This inquiry, however earnestly I endeavour to keep it thoroughly scientific in tone and method, must inevitably be not something less but something more."[54] But out of necessity, parapsychology has had to become something less: it must exist in a theoretical vacuum, outside the context of human experience, in a void of philosophical significance. Because of its profound threat to the physicalist paradigm, it must be silent and let the "facts" speak for themselves. And yet, these facts themselves have been for the most part inaudible to science. In the rather decisive words of parapsychology's most prominent defector, Susan Blackmore: "Psychical research has failed. It has failed not only to fulfill the hopes of its founders of one hundred years ago, but it has failed to establish itself as a respected area of scientific enquiry and, more generally, to contribute substantially to our understanding of human nature."[55] That may be so, but it is not entirely clear to what extent that pertains to the researchers or the phenomena or the fixed understanding of human nature to which it failed to contribute.

BEYOND THE DESPOND

On a "starlight walk" in 1869, Sidgwick and Myers "caught together the distant hope that Science might in our age make progress to open the spiritual gateway which had been thought to close."[56] This was the heroic project that became psychical research in 1882, and then began to dramatically pull in its sails a few decades later. Nevertheless, their ship appears to have come in once again with the Templeton Foundation, established in 1987. Victorian to his fingertips, billionaire founder John Templeton believed in the sacred principle of progress: the spirit of advance that characterized modern science must now be allowed to accelerate the evolution of theology and make it once again "the queen of the sciences." To that end, the foundation gives up to $70 million a year in grants, much of it to major research universities pursuing work mainly in quantum physics, moral philosophy, psychology, and evolutionary biology.

Like the SPR, the Templeton Foundation generates a lot of controversy. But it is not just the nature of the controversy (the desire to import religious questions into scientific frameworks) that makes the two organizations so similar to nineteenth-century psychical research, but the location of the controversy. This debate is unfolding right in the heart of academia itself. By passively funding, rather than initiating, sympathetic academic research, the foundation maintains this as a dialogue internal to science, prosecuted by elite actors. With its $2.1 billion endowment and soft sell, the foundation has a list of beneficiaries that is a veritable "who's who" of science and philosophy, reconstituting the kinds of social and intellectual networks not seen since the early days of the SPR.[57] (As we have seen, that was a cast that included a past and a future prime minister, archbishops, Royal Society presidents, Nobel laureates, Knights of the Realm, and so on.) Recent awards of the £1 million Templeton Prize, which honors "a living person who has made exceptional contributions to affirming life's spiritual dimension," have gone to academic celebrities: teleological cosmologists, such as Martin Rees (2011), George F. R. Ellis (2004), John Polkinghorne (2001), and Bernard d'Espagnat (2010); and advocates of theistic evolution, such as Francisco J. Ayala (2010) and Arthur Peacock (2001). The amount of the Templeton Prize is, notably and intentionally, larger than that of the Nobel Prize.

Templeton money is also behind the creation of the AAAS program "Dialogue on Science, Ethics, and Religion," established in 1995, as well as the founding of the Faraday Institute for Science and Religion at Edmunds College, Cambridge University. Such incursions into the citadel inspired the criticism that "the Templeton Foundation is sneakier than the creationists" for the manner in which it is insinuating religion into science.[58] But what does all this recent "talk of God" saturating the public discourse on science really mean?[59] By all appearance, the reconciling impulse of the last fin de siècle appears resurgent once more, endangering the boundary of science and religion. But is this a soft reconciliation or a hard one? Looking more closely, this rapprochement is much more in the spirit of Kelvin and Faraday than Sidgwick and Myers. The midcentury architects of modern secularization put in place a two-sphere model of accommodation that has been notably more successful than the French Revolution's "cult of reason," Comte's "religion of humanity," Tyndall's scientism, or, for that matter, psychical research itself. All this solicitation over religion is probably best understood as part of the routine maintenance of this "two-sphere" model coming from within science itself. The progress of globalization; the writings of Richard Dawkins; the terrorist attacks of September 11, 2001; the potency of naturalistic reductions of human nature coming out of genetics and neurobiology: all this potential friction requires some fine-tuning of the harmony between the separate spheres of religion and science. Even the seemingly metaphysical positions taken by Templeton-funded scholars are little more than a cordial nod to God. For all the concerns raised over anthropic cosmology and evolutionary theism, they keep "intelligent design" so far out at the cosmological fringes that it plays no causal role in biological evolution or theoretical physics. God remains right where Voltaire wanted him, but without being rudely escorted off the property.

It remains to be seen if this highbrow reconciliation will be able to countermand the conservative revival of a populist warfare thesis. Certainly, "religion as folk wisdom and natural law" versus "godless secular elitism" is not a winning cultural strategy for progressives. So, while all this attempt at dialogue is unsavory to purists, it is perhaps naive to think that science can remain fully aloof from the politics of knowledge and its corollaries of power. (Denial of climate change, for example, happens not so much at the level where the science is produced, within the university system, but at the level where it is received, attached to the agendas of nonscientists.) When the BAAS set about institutionalizing science in 1833, the goal was not just to be intellectually independent but also to create a guiding role for scientists in public life, one in which science would be respected rather than feared. This meant maintaining the intellectual status of theology in relation to (but emphatically, not within) scientific discourse.

Despite the concerns repeatedly raised by leading members of the AAAS, this reconciling impulse of the last two decades is more a continuation of the deep structures of the status quo than a disruption of them. Templeton Foundation money has funded not some growing influence of religion driven from below, but an elite, accommodationist counterplay managed from above. (Attention here is paid not to the political or theological allegiances of the foundation, but to where the money actually goes.) Perhaps the proper question regarding this issue is: did Templeton get his money's worth? So far, all he has financed appears to be little more than a feel-good dialogue between science and religion—one in which religionists are made to "feel good" about science rather than vice versa. Nothing like Templeton's hoped-for "giant leap forward in our spiritual understanding" has yet come to pass.[60] But when largesse flows to the establishment, it tends to reestablish itself. One blogger put it best when he accused the Templeton-funded Faraday Institute not of brainwashing children into religious faith, but rather of brainwashing kids into becoming accommodationists.[61] The model of harmony is as much a manufactured narrative of the nineteenth century as "the crisis of faith," the difference being that its mythology was more active in giving rise to present-day institutional realities.

However, this does not invalidate the observation made by some thinking Victorians that science and religion were now two and not one precisely because they were conflicted. This conflict could be buried and denied, but ultimately not gotten around. "Separate spheres," even harmonious ones, were intrinsically problematic because they left the project of knowledge divided. This was Balfour's point when he wrote *The Foundations of Belief* (1898). It was also the impetus behind the founding of the Metaphysical Society in 1869, and again behind the creation of the Synthetic Society in 1898. But only the Society for Psychical Research responded with a radical plan of action instead of a discussion group.

Unlike those who sought to bolster the claims of metaphysics by limiting the claims of science, psychical researchers pursued an opposing strategy: expand science into a theistic domain. This was the "hard reconciliation," a concept of secularization based on a single, synthetic sphere. And this required an atmosphere not of accommodation, but of crisis and catastrophe, the ideological prelude to psychical research. The original founders of the SPR (like John Templeton later on) hoped to advance the evolution of religion by attaching it to humanity's most wildly successful knowledge enterprise: Science. But psychical researchers were willing to break eggs to make that omelet. For all its politesse and status-consciousness, the SPR aimed to violate, not harmonize, accepted scientific and religious conventions. It was a full-out, empirical assault both on the physicalist paradigm and on revealed religion. The new evolutionary model included guidance by psychical force, making the "intelligence" in "design" our own; and this same psychical intelligence could likewise evolve our knowledge of an ultimate reality, creating a human-centered cosmology in which self-knowledge, science, and divine revelation converged in the unitary sphere of a theistic science. Recently, there have been a few developments that seem to suggest the reconciliation may be intensifying. These are the Templeton-funded Immortality Project at University of California Riverside and the AWAreness during REsuscitation study (AWARE) led by Dr. Sam Parnia, a specialist in cardiopulmonary resuscitation at the Stony Brook University School of Medicine. Both studies involve an inquiry into the phenomena colloquially called "near-death experiences," or NDEs. This research could signify the full extension of the current cycle of rapprochement, sending the pendulum swinging in the other direction. Or perhaps it means something more: a decisive break across the boundary of science and religion, the SPR's insurgent drive to synthesis asserting itself once again within academia.

The first AWARE study (2008–12) was remarkable for its willingness to explicitly entertain the reality of the NDE at the center of institutional medicine; it involved twenty-five major hospitals in Europe, Canada, and the United states. While such concerns as "the psychological benefits of the

NDE" and "the role of brain-blood perfusion in resuscitation outcomes" gave AWARE some medical cover, its core modality involved testing the audiovisual perceptions claimed by patients who died on the operating table.[62] Only patients in full cardiac arrest were considered, and they had to identify a visual target hidden somewhere in the operating room, viewable only when "floating from above" as described in the classic NDE. The image was generated at random to ensure double-blindness. ("Near death" is a bit misleading, since the significance claimed for these episodes is that they occur when the person is actually dead. Parnia preferred the term "actual death experience" for his study.) Since one's body would presumably be lying dead on the table at the time the target was acquired, such results would be difficult to sideline. The orthodoxy of the "null hypothesis" would be made vulnerable to challenge on a more legitimate basis, opening the floodgates on similar research currently walled off by universities, where it has remained unfundable and invisible for more than a century. As yet, there have been no positive hits on the images, but Parnia feels that auditory and emotional awareness have been established to a reasonable degree of certainty during clinical death. Rather than accepting this as evidence for mental awareness independent from the brain, the more conservative approach to these anomalies is to consider a new chronology for death, with consciousness unwinding more slowly as a series of microprocesses occurring even after major organ failure. Expectant fans awaiting AWARE's final results may be disappointed to find that this ends up being framed as a revolutionary study about dying rather than of an afterlife.

In 2012, after the initial AWARE study had been concluded, the Templeton Foundation cast a wider net at the question, tapping John Martin Fischer, an eminent scholar of free will and moral responsibility, to lead a $5.1 million interdisciplinary research effort under the masthead of The Immortality Project (2012–15). Fischer, whose research inclinations did not lean in this direction, undertook this duty under the somewhat strange misconception that "no one has taken a comprehensive and sustained look at immortality that brings together the science, theology and philosophy."[63] And yet, scientists, theologians, and philosophers have all converged upon this very topic in a manner both comprehensive and sustained since 1882, across a variety of research platforms explicated in thousands of scholarly publications. The effortless assertion that the Immortality Project involves groundbreaking novelty signals Fischer's insider sensibility by breezily discounting 130 years of unsanctioned research. Rejected knowledge thus creates an ironic criterion for arrogating academic authority in its taboo domain: total ignorance of the existing scholarship. The choice of a philosopher rather than a scientist to lead the effort set a certain tone, favoring the latitude of humanism over the empirical reduction to "true or false." So, despite its air of maverick investigation, with "Immortality" and "Near Death Experiences" blinking on the marquee, a little more than half of the funding went to talking about how we conceptualize immortality, rather than making a direct test of it. Theologians received grants to ponder immortality theologically, while humanists were funded to consider its temporal and civic dimension, its desirability, the role of memory and imagination in its description, and other cultural dimensions of its construction.

John Templeton (1912–2008) had hoped for something more concrete than a mere public display of institutional affection.[64] But there are limits to what his legacy foundation can achieve and still remain relevant. To work with academic stars and major universities, these projects must unfold in the framework of a soft reconciliation. To fund a direct challenge to the physical worldview, or, conversely, to conspicuously obstruct religious hope in actual "spiritual realities," threatens the fragile peace. A multidisciplinary study that encouraged scientists, theologians, and philosophers all to elaborate their own ideas of immortality would seem to inevitably cross those wires. And it did. More than $2 million went to fund empirical research, which conventionally meant looking for physical and psychological causes in lieu of spiritual ones. That seems more like a collision than a conversation. Prize money went to neuroscientists in Israel examining the NDE "life review," offering brain-based explanations for the way these narratives were stored and the adaptive psychology of deploying them in times of crisis. Researchers at the University of Barcelona used immersive virtual reality to simulate embodiment in another existence, allowing test subjects to survive their death (the perennial difficulty) and report on aspects of this experience. This included the tunnel leading to the white light, life review, and being out-of-body, suggesting that the NDE was a psychologically manufactured event, unfolding in the brains of people who only *thought* they were dying. Previous research came to similar conclusions from the other end, looking at spontaneously occurring NDEs, determining that in nearly one-third of these instances, people were not actually in danger of dying. Like the Immortality Project subjects, they only felt it to be true. Additional Templeton funding went to Dr. Sam Parnia to continue his NDE hospital study, but this time widening the pool beyond cardiac arrest patients to include all planned and unplanned losses of consciousness. This gave the new project a less controversial framing, since looking for NDEs in the brains of living persons would seem more in line with Fischer's thesis that these were brain-based hallucinations. Parnia, however, still understood the research as casting a net for paranormal phenomena that might occur under duress as well as death, any situation that caused the brain to "tune' the mind differently. (The metaphor often used for the nonphysical paradigm features the brain as a kind of receiver and consciousness as the transmission.) The flagship of the Immortality Project was Fischer's own book, coauthored with his research assistant Benjamin Mitchell-Yellin, *Near-Death Experiences: Understanding Our Visions of the Afterlife* (2016). Despite the lure of "immortality" and the winged creatures on the cover, the authors advanced their analysis of five NDE narratives against this backdrop of physical evidence that strongly implied there was no such thing.

The problem with the Immortality Project is not its scholarship, its leadership, or any given tributary of funded research, all interesting questions pursued by high-caliber scholars. But as an exercise in rapprochement, as Templeton projects are, it fails. Its difficulties inhered at the level of conception, but there were also unforeseen challenges that made a tactful execution of this cross-cultural conversation unlikely. Instead of parachuting into a deserted field of research, as Fischer had imagined himself doing, he found the topic already densely occupied with credentialed investigators, many with medical training. In addition to the AWARE study, there were multiple medical researchers working their own sustained lines of inquiry, such as Dr. Jeffrey Long (a radiation oncologist), Dr. Pim Van Lommel (cardiologist), Dr. Bruce Greyson (psychiatrist), and Dr. Edward Kelley (who earned a Ph.D. in psycholinguistics and neuropsychology from Harvard), among others attempting to elevate NDE research into something more disciplinary and data-driven. To the extent the Templeton Foundation had hoped to stake out a larger academic position in a burgeoning area of popular interest, the effort proved rather more complicated. The friendly foray into the afterlife became immediately more fraught now that it involved an explicit defense of science's own internal boundary. Even though Fischer and his coauthor, Benjamin Mitchell-Yellin, stated for the record that "no part of [their] view in any way denigrates religious belief or a belief in the afterlife," that forbearance does not apply to spiritualities that infringe on the framework of the physical sciences. As far as that goes, the authors admit that their "aim is to call into question a particular route to religious beliefs and beliefs about the afterlife"—namely, the NDE—precisely because it asserts these claims on the basis of direct experience and even additional support from scientific research.[65]

The awkwardly "respectful" preface to *Near-Death Experiences: Understanding Our Visions of the Afterlife* makes plain the difficulties of the authors' situation, made all the worse by their inability to fully grasp it. They write, "We have a deep respect for individuals who have had near-death experiences. Indeed, we hope that our serious attention to these remarkable experiences displays our respect. We also respect the deep religious convictions of the many who have thought about the significance of these experiences."[66] But to state one's respect (however many times) is not the same as conferring it or, at the very least, making others believe that one has done so. The authors' many protestations suggest they sense this disconnect but choose to deflect it as some misunderstanding arising from the other side. Is it? The upshot of their research is that NDEs are, at best, a productive hallucination, sincerely felt, "deeply attractive and even deeply inspiring." This humanistic, transformational NDE celebrates our psychological capacity to extract meaning and growth in the face of difficult experiences, and also the capacity of mind to give comfort in the face of death. And yet, those are not the grounds upon which those who have experienced NDEs value these experiences, which are felt as direct spiritual revelations of a life after death. Consequently, they are unlikely to feel honored by Fischer and Mitchell-Yellin's scholarly attention, no matter how many times the philosophers instruct them that they should. To deny the physical validity of these experiences is to turn them into psychiatric events and "respect" into something more akin to professional "sympathy," encouraging patients to make the best of an adverse medical experience. The disingenuousness of that "respect" is most clearly communicated in the way they discourage those who have not experienced them from investing in these stories themselves, characterizing such beliefs as a form of "false hope" and "wishful thinking." The implication of such a warning is, of course, that the NDE is an intellectual contagion that should not be allowed to spread. Instead of a world-redeeming truth to be evangelized on the internet, it is a regressive, superstitious credulity sourced in an episode of madness. The authors approach the subject as if they are walling off a dangerous meme, but they could just as easily get themselves shut out instead of the very conversation they wish to start.

My point is not to question the valor or validity of these academic arguments, but to look more deeply into the broken diplomacy in which they take part. For this, the Immortality Project seems uniquely illustrative, highlighting the limits of accommodation and the degree to which "wishful thinking" and "false hope" pervade the modern institutional sensibility. Fischer and Yellin at once wish to assert control over, and remain friendly with, so-called NDEers, attempting to close the distance between their positions with a few rhetorical flourishes of the soft reconciliation. They maintain that "the purpose of the book has not been to discredit those who share their near-death experiences," but only "to critically examine the purported implications of those experiences." This seems to imply that the latter has absolutely no bearing on the former: how we esteem its knowledge, accord status to the believer, and allocate resources to evaluate it and potentially benefit from it. In smoothing over what they take to be but a few, untroubling inconsistencies, the authors offer only: "In the end, we believe we can have our cake and eat it too."[67] This is a familiar kind of magical thinking to which authority is prone: erasing the power of the other, while dreaming up the logistically unimpeded, morally unobstructed exercise of its own advantage. In this scenario, scientists are greeted as liberators and showered with rose petals, as the masses recognize that the truth has set them free. This lack of insight into the other suggests overconfidence, as if one's enforcement strategy didn't require mutual understanding. And in this particular academic fantasy of power, knowledge flows in one direction. This is also a way of distracting from one's own vulnerability, exaggerating the success of one's own initiatives and ignoring those externalities beyond one's control. An article on NDEs that appeared in *Scientific American* (September 2011) assured its readership that "near-death experiences are often thought of as mystical phenomena . . . research is now revealing scientific explanations for virtually all of their common features."[68] Since then, interest in the topic has only gained ground.

One is reminded of Faraday's attempt to pin down table turning in 1853 as a muscular reflex, only to watch these magnetic circles spin out of control into the so-called "spirit craze" over the next few decades. There are similarities worth noting between the rising tide of NDE research and the popularity of Victorian spiritual investigation in terms of their cultural dynamics. Both are lay insurgencies that intentionally mix the idioms of science and spirituality, subverting the segregation and control of these narratives by institutional elites. The near-death experience and the séance encounter focus the power of knowledge directly into the hands of the individual, allowing them personal confirmation of a spiritual reality along with access to its store of wisdom. Both nineteenth-century spiritualists and twenty-first-century NDEers partner with scientists selectively, as participants rather than merely objects of investigation, creating their own institutional spaces and hierarchies, and controlling their own discourses through a range of publications. There is no fixed ideology; these are open-source systems, grown through the consent and collaboration of autonomous individuals. Unlike the separatist impulse of the cult with its power focused on a leader, spiritualism and the NDE movement spread horizontally into the world, networking across groups to form new kinds of alliances. Belief in séance spiritualism or the near-death experience, does not itself constitute a religion, or a politics, or any one specific social reform movement. They are instead powerful amplifiers: potentially deepening one's Christian faith, one's investigation of nature, one's utopian hopes for a better world.

Both movements arose in the wake of a significant media revolution, a century and a half apart: the penny press taking off in the 1840s, and the modern internet. Both lowered not only the bar to accessing information, but also, and more significantly, the bar to publishing and promulgating it. The mid-Victorian period was a watershed moment when the once passively instructed believers began to realize their power as economic consumers, and even makers, of knowledge, setting in motion the proliferation of competing truths that continues to roil the marketplace of ideas, where the threshold of entry seems, increasingly, to be: "is anyone buying it?" At the same time, official knowledge was becoming more disciplined, not less, maintaining intellectual hygiene by filtering out lay opinion. But Victorian spiritualists were looking to be part of that public process, as justified by the moral and intellectual authority of their own empirical investigation. They dealt themselves into the public discourses of science and religion, promising to transcend and perhaps even reconcile these divided spheres of knowledge. The truth they were after was meant to be shared and refused to be shamed, pulling in, under its umbrella of respectability, millions of ordinary people going about their daily lives in an unremarkable manner even as the knowledge they touched trembled on the brink of some extraordinary, civilizational transformation.

There is something here that resonates deeply with the NDE narrative, especially as it evolves a more public character. Hundreds of near-death experiences have already been archived on dedicated websites like the Near-Death Experience Research Foundation (NDERF) and the International Association for Near-Death Studies (IANDS), available to professional researchers and individual seekers alike. Not unlike the SPR's *Phantasms of the Living* (1886) or "Census of Hallucinations" (1894), these compendiums take the cultural scattershot of rumor and anecdote, and turns it into something more akin to a database amenable to a degree of quantitative analysis. Adding to this catalogue of lay testimony are the growing number of medical professionals willing to contribute their own anecdotal support for patient experiences, giving the NDE a layer of professional verification. When these pools converge, and the doctors testify to their own near-death experiences, it can constitute a seismic cultural event, as with Dr. Eban Alexander's *Proof of Heaven* (2012) or Dr. Mary Neal's *To Heaven and Back* (2012). These elevated the NDE narrative to national attention just as the Immortality Project was getting under way. But it is the formal studies, undertaken with institutional support and finding their way into medical journals, that threaten to impose the NDE onto professional science. The NDE's scientific encounter has already proven much more robust than anything achieved by Victorian spiritualism. The spiritualism movement's institutional legitimacy began to crest in the early 1870s with the Dialectical Society and Crookes's "scientific spiritualism," but collapsed a few years later, hardly halfway through the decade, under the weight of its own improbability. By contrast, scientific interest in NDEs has only grown through efforts to conform to scientific standards. As a research object, it has neither vanished under scrutiny nor blown up in scandal, as might be expected under the glare of this attention.

But the real, evangelizing power of the NDE comes not from this credentialed research as a form of disembodied data, but from each given incident of personal testimony that manages to deeply touch its listener. Many near-death accounts are catalogued as video interviews or recorded live before an assembly of seekers, an unmediated transaction between the teller and the told. This is a human-to-human communication that recruits the kinds of judgments social creatures are good at making, boiling the factual analysis to its critical essence: is this person credible? If the answer is yes, then the power of the testimony takes on a resilience difficult for science to touch. It is not an abstract "belief" that has been seeded, but a sense of trust. It allows that reality to be shared even where it cannot be verified or understood, forging a bond of common interest. By contrast, all the core samples and ocean buoy readouts in the world are not enough to convince persons who feel climate science is a scheme to defraud them. The accuracy of facts hardly matters if they are to be twisted against one. Near-death experiencers travel to the source where the moral and factual implications of knowledge are one, conveyed as a sweeping revelation and not as bits of data with decimal points. That's the advantage of the supernatural, which deals in certainties and totalities that are difficult to hold back. There is sense to urging caution. Skepticism is the discipline of resisting such attractive tales and the human tendency to prefer error to ignorance. We are "storytellers," Fischer advises, and we need to impose our sense on the world.[69] The prescription for this mythic tendency: trust the facts, not each other.

It is not unreasonable to present the public with alternative physical explanations for NDEs or to urge fellow researchers to impose the highest standard of proof for an improbable hypothesis. The rules of science are not what triggers institutional resentment, so much as the appearance of scientific rule. Yet, Fischer and Mitchell-Yellin impose an insider's geography from the start by characterizing NDE research as a form of supernaturalism. Applying that term is inherently delegitimizing, placing NDE research outside of nature and outside the scope of empirical inquiry. It elevates Newtonian physics to the status of reality, when it merely occupies the foreground of our phenomenological perspective and therefore most reliably describes our experience. The authors fail to reflect on the politics of knowledge in which they play a part, even as they ably walk others through the fallacies of their reasoning. This binary view of natural/supernatural is a specific strategy employed by secularists to order the social arrangements of knowledge in a certain way. Frederic Myers, who sought an alternative paradigm for secularization, rejected that language and constituted a new category for psychical phenomena, "supernormal." The term is still in use by parapsychologists to denote phenomena that operate outside the understanding of nature, but not outside of nature itself. It is not enough for Fischer and Mitchell-Yellin to state that they "agree that supernaturalism should get a fair hearing," if they undercut that intention at the same time. As bridge building is among their avowed aims, then it is not enough to be "fair," there must also be the appearance of fairness. What does this bridge look like from the other side? Conventionally, disciplinary science is peer-reviewed. But as this "supernatural" topic has been placed somewhere outside of science, the ability to participate in the production of knowledge is significantly lowered. The authors advance their own interpretation of experimental results in ways that supersede the judgments of the primary researchers. They justify this by separating out the technical expertise involved in establishing the facts from the critical insight used to interpret them. But that intellectual division of labor doesn't quite work. The significance of facts may require deeper knowledge of the brain and body. One might connect the dots differently, depending on one's own disciplinary logic.

The authors bring this same psychophysical parallelism to the NDE, which they see as unfolding in two distinct frames of reality. In one of these, the inner realm, the NDE gets to be entirely genuine, while in the external world, physical laws are allowed to prevail. It is the rapprochement redux: between the authors and the NDEers, between the physicists and the philosophers, each taking charge of their allotted sector of truth, each contributing to the summary of meaning, all in harmony. And yet, on closer examination, there is a hierarchical shuffle that cannot satisfy. The neurologist must be deferred to, the NDE researcher is overruled, and the NDEer is almost entirely ignored, making it into this study only as a disembodied narrative. The authors conducted no additional interviews of their subjects because seeking more information from the source was not warranted. Since these events all unfolded in the imagination, reconstructing that hallucination only put its accuracy at risk. The authors thus rendered the person of the NDE narrator irrelevant to the full interpretation, which now rested entirely in their hands as masters of meaning. But this disembodying of the NDE story is strangely at odds with its own physicalist assumption: mental states are always conditioned by neurological correlates. So, even the imaginative features of the NDE implicate certain brain functions both at the level of the experiential qualia (the tunnel, temporality, emotional valence, egoic experience, sense of significance) and also, more practically, at the level of the encoding of memories, their stability over time, their clarity and depth of detail. Other aspects of the NDE, like the simultaneous unfolding of events described by Jung, seem so far outside human phenomenology that they seem to defy the powers of imagination in ways that tunnels, angels, and bright lights do not, and thus add a level of neural significance. To let the NDE story float like a fantasy in a jar seems like a philosopher's retreat from reality, capturing the subject on one's own intellectual terms. While it is unfair to expect two scholars to comprehensively survey this cross-disciplinary field, I do think an opportunity was lost to demonstrate a genuine willingness to do so. That is what the fragility of this epistemic moment most requires: allowing people to peer into the rigors of the academic process in a way they could fully appreciate, while feeling that their own perspectives were given consideration as well.

I have noted the similarities between Victorian spiritualism and the NDE movement, but there are differences as well. By the time spiritualism began to draw formal attention from psychologists and physiologists, it was as physical mediumship, in the 1890s. Unlike Crookes's more open-ended scientific spiritualism, this research was driven by various academic agendas and placed within theorists' own interpretive schemes. Parapsychology stripped down what remained of the Sidgwick Group's original séance curiosity to a quantifiable mental effect, devoid of any spiritual or even theoretical content. All this diminished the public stake in a paranormal narrative that had become somewhat dehumanized and technical. But the arc of the NDE movement seems to be running a different course, growing in concert with sympathetic research interest, while keeping its spiritual core intact, with the NDEer still occupying the center. Even if this research makes no further progress into mainstream science, it promises to grow at the edges, and it is substantial enough ground for NDEers to hold and defend. To slight or ignore the stable of researchers from elite universities who pursue this research will only enhance the credibility of NDE science in the light of a perceived institutional bias. The retreat made by spiritualism was facilitated by scandal and fraud, but NDE groups operate as not-for-profit resources for experiencers and as centers of study. While spiritualism expanded commercially, adding more and more exotic fees-for-service, the NDE movement still remains focused on its own exploration, occasionally monetized the old-fashioned way when an individual turns psychic or sells their story. To the extent that NDE research remains outside of a rigorous back-and-forth with the rest of science, it will still exercise authority over this narrative, splintering the sense of science as universal knowledge.

Whether this more public conversation will result in a deepening mainstream engagement of the NDE or, conversely, galvanize its more determined and organized expulsion by science remains to be seen. But it does seem clear that it will further degrade the equilibrium of these "separate spheres," which must also find their balance against the weight of public opinion. For many, the withholding of scientific support for religious narratives does not feel like academic neutrality but like an intellectual threat. This has been a feature of secularization ever since popular literacy, public opinion, and the commercial press came together in the nineteenth-century private sphere. The authority of science to define physical reality cast a shadow pronouncement over the faithful, unspoken but heard anyway: everything not so defined by science became suddenly *less* real. This was the nihilistic power Frederic Myers recognized, coiled up inside even the most harmonious exercise of empirical authority, when he wrote: "whatever science does not tend to prove, she in some sort tends to disprove."[70] That an academic study of immortality would raise certain hopes that could potentially turn to anger seems rather obvious, especially when it discouraged the very studies it might be expected to engage. But anger would at least imply a degree of relevance, granting academia its share of deciding power. The response has been something more akin to dismissive. Those seeking scientific support for NDEs have other sources to turn to. They do not stop to wrangle concessions from those they can simply choose to make irrelevant.

In the past five years or so, a note of triumphalism has crept into the rhetoric of those fighting their popular insurgency against "materialistic science," consuming books and streaming lectures on alternative archaeology, alternative physics, alternative psychology, and so on. This sentiment doesn't just pervade the chatrooms, the podcasts, or the comment sections where personal opinion goes to congregate. This shift in attitude also appears in the rhetoric of those more formally involved in the construction of forbidden knowledge. The biologist Dr. Rupert Sheldrake, known for his theory of morphic resonance, and Dr. Dean Radin, the lead experimental investigator of psi phenomena, have acquired a less embattled posture. Where in the past they narrowed their language to fit as tightly as possible within academia's constraints, they now speak with an air of vindication that allows them to take a larger view. Sheldrake's most recent book was titled *Science and Spiritual Practices* (2018) while Radin's book, *Real Magic: Ancient Wisdom, Modern Science, and a Guide to the Secret Power of the Universe* (2018), makes no bones about the consilience he sees between the "reality of psi" and the historical conceit of magic. Even spiritual curmudgeons like Sam Harris and Robert Wright now have books touting their meditation practices and mystical insights, while Michael Pollan has turned his scholarly attention to magic mushrooms in an effort to become more spiritually open. What's next, Richard Dawkins in the ashram? But even while this receptivity feels of the moment, a mix of popular zeitgeist and leading cultural opinion, the scientific component on which it draws continues to come from the institutional edges. The center, for now still holds. As when Daryl Bem tried to advance proof of psi inside the castle keep, psychology was mobilized against him. Having a Division of Perceptual Studies at the University of Virginia does not necessarily imply more success for the NDE than parapsychology attained through the Rhine Institute (although, admittedly, it gets a more confrontational hypothesis through the gate). If anything, such adjunct institutional spaces can be ways of keeping research out of one's department, more like a ghetto than a true concession of ground. Of the millions of dollars spent on the Immortality Project, little of it went to open up an internal debate about the nature of consciousness within science. Rather, activity flowed in the other direction, reinforcing the physical orthodoxy in the popular sphere.

For most neuroscientists, the NDE remains the ischemic hallucination of a desperate mind struggling to "survive" its own death per the dicta of evolution. The tunnel thought to be a passage out of this world merely symbolizes the mental struggle to hold open our place in it, something more akin to Thomas Metzinger's "ego tunnel." We are not *in* that tunnel, we *are* that tunnel, a corridor through reality carved by perception, as the brain converts the constant flow of stimulus into the ephemera of experience. To follow that tunnel in hopes of an eternal existence is only to discover we were never even here in the first place. Metzinger's metaphor is a poetic update of the phenomenological view of the self laid down by psychology in the last decades of the nineteenth century. While this reframing of reality as a largely mental rather than physical experience marks the end of the Victorian age and the beginning of the modern, that should not obscure the fact that these sleek philosophical insights were still underlain by an industrial science. And that original disciplinary framework remains in place. While today's neural tomography gives a far better picture of the brain than an autopsy, and a complex synaptic ballet has replaced Charcot's crude cerebral energies, this is the same neurochemical machinery worked out by Hermann von Helmholtz over a century ago. No real additional understanding of how this physical scheme gives rise to a subjective one has since been acquired. (Of course, significant knowledge has been gained about how these two things correlate, but that is not the same thing.) The "hard problem of consciousness" was an enormous obstacle immediately recognized by Wilhelm Wundt, Helmholtz's protégé, and other experimental psychologists trying to correlate stimulus and perception.[71] Some one hundred fifty years later, the failure to resolve the issue has normalized it instead. What should be a glaring insufficiency is instead more like a bother to be worked around. The "hard problem" is simply too big to fit in any research proposal; there is no related hypothesis, at least in Newtonian physics, that could be levered into some testable theory.

While progress has piled up on either side of the divide, mapping the advances of neurology to psychology and *vice versa*, the wormhole connecting these into a single ontological system remains a mystery. The phenomenological view holds consciousness itself to be an emergent phenomenon, an object rendered by the brain to manage the chaos of the perceptual field. As a theoretical assertion, it sounds like a suitably cautious take on a singular mystery. But I'm not so sure. Biologizing the mind is a substantial assertion of ontology, the boldest form of explanation possible. It tells us what something is and, by extension, what it is not. This is despite the fact that none of the antecedent qualities known to neurochemistry comes in any way close to explaining mental experience. And it gets a bit circular when, from the standpoint of those who study it, the brain is also a phenomenological object, giving rise to the consciousness from which it derives. To say this is not to advocate for some radical departure from proven hypotheses or to abandon conservative methods of induction. But as the Victorian "supernormal" reminds us, the boundaries of science are also drawn according to what we are willing to imagine as possible. That consciousness exists as an effect without a cause would seem to hold open the boundaries of conjecture, but it remains locked inside the closed system of Victorian thermodynamics. Thus, even as modern physics ponders the possibility of a world composed at its basis of bits of information, the study of the mind still proceeds mainly in reference to matter. Far from succumbing to the temptations of indeterminacy, psychology became more mechanistic, not less, with the advent of quantum mechanics, doubling down with Skinner's radical behaviorism in the 1930s (objectifying and measuring the conditions of innermost experience).

There is an underlying utility to empirical knowledge that resists the importation of quantum thinking into human affairs as an impractical, unproductive addition. Instead of the foundations of reality, it becomes a kind of fanciful math that plays no real part in the human equation. Whatever value such knowledge might have for understanding the pathway of an electron, letting such acausal, "magical" thinking reenter the personal sphere of cause and effect is regressive from the standpoint of human agency, where secular knowledge lives. This moral suspicion is expressed in the favored term of disparagement, "quantum woo." "Woo" prefers to believe, to intend, to consciously direct its affairs, particularly in medical and financial matters where the stakes are high, the procedures arduous, and the outcomes uncertain. But it is also a practical defense of disciplinary knowledge, which would become too unmoored, too open to interpretation, too subject to lay incursion if removed from solid ground. Psychology's physical reduction of the mind to the brain, however it may be warranted in theory, must also be seen as caught up in the politics of knowledge. Its own, and that of the rest of empirical science. Given psychology's foundational role in explaining the human mind, it was not just an instrument of scientific investigation; it was also its establishing premise. Introspecting the processes of perception was the first order of business for both Descartes and Locke, and reaffirmed by Edmund Husserl in the early twentieth century, updating the philosophy of knowledge with psychology's latest laboratory advances.[72]

But whatever the philosophical overreach of scientific materialism, it has served to hold the line at sensory cognition and physical cause, defending skepticism against the gnostic forces working in reverse. This was the perennial difficulty, keeping knowledge doubtful. The sovereignty of the human mind relied on this embrace of its own ignorance: the right *not* to know, *not* to believe, the right to dispute, the right to consent only to those facts subject to verification. Locke's epistemology was never about setting limits on personal beliefs, such as faith in God. The aim was to prevent public monopolies on truth that let one person set him- or herself up like a god over another. Some three and a half centuries after the publication of *On Tyranny* (1689), the cosmologist Lawrence Krauss draws the same line that Locke did between individual rights and natural law, writing: "I see a direct link, in short, between the ethics that guide science and those that guide civic life."[73] (His essay was in response to the religious justification a country clerk used to deny a marriage license to a same-sex couple.) Krauss takes a classically Comtean view: the price of liberty is religious certainty. To the extent that the arc of secularization bends toward justice, it is as freedom *from* rather than freedom *of* religion. That may be true when it comes to religion's social prescriptions, but not its inner life, which has been a significant factor in reform movements. But Krauss sees the danger as inherent in the conceptual displacement of humanity by "something greater," a writ of entitlement waiting to be exercised. In his essay titled "Why Scientists Should Be Militant Atheists," Krauss urges the duty to challenge religious belief before it gives rise to discriminatory behavior. While most scientists would distance themselves from such a remark, Krauss gets at truths others would rather deny: science involves social values and relations of power. Its ability to project its knowledge into the public sphere rests on a certain degree of political will that it should do so.

But getting the consent of the governed is turning out to be a problem in democracies across the globe. The skeptical epistemology that gave rise to these arrangements imposed no obligation to believe in them. Nowhere is the relationship between science and civil society more strained than in the American context, where political agreements seem to be rapidly unraveling. Part of this goes back to the sectarian nature of the country's founding freedoms. The "live and let live" approach that shaped Europe's religious attitudes was forged in populations imploding into violence, willing to surrender passion in the name of peace. But the original American dream was to be uncompromising in matters of faith, to live in the spiritual community of one's choosing, and to bring worship into daily life. As a cultural trajectory, it found itself at a disadvantage. Whatever remains of this expectation of unfettered religious agency, it seems to be enough to have created conflicting schemes of American secularization. Recently it has devolved into an all-out culture war. It's not a conflict best prosecuted through religious legislation (though that is occasionally tried). Instead of proscriptions, the aim is to seize the apparatus of rights. For every social media app providing its users with multiple gender options (Facebook rolled out fifty-two in 2014), there is an equal enlargement of "traditional" rights on the other side, with "all the guns you can open-carry." The very extravagance of these exercises in individual fulfillment seems to deliberately exceed the capacity for social containment. This is a cultural divide so deeply split along the epistemic seam between faith and reason that it appears it would rather rend itself than be repaired. The battle cry of "fake news" and "alternative facts" is not a demand for better data, but a deliberate rejection of facticity itself. The way to triumph over the expertise of "liberal elites" is not to be "better informed," but to be impervious to information.

That modern civil societies are stumbling backward into more traditional or even atavistic forms of community (racism, ethnic nationalism, religious sectarianism) is at heart a failure to make the case for their own creedal communities. Trust has not been engendered in their bonds of duty, practical solutions, or technocratic leadership. It does not help that the natural philosophy that gave rise to this politics has retreated from the public sphere, projecting only the power of its technology and not its ideals. The interlocutors that do remain—skeptics like Krauss, Richard Dawkins, and Michael Shermer—seek to correct the faults of accommodation by evangelizing reason as a way of knowledge and a way of life. They put the squeeze on theology, trying to gain for science the rule of influence commensurate with its contribution. This is more likely to prevent the exercise of scientific authority than defend it, as accommodationists figured out in the nineteenth century. People do not trust that which threatens them. The United Nations Intergovernmental Panel on Climate Change is tossed in the garbage like a bad script, while belief in chemtrails becomes an article of faith. The idea of scientific "objectivity" has proven difficult to immunize against the perception that such knowledge rewards some more than others. There is a rebellious *jouissance* in the popular rebuff of expertise; those whom the truth set free (though they never asked) set themselves free of the truth as the ultimate expression of liberty.

Given the sense of derailment overtaking "the liberal world order," it is worth revisiting the psychical moment for the insight of its philosophical intervention. While the harmony of "separate spheres for science and religion" seems to have proven itself a naive fiction dreamed up by out-of-touch elites, the discarded concerns of the Victorian crisis seem almost prescient. Its founding generation—thinkers such as Sidgwick, Myers, Gurney, Lodge, Balfour, and Lang—gave deep consideration to the mythic functions of science, contextualizing their empirical research in writings that circled back to the social and spiritual aspirations of knowledge. Psychical research put universal questions to nature on behalf of humanity. These were questions without boundaries, regarding a nature without limits, but still inscribed within a science that retained allegiance to modernity.[74] While it could not satisfy the longings for traditional faith, it did not square off against them, remaining deeply curious regarding all manner of sacred and extraordinary experiences, always seeking some more universal language to return these religious ideas to nature and humanity. Most important, psychical research retained the necessary sense of wonder in the world to truly engage with it.

Notes

INTRODUCTION

1. By the 1870s, elements of the spiritualistic and mesmeric trance had become entangled with the psychiatric diagnosis of hysteria, opening up new lines of inquiry into the subconscious and restoring some of the lost adjacency between madness and religion. For more on these developments in psychiatry as they unfolded within and around Charcot's clinic at the Pitié-Salpêtrière, see M. Brady Brower, *Unruly Spirits: The Science of Psychic Phenomena in Modern France* (Urbana: University of Illinois Press, 2010), xv–92; Adam Crabtree, *From Mesmer to Freud: Magnetic Sleep and the Roots of Psychological Healing* (New Haven, CT: Yale University Press, 1993), 333–417; Henri F. Ellenberger, *The Discovery of the Unconscious: The History and Evolution of Dynamic Psychiatry* (New York: Basic Books, 1981), 110–346; Alan Gauld, *A History of Hypnotism* (Cambridge: Cambridge University Press, 1995), 297–536; and Ian Hacking, *Rewriting the Soul: Multiple Personality and the Sciences of Memory* (Princeton, NJ: Princeton University Press, 1998), 142–97; Renee Haynes, *The Society for Psychical Research, 1882–1982: A History* (London: MacDonald, 1982), 1–240; Roger Luckhurst, *The Invention of Telepathy* (Oxford: Oxford University Press, 2002), 1–252.

2. Janet Oppenheim's breakthrough academic study of the Victorian supernatural, *The Other World: Spiritualism and Psychical Research in England* (New York: Cambridge University Press, 1985), reckoned psychical research as a "pseudo-science," and undertook its study as part of a continuum including spiritualism and theosophy. While her book did document the many significant ties between orthodox science and the SPR in terms of institutional actors and even some theoretical speculation, it never fully transcended the bias of its own assumptions to push deeper into the merits of the society's scientific self-conceit, setting the tone for many such cultural studies that followed.

3. Charles Taylor, *Sources of the Self: The Making of Modern Identity* (Cambridge: Cambridge University Press, 1989), 159–76.

4. In *Console and Classify: The French Psychiatric Profession in the Nineteenth Century* (Chicago: University of Chicago Press, 2001), Jan Goldstein interprets the rise of psychiatry and the asylum as part of the larger process of French secularization, transferring the traditionally religious domain of madness into the hands of the state and its academic deputies. In the 1880s, it seems that the brain briefly stood in danger of being "repossessed" by spirit even in Charcot's neurological sanctuary.

5. In reality, the vast majority of people were eager to reconcile scientific progress with Christian purpose, refusing this proposition of either/or imposed by the warfare thesis. Even polemicists like Herbert Spencer and John Tyndall were mainly fighting an institutional battle over domains

of knowledge and never meant to divide the constituencies of science and religion into atheists versus theists. This revision has been under way in earnest since 1970 with Anthony Symondson's *The Victorian Crisis of Faith* (London: SPCK, 1970), giving rise to various theories ranging from economic, gender, social, and industrial anxieties to Anglicanism's failed theodicy and the crisis as a work of academic self-projection. For an overview of this historiography, see John Brooke and Geoffrey Cantor, *Reconstructing Nature: The Engagement of Science and Religion* (New York: Oxford University Press, 2000), 1–74; as well as Gary B. Ferngren, ed., *Science and Religion: An Historical Introduction* (Baltimore: Johns Hopkins University Press, 2002), 1–30. These works adequately address the complexity of the interaction between science and religion, and the need to avoid any particular master narrative. My focus on elite unbelief somewhat revives the importance of the original ideological dimensions of the crisis abandoned by revisionists, but in the interest of adding to rather than rejecting that historiographic framework.

6. Timothy Larsen correctly interprets this language of doubt as an expression of fundamentally religious concerns in *Crisis of Doubt: Honest Faith in Nineteenth-Century England* (New York: Oxford University Press, 2007), 1–17, 239–53, extending "crisis" to include anxiety about other people's faith, if not one's own. Certainly, in the personal trope of "lost faith" there is also this sense of looming social peril, should doubt become a contagion that endangers religious community and cultural character. In *Modern Spiritualism and the Church of England, 1850–1939* (Woodbridge, UK: Boydell Press, 2010), 1–252. Georgina Byrne examines how spiritualism entered the common culture as a venue for fashioning a more emotionally satisfying and immanent theology, demonstrating how "doubt" concerning traditional Anglican theology could drive the intensification of spirituality as much as diminish it. For an excellent survey of Victorian religious subcultures and the extent to which religion permeated everyday life across denominations and amid scientific developments, see Julie Melnyk, *Victorian Religion: Faith and Life in Britain* (Ann Arbor: University of Michigan Press, 2008), 1–51; as well as James C. Livingston, *Religious Thought in the Victorian Age: Challenges and Reconceptions* (New York: Continuum, 2007), 1–37.

7. The "crisis of meaning" here specifically references the epistemological problem posed by phenomenology. See Art Berman, "The Dilemmas of Empiricism," in *Preface to Modernism* (Urbana: University of Illinois Press, 1994), 120–98; Mark S. Micale, "The Modernist Mind: A Map" and "The Modernist Mind: A Timeline" (with Jesse Wegman) in Mark Micale, ed., *The Mind of Modernism: Medicine, Psychology, and the Cultural Arts in Europe and America, 1880–1940* (Stanford: Stanford University Press, 2004), 1–70.

8. After Crookes's infamous investigation of Florence Cook was concluded in 1874, even skeptical or overtly hostile investigations of séance spiritualism were seen to confer too much legitimacy upon their object and, by extension, to delegitimize the investigator.

9. In *Natural Supernaturalism: Tradition and Revolution in Romantic Literature* (New York: W. W. Norton, 1971), 65–71, M. H. Abrams puts the reenchantment of nature, undertaken in the secular framework of literary aesthetics, at the heart of the romantic project. The psychical notion of the "supernormal" attempted to render this impulse in the context of philosophy and science, allowing romanticism to unite with these hitherto rejected discourses in the late nineteenth century.

10. The mid-nineteenth-century "ether" was a universal, imponderable, hydrodynamic "fluid," thought to structure force, transmit energy, and even give rise to the vortex atom. It was elaborated within the context of thermodynamics, and despite its seemingly fanciful nature, it theoretically grounded a more common-sense, mechanical physics. The failure to confirm its existence toward the end of the century threatened to throw the colossal achievements of mid-Victorian energy physics into disarray.

11. A caveat is here necessary for William Crookes, who did willingly perpetuate a falsehood. However, he was not a casual or career liar by any means. If the scandalous testimony he asserted on behalf of Florence Cook can be attributed to a fault of character, it tends principally to pride more than to deceit.

12. Excellent cultural histories exploring the interconnected intellectual and symbolic dimensions of mesmerism, spiritualism, and psychical pursuits in the Victorian period include Nicola Bown, Carolyn Burdett, and Pamela Thurschwell, eds., *The Victorian Supernatural* (Cambridge: Cambridge University Press, 2009); Brower, *Unruly Spirits*; John J. Cerullo, *The Secularization of the Soul: Psychical Research in Modern Britain* (Philadelphia: Institute for the Study of Human Issues, 1982); John Warne Monroe, *Laboratories of Faith: Mesmerism, Spiritism, and Occultism in Modern France* (Ithaca, NY: Cornell University Press, 2007); Oppenheim, *The Other World*; Pamela Thurschwell, *Literature, Technology and Magical Thinking, 1880–1920* (Cambridge: Cambridge University Press, 2001), 1–64; and Alison Winter, *Mesmerized: Powers of Mind in Victorian Britain* (Chicago: University of Chicago Press, 1998). For a more technically and theoretically detailed account of scientific spiritualism, see Richard Noakes, "The Sciences of Spiritualism in Victorian Britain: Possibilities and Problems," in Tatiana Kontou and Sarah Willburn, eds., *The Ashgate Research Companion to Nineteenth-Century Spiritualism and the Occult Reader* (New York: Ashgate, 2012); Noakes, "Cromwell Varley FRS, Electrical Discharge and Victorian Spiritualism," *Notes and Records of the Royal Society of London* 61, no. 1 (January 22, 2007), 5–21; Noakes, "Telegraphy Is an Occult Art: Cromwell Fleetwood Varley and the Diffusion of Electricity to the Other World," *British Journal for the History of Science* 32, no. 4 (December 1999): 421–59; Noakes, "Instruments to Lay Hold of Spirits: Technologizing the Bodies of Victorian Spiritualism," in Iwan Rhys Morus, ed., *Bodies/Machines* (Oxford: Bloomsbury Academic, 2002), 125–64.

13. The American SPR was the first such organization outside Britain and a sister society of the SPR, established in 1885 under the auspices of William James, but it quickly succumbed to an influx of spiritualists who altered the original character of the SPR. Other offshoot societies that fanned out across Europe over the next two decades never achieved the status of the SPR, although they emulated such scholarly features as its journal.

14. Arthur Balfour, prime minister from 1902 to 1905, served as president in 1893. Other notable presidents include Henry Sidgwick (1882–84 and 1888–92), William James (1894–95), Sir William Fletcher Barrett (1904), Nobel Prize–winner Charles Richet (1905), Henri Bergson (1913), John William Strutt, 3rd Baron Rayleigh (1919), and all four of my subjects: Frederic William Henry Myer (1900), Oliver Lodge (1901–3 and 1932), Sir William Crookes (1896–99), and Andrew Lang (1911). E. W. Benson, archbishop of Canterbury, and Alfred, Lord Tennyson, poet laureate, were also members, as was the statesman William Gladstone.

15. This was Oliver Lodge's personal observation, marking the growing openness to vitalistic considerations in biology and monistic thinking in physics, which began finding more favor in physics starting in the late 1880s. For more on this, see Peter Bowler, *Reconciling Science and Religion: The Debate in Early-Twentieth-Century Britain* (Chicago: University of Chicago Press, 2001), 87–188.

16. Thermodynamics, as it was variously elaborated by British and continental physicists over the course of the 1830s, 1840s, and early 1850s, gave all energy (be it gravitational, magnetic, kinetic, or even the biochemical processes that "animated" life) an essentially quantitative character, revealing the interconnected, interconvertible nature of all physical forces and uniting their branches of study into a single, powerful mechanical orthodoxy.

17. Oliver Lodge speculates about this problem of conservation in "Experiences of Unusual Physical Phenomena Occurring in the Presence of an Entranced Person (Eusapia Palladino)," *Journal of the Society for Psychical Research* 6 (November 1894): 306–60.

18. For more on thermodynamics and the Victorian imagination, see Barri J. Gold, *Thermopoetics: Energy in Victorian Literature and Science* (Cambridge, MA: MIT Press, 2010), which explores how the implications of entropy, such as "cosmic heat death," and the thermodynamic world of wasted energy and work challenged the optimism of the era.

19. Bowler, *Reconciling Science and Religion*, 1–56, 362–93.

20. For more on Romantic-era brain science, see Alan Richardson, *British Romanticism and the Science of Mind* (Cambridge: Cambridge University Press, 2005), 1–92; as well as Martin Halliwell,

Romantic Science and the Experience of Self: Transatlantic Crosscurrents from William James to Oliver Sacks (Brookfield, VT: Ashgate, 1999), 1-35.

21. John Tresch lays out the technological subculture of "romantic mechanism," flourishing in and around Paris in the 1820s and 1830s, crossing the physical mechanics of the French Enlightenment with the theoretical metaphysics of *Romantische Naturphilosophie*. See John Tresch, *The Romantic Machine: Utopian Science and Technology after Napoleon* (Chicago: University of Chicago Press, 2012). Both Tresch and Richardson give an excellent sense of what that transition entailed in the 1830s as the BAAS and the Berlin Physical Society began to press for more rigorous oversight of the practice of and intellectual participation in science, creating a far more hostile milieu for any such secular metaphysics both in theory (conservation) and principle (a "gentlemanly science" based on consensus, decorum, and elite leadership).

22. Freud formalized this view in 1893-95 with the concept of "cathexis," explaining the overloading of neural networks with potential energy, $Q(n)$, which was subject to both the laws of inertia and the need to achieve equilibrium through periodic discharge. This provided a medical, histological view of conscious behavior cycling through periods of repression and compulsion, which was itself mapped onto a purely quantitative understanding of the underlying collection and discharge of the energies "fueling" such urges. Thus Freud not only merged the dominant trends in neuropsychiatry and psychophysics for over two decades to assert a rival to the problematic "subliminal self," he likewise hoisted psychology onto the concrete foundations of evolutionary biology and thermodynamics, ensuring the future prestige of the discipline as well as his own eventual ascendancy. (See Sigmund Freud, "Project for a Scientific Psychology," in *The Standard Edition of the Complete Psychological Works of Sigmund Freud, Volume I (1886-1899): Pre-psychoanalytical Papers and Unpublished Manuscripts* (New York: Vintage, 2001), 298-357, as well as chapter 4 of this volume.)

23. Behavioral psychology is something of a catch-all term to describe efforts beginning in the early twentieth century to pursue a more concrete, "cause and effect" study of human behavior, most often associated with Pavlov's dogs in the late 1890s and later with the behaviorism of John B. Watson and B. F. Skinner in the twentieth century.

24. While the term "parapsychology" had been coined in the 1880s, it was not adopted until later, when J. B. Rhine established his Parapsychology Laboratory as part of Duke University's psychology department in 1930. However, the statistical methodology and strictly laboratory settings that defined its practices had been pioneered earlier by Charles Richet in the mid-1880s under the aegis of the SPR and provided the platform for earlier academic initiatives like that of ex-psychical researchers John Edgar Coover at Stanford in 1911.

25. This concept of NOMA was laid out by Stephen Jay Gould in "Non-overlapping Magisteria," *Natural History* 106 (March 1997): 16-22. It has come to designate the notion of separate spheres of authority for science and religion, in which each is according its proper preserve and dignity.

26. See Theodore Porter, "The End of Naturalism: From Public Sphere to Professional Exclusivity," in Gowan Dawson and Bernard V. Lightman, eds., *Victorian Scientific Naturalism: Community, Identity, Continuity* (Chicago: University of Chicago Press, 2014), 265-87. Porter sees in the philosophical retreat from naturalism a parallel disengagement with the public sphere as elite scientists (even former rabble-rousers like Huxley) sought a more irenic and respectable public face for institutional science, abandoning the substantive basis for any such ethical commentary by adopting an epistemology of descriptivism and affirming its status through public utility and technical expertise. In *Huxley's Church and Maxwell's Demon: From Theistic Science to Naturalistic Science* (Chicago: University of Chicago Press, 2015), Matthew Stanley historicizes this aspirational naturalism by showing its earlier strategic appropriation of the social and moral programs of traditional theistic science. Secularists like Huxley succeeded because they understood that liberating science required replacing the entire framework in which natural knowledge was traditionally held (with all its educational, epistemological, and social burdens), not just its content.

CHAPTER ONE

1. Heb. 11:1.

2. For a more complex view of the various moral, cultural, and political agendas pursued through the elite positivism of Lewes and Eliot, and the response of critics like Sidgwick, see T. R. Wright, *The Religion of Humanity: The Impact of Comtean Positivism on Victorian Britain* (Cambridge: Cambridge University Press, 2008), especially 125–202.

3. As used here, "popular science" designates the ways in which the nonprofessional public consumed, appropriated, and deployed scientific narratives in ways meaningful to themselves. For more on its changing meaning over time, see David Allen, "Amateurs and Professionals," in *The Cambridge History of Science*, vol. 6: *Modern Life and Earth Science* (Cambridge: Cambridge University Press, 2009), 13–33; and Bernard Lightman, *Victorian Popularizers of Science: Designing Nature for New Audiences* (Chicago: University of Chicago Press, 2007), 1–38, 353–422. "Popular" acquired its pejorative connotation as institutional science fought to elevate the social status and cultural authority of its professionally trained practitioners, which remained a somewhat open contest until the 1880s.

4. See Lightman, *Victorian Popularizers of Science*, 39–94; and James Moore, *The Post-Darwinian Controversies: A Study of the Protestant Struggle to Come to Terms with Darwin in Great Britain and America 1870–1900* (Cambridge: Cambridge University Press, 1979), 217–352, for efforts to adapt Darwinian narrative to a liberalized Anglican theology in the latter half of the nineteenth century.

5. Oliver Lodge, "The Outstanding Controversy between Science and Faith," *The Hibbert Journal* 1 (October 1902–July 1903): 60.

6. Timothy Larsen, *Crisis of Doubt: Honest Faith in Nineteenth-Century England* (Oxford: Oxford University Press, 2007), 1–17, 239–53.

7. One of the two girls, Margaret Fox, confessed years later, in 1888, that she had merely been cracking her toes under the table and that her subsequent spiritual celebrity was based on fraud. Charles Richet, *Thirty Years of Psychical Research: Being a Treatise on Metapsychics* (New York: Macmillan, 1923), 26–36.

8. Joseph McCabe placed the date earlier, as 1847 in France and by 1848 in England, before the arrival of spiritualism. See Joseph McCabe, *Spiritualism: A Popular History from 1847* (London: T. F. Unwin, 1920), 15. Alison Winter emphasizes that the practice of tipping itself arrived with spiritualism. See Winter, *Mesmerized*, 20. Table moving was probably was an established phenomenon, as noted by McCabe onto which tipping communication was grafted with the arrival of spiritualism.

9. Mary Botham Howitt, *Margaret Howit: An Autobiography*, vol. 2 (Cambridge: Cambridge University Press, 2010), 99.

10. Cables connecting England and France were laid in 1847 and commonly referred to as "the railway of thought," capturing some of spiritualism's own disembodied intellectual mode of transmission. This defining phrase can first be found in *Herapath's Journal and Railway Magazine* 8, no. 368 (July 4, 1846): 863.

11. *The Zoist* was already in publication by 1842, several years before the arrival of spiritualism on English shores, having absorbed French mesmerism and *séance magnétiques* into the framework of phrenology. This early British reception of France's alternate consciousness paradigm had a decidedly psychological rather than mystical spin, although the idea of an actual force or some mysterious rapport persisted as a strain of this intellectual interest. Spiritualism readily took over this magnetic infrastructure, adding back into it the exotic metaphysics that early British mesmerists saw fit to remove. Edward William Cox, the founder of the Psychological Society in 1875, established a national reputation in the 1830s as one of the leading expositors of phrenology and later as an investigator of spiritual phenomena (though not precisely a spiritualist). He provided an important intellectual context for Frederic Myers in the 1870s, bringing together in his nearly half-century of research the shadow history of consciousness that fell between spiritualism and

academic psychology. For more on the early British contexts of mesmerism, magnetism, and phrenology, see Winter, *Mesmerized*, 109-62.

12. Thomas Lake Harris, *Modern Spiritualism, Its Truths and Its Errors: A Sermon* (New York: Church Publishing Association, 1860), 31.

13. W. H. Ferris, "The Theology of Modern Spiritualism: Its Infidelity," *The Ladies Repository* 16 (January 1856): 297-300.

14. William Ramsay, *Spiritualism: A Satanic Delusion and a Sign of the Times* (Rochester, NY: H. L. Hastings, 1857); and Charles Cowan, *Thoughts on Satanic Influence; or, Modern Spiritualism Considered* (London: Seelys, 1854).

15. Catherine Crowe, *Spiritualism and the Age We Live In* (London. T. C. Newby, 1859), 48.

16. Joel Tiffany, *Lectures on Spiritualism: Being a Series of Lectures on the Phenomena and Philosophy of Development, Individualism, Spirit, Immortality, Mesmerism, Clairvoyance, Spiritual Manifestations, Christianity, and Progress, Delivered at Prospect Street Church, in the City of Cleveland, during the Winter and Spring of 1851* (Cleveland: J. Tiffany, 1851), 3.

17. Crowe, *Spiritualism and the Age We Live In*, 52.

18. Tiffany, *Lectures on Spiritualism*, 64.

19. For more on magic, literary modernism, and identity formation, see Timothy Materer, *Modernist Alchemy: Poetry and the Occult* (Ithaca, NY: Cornell University Press, 1995); Leigh Wilson, *Modernism and Magic: Experiments with Spiritualism, Theosophy and the Occult* (Edinburgh: Edinburgh University Press, 2013); Alex Owen, *The Places of Enchantment: British Occultism and the Culture of the Modern* (Chicago: University of Chicago Press, 2004).

20. For an excellent biography of Aleister Crowley, see Lawrence Sutin, *Do What Thou Wilt: A Life of Aleister Crowley* (New York: St. Martin's, 2002). Crowley, perhaps more than any other contemporary, dragged the public spirit of the Victorian occult back into the Faustian shadows. He began to formulate his cult of modern magic, Thelema, in the early twentieth century, claiming he could draw supernatural power into human hands by summoning those shadow forces under eclipse by both Christianity and science. The goal was to live grandly and even transgressively in this life. He was, in that sense, jumping *off* the bridge between naturalism and theology that Victorian spiritualists and even psychical researchers thought they could mend.

21. Aldous Huxley revived the romantic connection between drugs, spirituality, and consciousness in *The Doors of Perception* (London: Chatto & Windus, 1953), but it was the efforts of Timothy Leary, Ralph Metzner, and Ram Dass (see *The Psychedelic Experience: A Manual Based on the Tibetan Book of the Dead* [New York: Citadel Press, 1964]) that fully entwined these themes within an Eastern religio-mystical platform to power a revolutionary social movement.

22. Linda M. Lewis, *Elizabeth Barrett Browning's Spiritual Progress: Face to Face with God* (Columbia: University of Missouri, 1998), 139.

23. "Science and Spiritualism," *Pharmaceutical Journal: A Weekly Record of Pharmacy and Allied Sciences*, January 11, 1873, 545.

24. Michael Faraday, "The Table-Turning Delusion," *New-Hampshire Journal of Medicine* 4-5 (January- December 1854). See also Winter, *Mesmerized*, 276-305.

25. Agénor Étienne de Gasparin, *Science versus Modern Spiritualism: A Treatise on Turning Tables*, vol. 1 (New York: Kiggins & Kellog, 1856), xxi.

26. Chau H. Wu, "Electric Fish and the Discovery of Animal Electricity: The Mystery of the Electric Fish Motivated Research into Electricity and Was Instrumental in the Emergence of Electrophysiology," *American Scientist* 72, no. 6 (November-December 1984): 598-607.

27. Lightman, *Victorian Popularizers of Science*, 353-422. While the aim was to defend the integrity of scientific ideas in this open commercial forum, Tyndall and Huxley's strident naturalism created blowback for the scientific institutions they wished to defend. These so-called "execrable professors of physical science" could be used to rile up a more general anti-establishment sentiment, conflating science with scientism. As much as threats of clerical interference or judgment

from "the outside," it was the failure of fellow professionals to rigorously defend "unfettered science" that gave rise to Huxley's X-Club and solidified the naturalist's rhetorical position, drawing attention to a debate within science that most of the leadership wished to quell. See J. F. M Clark, "John Lubbock, Science and the Liberal Intellectual," *Notes and Records: The Royal Society Journal of the History of Science* 68 (2014): 65-87.

28. The locus of this effort at rapprochement was in North Britain; it was spearheaded by physicists like William Thomson, James Clerk Maxwell, and Peter Guthrie Tait, who were alarmed by some of the needless lack of deference on display in the more secular urban South. Scientific freedom, in their eyes, could still proceed on a friendly footing with Christian ideals, but that required the profession make a point of being more politic with regard to the public. Crosbie Smith, *The Science of Energy: A Cultural History of Energy Physics in Victorian Britain* (Chicago: University of Chicago Press, 1998), 150-69.

29. William Howitt gives that figure in "Spiritualism Defended," an open letter published in the *Dunfermline Press*, July 4, 1868. The quote appears in *Pharmaceutical Journal: A Weekly Record of Pharmacy and Allied Sciences* 3 (1872-73): 545, written by a staff author. A more conservative estimate made by spiritualists themselves put this number at 1 million to 2 million by 1852, the year of their first convention. This more exaggerated figure of 11 million was suggested by a speaker from the American Catholic Congress in 1854, and became widely circulated by those both panicked and enthused by the movement. Byrne, *Modern Spiritualism and the Church of England*, 19.

30. This distribution is from Proquest's *C19: The Nineteenth Century Index*, which indicates a peak in 1876-77, with more than three hundred published articles concerning spiritualism appearing in the mainstream press alone. While these were not necessarily supportive, this level of attention, even from detractors, signifies the degree of public concern.

31. Alex Owen, *The Darkened Room: Women, Power, and Spiritualism in Late Victorian England* (Chicago: University of Chicago Press, 1989), 42—75.

32. Allen, "Amateurs and Professionals."

33. Edward William Cox, *Spiritualism Answered by Science: With Proofs of a Psychic Force* (London: Longman & Co., 1872), 10.

34. Ruth Richardson, "Why Was Death So Big in Victorian Britain," in Ralph Anthony Houlbrooke, ed., *Death, Ritual and Bereavement* (London: Routledge & Kegan Paul, 1989), 105-23.

35. Martha McMackin Garland, "Victorian Unbelief and Bereavement," in Ralph Anthony Houlbrooke, ed., *Death, Ritual and Bereavement* (London: Routledge & Kegan Paul, 1989), 151-67.

36. Alfred Russel Wallace, *On Miracles and Modern Spiritualism: Three Essays* (London: James Burns, 1875), 160.

37. See Paul Hazard, *Crisis of European Mind: 1680-1715* (New York: Meridian Books, 1963), for the early Enlightenment challenge to the religious and Platonic entrenchment of natural philosophy. For more on the nineteenth-century Anglican confrontation with philosophical naturalism and evolutionary theories, see Peter J. Bowler, *Evolution: The History of an Idea*, 25th ed. (Berkeley: University of California Press, 2009); C. C. Gillespie, *Genesis and Geology: A Study in the Relations of Scientific Thought, Natural Theology, and Social Opinion in Great Britain, 1790-1850* (New York: Harper, 1959); James Secord, *Victorian Sensation* (Chicago: University of Chicago Press, 2000); Frank Turner, *Contesting Cultural Authority: Essays in Victorian Intellectual Life* (Cambridge: Cambridge University Press, 1993); Richard Yeo, *Science in the Public Sphere* (Aldershot, UK: Ashgate/Variorum, 2001); Robert Young, *Darwin's Metaphor: Nature's Place in Victorian Culture* (Cambridge: Cambridge University Press, 1985).

38. Lorraine Daston and Katharine Park, eds., *The Cambridge History of Science*, vol. 3: *Early Modern Science* (Cambridge: Cambridge University Press, 2006), 401.

39. Margaret Jacob, *The Newtonians and the English Revolution 1689-1720* (Brighton, UK: Harvester Press, 1976), 162-200.

40. Turner, *Contesting Cultural Authority*, 131-50.

41. For a more pluralistic understanding of this intellectual milieu see Nicola Bown and Carolyn Burdett, eds., *The Victorian Supernatural* (Cambridge: Cambridge University Press, 2009); Roger Cooter, *The Cultural Meaning of Popular Science: Phrenology and the Organization of Consent in Nineteenth-Century Britain* (New York: Cambridge University Press, 2005); Hilary Grimes, *The Late Victorian* (Aldershot, UK: Ashgate, 2011); Luckhurst, *The Invention of Telepathy* (Oxford: Oxford University Press, 2002); Monroe, *Laboratories of Faith*; Richard Noakes, "Cromwell Varley FRS, Electrical Discharge and Victorian Spiritualism," *Notes and Records of the Royal Society of London* 61, no. 1 (January 22, 2007), 5-21; Noakes, "Telegraphy Is an Occult Art"; Oppenheim, *The Other World*; Owen, *The Darkened Room*; Roy Wallis, *On the Margins of Science: The Social Construction of Rejected Knowledge* (Keele, UK: University of Keele, 1979); and Winter, *Mesmerized*.

42. See Richard J. Helmstadter and Bernard Lightman, eds., *Victorian Faith in Crisis: Essays on Continuity and Change in Nineteenth-Century Religious Belief* (Stanford: Stanford University Press, 1990); David Lindberg and Ronald L. Numbers, eds., *When Science and Christianity Meet* (Chicago: University of Chicago Press, 2003); Colin A. Russell, "The Conflict Metaphor and Its Social Origins," *Science and Christian Belief* 1 (1989): 3-26; Turner, *Contesting Cultural Authority*.

43. Turner, *Contesting Cultural Authority*, 73-100.

44. While not all of these names are on the membership rolls, they are still in the mix. Leslie Stephen was a member of Sidgwick's Cambridge group, the Metaphysical Club, Gurney's Scratch Eight club, and Balfour's Synthetic Society, and was a deeply interested amateur in the study of the unconscious, a frequent topic of his Sunday Tramps walking club. James Ward was a member of the Synthetic Society and one of Henry Sidgwick's closest friends, a psychologist who advocated for "empirical idealism" and recognized the possibility of "a rapport or telepathy." George Romanes was on the SPR planning committee but dropped out before becoming a member. He is, however, listed as a member of Charcot's Société de Psychologie Physiologique, and he considered a role for consciousness in evolution. Samuel Butler was a friend of Gurney and Myers, and interested in psychical and psychodynamic questions; his book *Unconscious Memory* (1880) put him at the forefront of this opening-up of psychology.

45. Jack Morrell and Arnold Thackray, *Gentlemen of Science: Early Years of the British Association for the Advancement of Science* (Oxford, UK: Clarendon Press: 1981), 224-41.

46. Peter Bowler, *The Invention of Progress: The Victorians and the Past* (New York: B. Blackwell, 1989), 35-36; and Bowler, *Evolution: The History of an Idea*, 141-76.

47. There are some qualifications to be made. George Romanes actively investigated spiritual phenomena alongside his Cambridge colleagues in the 1870s and accepted an invitation to be among the SPR's founding members. Other obligations kept him from attending that meeting in 1882, though he did participate in some later thought experiments in 1883, after which point his association with the SPR seems to have come to an end, probably due to the detection of fraud and his need, as an evolutionary physiologist, to distance himself from potential contamination. (See Oppenheim, *The Other World*, 286.) Sidgwick, Stephen, and Butler were all born in the 1830s, and their respective Cambridge enrollments began in the 1850s. Butler left in 1859, but Stephen and Sidgwick continued on into the 1860s and 1870s, respectively.

48. Frederic Myers, *Fragments of Inner Life: An Autobiographical Sketch* (London: Society for Psychical Research, 1961), 21-29.

49. See William C. Lubenow, *"Only Connect": Learned Societies in Nineteenth-century Britain* (Suffolk, UK: Boydell & Brewer, 2015), 129-60, for more on the scientific and social agenda of the X-Club and its stand against classism and clericalism in science.

50. Herbert Spencer, *First Principles*, 4th ed. (London: Williams & Norgate, 1880), 8-24, 98-126.

51. Arthur Sidgwick and Eleanor Sidgick, *Henry Sidgwick: A Memoir* (London: Macmillan, 1906), 357.

52. William M. Reddy, "Against Constructionism: The Historical Ethnography of Emotions," *Current Anthropology* 38, no. 3 (1997): 327-51; Reddy, "Emotional Liberty: Politics and History in the Anthropology of Emotions," *Cultural Anthropology* 14 (1999): 256-88; and Reddy, *The*

Navigation of Feeling: A Framework for the History of Emotions (Cambridge: Cambridge University Press, 2001), 3-138.

53. See David Hempton, *Evangelical Disenchantment: Nine Portraits of Faith and Doubt* (New Haven, CT: Yale University Press, 2008); and Frank Turner, *Between Science and Religion* (New Haven, CT: Yale University Press, 1974), for biographical arcs that return Victorian renunciates to positions of faith.

54. Reddy, "Emotional Expression as Type of Speech Act," in *The Navigation of Feeling*, 63-111.

55. Sidgwick is a central figure here because of his high standing in the academic community. There was a rash of such gestures at Cambridge between 1869 and 1871, with Sidgwick's being among the first. The acts were repealed in 1871 largely as a result of his efforts, and he resumed his association with the university as a lecturer in moral philosophy. J. P. C. Roach, ed., "The University of Cambridge: The Age of Reforms (1800-1882)," in *A History of the County of Cambridge and the Isle of Ely*, vol. 3: *The City and University of Cambridge* (London: Victoria County History, 1959), 235-65.

56. Taylor, *Sources of the Self: The Making of Modern Identity*, 404.

57. Myers, *Fragments of Inner Life*, 13.

58. William K. Clifford, *Lectures and Essays*, ed. Leslie Stephen and Frederick Pollock (London: Macmillan, 1886), ix.

59. Frederic Myers, "Charles Darwin and Agnosticism," in *Science and a Future Life* (London: Macmillan, 1901), 64.

60. Hempton, *Evangelical Disenchantment*, 1-40.

61. Frederic Myers, *Fragments of Inner Life*, 14.

62. This is the famously provocative statement made in Huxley's review of Darwin's *Origin of Species* just weeks before his debate with Bishop Wilberforce at the annual BAAS meeting of 1860.

63. Leslie Stephen, "Some Early Impressions," *National Review* 42 (1903): 532.

64. Myers, *Fragments of Inner Life*, 15.

65. Frank Turner, "The Victorian Conflict between Science and Religion: A Profession Dimension," in Gerald Parsons, ed., *Religion in Victorian Britain*, vol. 4: *Interpretations* (Manchester, UK: Manchester University Press, 1992), 170-97.

66. Edmund Gurney, *Tertium Quid* (London: K. Paul, Trench & Co., 1887), 137-38.

67. The term "subconscious" was originally coined by Pierre Janet in 1886 to improve upon the designation "unconscious," which failed to convey the appropriate sense of intelligent activity. The notion itself, however, entered psychology nearly a decade earlier with Charcot's consideration of trance consciousness as its own variety of sentience. See Ellenberger, *The Discovery of the Unconscious*, 372-73.

CHAPTER TWO

1. While the conflict between the truth-claims of science and the revealed truth of religion was at the core of the natural philosophy of the Enlightenment, the intensity of this debate, as well as its extension into the clerical and popular realm, was indeed new. For more on this, see Frederick Gregory, "Intersections of Physical Science and Western Religion in the Nineteenth and Twentieth Centuries," in *The Cambridge History of Science*, vol. 5: *The Modern Physical and Mathematical Sciences* (Cambridge: Cambridge University Press, 2002), 36-39.

2. Frederick Gregory explains that the willingness to surrender nature to science was the strategy of elite German theologians but met with pervasive resistance in other walks of life. See Frederick Gregory, *Nature Lost? Natural Science and the German Theological Traditions of the Nineteenth Century* (Cambridge, MA: Harvard University Press, 1992). The dénouement of strictly separate spheres did not fall fully into place until the early twentieth century.

3. Gould's NOMA assumes a consensus that faith is a matter of private conscience, with a scope of action subordinate to the demands of science and civility. To the extent one rejects those priorities, "harmony" can seem less than ecumenical.

4. Christians in the latter half of the nineteenth century struggled to find some theistic accommodation with debates over Darwinian evolution in order to maintain the relevance of religious ideas in a modernizing world. For a survey of these theological debates regarding evolutionary theory, see James R. Moore, *The Post-Darwinian Controversies: A Study of the Protestant Struggle to Come to Terms with Darwin in Great Britain and America 1870-1900* (Cambridge: Cambridge University Press, 1981), 193-298.

5. For more on how the idea of violent competition within nature became a pervasive, pernicious metaphor normalizing economic competition, imperial rivalry, and the march toward war beginning in the 1880s and 1890s, see Robert Bannister, *Social Darwinism: Science and Myth in Anglo-American Social Thought* (Philadelphia: Temple University Press, 2010).

6. For more on the breakout materializing mediums Florence Cook and Mary Rosina Showers and their champion Florence Marryat, see Georgina O'Brien Hill, "'Above the Breath of Suspicion': Florence Marryat and the Shadow of the Fraudulent Trance Medium," in Tatiana Kontou, ed., *Women and the Victorian Occult* (New York: Routledge, 2010), 59-73.

7. For more on the context in which Crookes came up through the ranks (1850s to 1870s), see Lightman, *Victorian Popularizers of Science*, 1-38.

8. For more on this intensifying professional censure in the 1870s, see Peter Lamont, "Spiritualism and a Mid-Victorian Crisis of Evidence," *Historical Journal* 47, no. 4 (December 2004): 897-920.

9. Elana Gomel, "Spirits in the Material World: Spiritualism and Identity in the Fin de Siècle," *Victorian Literature and Culture* 35, no. 1 (2007): 189-213.

10. Alfred Russel Wallace, *In Defense of Modern Spiritualism* (Boston: Colby & Rich, 1874), 39.

11. Roger Smith, "The Physiology of the Will: Mind, Body and Psychology in the Periodical Literature from 1855-1875," in G. N. Cantor and Sally Shuttleworth, eds., *Science Serialized: Representation of the Sciences in Nineteenth-Century Periodicals* (Cambridge, MA: MIT Press, 2004), 81-110.

12. This was the complaint leveled against the twentieth-century behaviorists.

13. Mark S. Micale, "The Salpêtrière in the Age of Charcot: An Institutional Perspective on Medical History in the Late Nineteenth Century," *Journal of Contemporary History* 20, no. 4 (October 1985): 703-31.

14. Smith, "The Physiology of the Will," 100.

15. For more on these interpretive narratives, see Sarah Willburn, "Viewing History and Fantasy through Victorian Spirit Photography," in Kontou and Willburn, eds., *The Ashgate Research Companion to Nineteenth-Century Spiritualism and the Occult Reader*, 359-82; and Jennifer Tucker, *Nature Exposed: Photography as Eyewitness in Victorian Science* (Baltimore: Johns Hopkins University Press, 2006).

16. Willburn, "Viewing History and Fantasy through Victorian Spirit Photography," in Kontou and Willburn, eds., *The Ashgate Research Companion to Nineteenth-Century Spiritualism and the Occult Reader*, 359-82.

17. Tucker, "Photography of the Invisible," in *Nature Exposed*, 159-93.

18. A good introduction to commercial spirit photography of the mid-nineteenth century can be found in essays by Crista Cloutier, Andreas Fischer, and Clément Chéroux, in the volume they coedited: *The Perfect Medium: Photography and the Occult* (New Haven, CT: Yale University Press, 2005), 1-71.

19. "Katie King" was, according to the biography circulated among spiritualists, the ghost of Annie Owen Morgan, daughter of Captain Morgan, whose spirit name was "John King." See Trevor Hall, *The Medium and the Scientist: The Story of Florence Cook and William Crookes* (Buffalo, NY: Prometheus Books, 1985), 96-97.

20. W. F. Barrett, "In Memory of Sir William Crookes, O.M., Etc.," *PSPR* 31 (1920-21): 26-27.

21. William Crookes, *Researches in the Phenomena of Spiritualism* (London: J. Burns, 1874), 108. (Crookes never authorized this collection of his articles.)

22. William H. Brock, *William Crookes and the Commercialization of Science* (Farnham, UK: Ashgate Publishing, 2008), 179-94.

23. *Psychic Science* 13, no. 1 (April 1934): 25-30.

24. Tucker, "Constructing Science and Brotherhood in Photographic Culture," in *Nature Exposed*, 17-64.

25. Willburn, "Viewing History and Fantasy through Victorian Spirit Photography," in Kontou and Willburn, eds., *The Ashgate Research Companion to Nineteenth-Century Spiritualism and the Occult Reader*, 359-82.

26. Crookes to Captain T. D. Williams, August 4, 1874, published in *Psypioneer Journal* 8, no. 7 (July 2012): 225.

27. Tucker, *Nature Exposed*, 159-93.

28. Oliver Lodge, "The Life of Crookes," *PSPR* 34 (1924): 313.

29. George M. Board, "The Psychology of Spiritualism," *North American Review* 129 (July 1879): 70-87; W. A. Hammond, "Spiritualism and Allied Causes and Conditions of Nervous Derangement" (New York: G. P. Putnam, 1876); W. B. Carpenter, *Mesmerism, Spiritualism, Historically and Scientifically Considered* (New York: D. Appleton. 1877).

30. Lamont, "Spiritualism and a Mid-Victorian Crisis of Evidence," 897-920.

31. Hudson Tuttle, *Arcana of Spiritualism: A Manual of Spiritual Science and Philosophy* (Boston: Adams & Co., 1871), 40.

32. Richard Noakes, "The Sciences of Spiritualism: Possibilities and Problems," in Kontou and Willburn, eds., *The Ashgate Research Companion to Nineteenth-Century Spiritualism and the Occult Reader*, 25-54.

33. Emma Hardinge Britten, in London Dialectical Society, *Report on Spiritualism of the Committee of the London Dialectical Society* (London: J. Burns, 1873), 412.

34. *Pharmaceutical Journal: A Weekly Record of Pharmacy and Allied Sciences* (January 11, 1873): 545.

35. William Howitt, "Spiritualism Defended," *Dunfermline Press*, July 4, 1868, 1-4.

36. Charles Bray, *Manual of Anthropology* (Oxford: Oxford University, 1871), 308.

37. Michael Faraday quoted in Crookes, *Researches in the Phenomena of Spiritualism*, 4.

38. T. H. Huxley, "Possibilities and Impossibilities" (1891), in *Science and Christian Tradition: Essays* (New York: D. Appleton, 1896), 192-208.

39. T. H. Huxley, letter excerpted from *Report on Spiritualism: Of the Committee of the London Dialectical Society*, 229.

40. Tuttle, *Arcana of Spiritualism*, 40.

41. Tuttle, *Arcana of Spiritualism*, 40.

42. Crookes, *Researches in the Phenomena of Spiritualism*, 6-7, 45-80.

43. For a good overview of how science was commercialized as a leisure activity and informative entertainment in the nineteenth century, see the introduction to Aileen Fyfe and Bernard Lightman, eds., *Science in the Market Place* (Chicago: University of Chicago Press, 2007), 1-22.

44. Luckhurst, *The Invention of Telepathy*, 32; and Noakes, "The Sciences of Spiritualism," 44.

45. Gregory, "Intersection of Physical Science and Western Religion in the Nineteenth and Twentieth Centuries," 47.

46. Bowler, *Reconciling Science and Religion*, 122-59.

47. Crookes, *Researches in the Phenomena of Spiritualism*, 24.

48. Jon Palfreman, "Between Scepticism and Credulity: A Study of Victorian Scientific Attitudes to Modern Spiritualism," in Roy Wallis, ed., *On the Margins of Science: The Social Construction of Rejected Knowledge* (Keele, UK: University of Keele, 1979), 201-36.

49. Richard Hodgson, "Report of the Committee Appointed to Investigate Phenomena Connected with The Theosophical Society," *PSPR* 2 (1884): 201-369.

50. Lamont, "Spiritualism and a Mid-Victorian Crisis of Evidence," 919.

51. George W. Stocking, Jr., "Animism in Theory and Practice: E. B. Tylor's Unpublished 'Notes on Spiritualism,'" *Man*, n.s. 6, no. 1 (March 1971): 91.

52. Stocking, Jr., "Animism in Theory and Practice," 91. Thus, spiritualism threatened not only the future of cultural evolution, but likewise Tylor's theory of cultural evolution.

53. Edmund Edward Fournier d'Albe, *The Life of Sir William Crookes* (London: T. Fisher Unwin, 1924), 24.

54. Frank A. J. L. James, "Of 'Medals and Muddles': The Context of the Discovery of Thallium: William Crookes' Early Spectro-Chemical Work," *Notes and Records of the Royal Society of London* 39, no. 1 (September 1984): 65–90.

55. Fournier d'Albe, *Life of Sir William Crookes*, 66.

56. Hannah Gay, "Invisible Resource: William Crookes and His Circle of Support, 1871–81," *British Journal for the History of Science* 29, no. 3 (September 1996): 311–36.

57. Quoted in Fournier d'Albe, *Life of Sir William Crookes*, 88–89.

58. Fournier d'Albe, *Life of Sir William Crookes*, 34.

59. Laurel Brake and Marysa Demoor, *Dictionary of Nineteenth-Century Journalism: In Great Britain and Ireland* (London: Academia Press, 2009), 110–11, 155.

60. Excerpted from Crookes's article for the *Popular Science Journal*, written in the wake of his discovery of thallium, in 1861, quoted in Fourner d'Albe, *Life of Sir William Crookes*, 66.

61. London Dialectical Society, *Report on Spiritualism of the Committee of the London Dialectical Society Together with the Evidence, Oral and Written and a Selection from the Correspondence* (London: Longmans, Green, Reader & Dyer, 1871). Though published in 1871, the correspondence is largely dated to 1869, indicating Crookes's rising profile as a serious spiritual investigator at least by that year.

62. London Dialectical Society, *Report on Spiritualism* (1871), 263.

63. London Dialectical Society, *Report on Spiritualism* (1871), 265.

64. Quoted in Fournier d'Albe, *Life of Sir William Crookes*, 137–38.

65. Richard Noakes, "Spiritualism, Science and the Supernatural in Mid Victorian Britain," in Bown, Burdett, and Thurschwell, *The Victorian Supernatural*, 23–43.

66. *The Spiritual Magazine* 6 (December 1871): 534.

67. Crookes, *Researches in the Phenomena of Spiritualism*, 6.

68. Crookes, *Researches in the Phenomena of Spiritualism*, 9–43.

69. Crookes, *Researches in the Phenomena of Spiritualism*, 6.

70. Crookes, *Researches in the Phenomena of Spiritualism*, 6–7.

71. For more on the defense of mind-reading in the public forum, see Roger Luckhurst, "Passages in the Invention of the Psyche: Mind-Reading in London, 1881–1884," in his *Transactions and Encounters: Science and Culture in the Nineteenth Century* (Manchester, UK: Manchester University Press, 2002), 117–50.

72. Allen Thomson, "Opening Address, Section D (Edinburgh, 1871)," *Report of the Meeting of the British Association for the Advancement of Science* 41 (1872): 121.

73. Crookes, *Researches in the Phenomena of Spiritualism*, 65.

74. Crookes, *Researches in the Phenomena of Spiritualism*, 65.

75. Richard Noakes, "Cromwell Varley FRS, Electrical Discharge and Victorian Spiritualism, *Notes and Records of the Royal Society of London* 61, no. 1 (January 22, 2007): 5–21; and Noakes, "Telegraphy Is an Occult Art."

76. Crookes, *Researches in the Phenomena of Spiritualism*, 58, 101.

77. Trevor Hall, *The Spiritualists* (London: Duckworth, 1962), 29.

78. Hall, *The Spiritualists*, 31.

79. Crookes, *Researches in the Phenomena of Spiritualism*, 102.

80. Hall, *The Spiritualists*, 54–73.

81. For details on this confession, see Hall, *The Spiritualists*, 99–108. Anderson's claims were investigated and supported by the SPR.

82. See Alex Owen's analysis of the Cook/Crookes relationship in the context of gender and power in *The Darkened Room*, 42–49.

83. Crookes, *Researches in the Phenomena of Spiritualism*, 111.

84. Crookes, *Researches in the Phenomena of Spiritualism*, 111.

85. The "shocking" allegation of an affair, contained in Hall's book, did not go over well in 1962, triggering an institutional defense of Crookes. The Anderson testimony came under attack some forty years after it was entered by deposition into the SPR's archives, and then that "damage"

to his testimony was used as evidence of Hall's shoddy scholarship. This crusade was carried out in the main by Richard Medhurst and Kathleen Goldney, whose objections to Hall consisted mainly of calling Anderson unreliable, the past unknowable, and Cook and Crookes victims of slander because they were no longer alive to defend themselves. Crookes's testimony was rehabilitated by making the case that Cook's phenomena could well have been real, eliminating the sexual quid pro quo Hall ascribed. (See Richard Medhurst and Kathleen Goldney, "The Anderson Testimony," *JSPR* 41 (June 1963): 93–97; and Medhurst and Goldney, "Sir William Crookes and the Physical Phenomena of Mediumship," *PSPR* 54 (1964): 25–157.) I do not quarrel with the right to make such an argument, only with the assumption that Hall's book now stands discredited. That was not the position of psychical researchers Eric Dingwall and Ruth Brandon, who seemed better able to assess the explanatory value of the affair rather than panicking at the sight of it. The idea of an affair is also supported by Marlene Tromp in *Altered States: Sex, Nation, Drugs, and Self-Transformation in Victorian Spiritualism* (Albany: SUNY Press, 2006), 21–48.

86. Crookes claimed that the book *Researches in the Phenomena of Spiritualism*, published in 1874, had been compiled and published without his knowledge or permission. Years later, when Sir Arthur Conan Doyle tried to revive the topic of Katie King in a very supportive book, a *contretemps* ensued upon Crookes's refusal. However, Crookes did not remain silent on the Home investigation, recounting the events for the *PSPR* (vol. 6) and allowing his reports to be republished elsewhere.

87. Fournier d'Albe, *Life of Sir William Crookes*, 248.

88. In a letter to Oliver Lodge on October 4, 1894, William James warned his friend not to publish his own researches into spiritualism or he "will be but another Crookes case to be deplored." William James Papers, Houghton Library, Harvard University (File 62, bMS Am 1092.1).

89. Fournier d'Albe, *Life of Sir William Crookes*, 248.

90. Fournier d'Albe, *Life of Sir William Crookes*, 248.

91. Palfreman, "Between Skepticism and Credulity," 89.

92. Randal Keynes, *Darwin, His Daughter, and Human Evolution* (New York: Riverhead, 2002), 280–82; and Palfreman, "Between Skepticism and Credulity," 89.

93. Keynes, *Darwin, His Daughter, and Human Evolution*, 280–82.

94. What Crookes observed was the flow of electrons off the cathode wire. He was never able to make the leap from atoms to electrons, but his apparatus was at the heart of J. J. Thomson's subsequent discovery in 1897.

95. William Crookes, *Radiant Matter, A Lecture Delivered to the British Association for the Advancement of Science at Sheffield Friday, August 22, 1879* (London: Davey, 1879), 30.

96. Robert K. DeKosky, "William Crookes and the Fourth State of Matter," *Isis* 67, no. 1 (March 1976): 36–60.

97. Fournier d'Albe, *Life of Sir William Crookes*, 179.

98. William Crookes, "Part of the Presidential Address Delivered to the British Association at Bristol, Sept., 1898," *PSPR* 14 (1898–99): 2–3.

99. Crookes, *Radiant Matter*, 30.

CHAPTER THREE

1. Myers, *Fragments of Inner Life*, 12.

2. Frederic Myers, "Account of Friendship with Henry Sidgwick. Oct 18, 1873," Box 13, Myers Collection, Special Collections, Wren Library, Cambridge University.

3. Myers, *Fragments of Inner Life*, 9.

4. Myers, *Fragments of Inner Life*, 12.

5. Turner, *Between Science and Religion*, 108.

6. Robert Goldstein, "Inclined toward the Marvelous: Romantic Uses of Clinical Phenomena in the Work of Frederic W.H. Myers," *Psychoanalytic Review* 79, no. 4 (Winter 1992): 579; and Alan Gauld, *The Founders of Psychical Research* (New York: Schocken Books, 1968), 275–99.

7. This term, as it emerges out of the midcentury debate spurred by Lionel Trilling, is often associated with that branch of modern philosophy dissenting against liberal values (that is, reason, political self-representation, individuality, scientific progress, and so on), but it should not be taken to signify something that is purely reactionary. Such antirationalists as Rousseau, Schelling, Blake, Burke, Rousseau, Kierkegaard, Schopenhauer, and Nietzsche, among others, originated their own version of an evolving humanism that can also be seen as supplementing, cautioning, or redirecting the Enlightenment, rather than "countering" it in the regressive sense. For more on the genealogy of this term, see Richard Schmidt's research blog *Persistent Enlightenment*, https://persistentenlightenment.wordpress.com. See also Schmidt, "The Counter-Enlightenment: Historical Notes on a Concept Historians Should Avoid," *Eighteenth-Century Studies* 49, no. 1 (2015): 83-86.

8. Jacques Barzun, *Classic, Romantic, and Modern* (Chicago: University of Chicago Press, 1975); René Wellek, "The Concept of 'Romanticism' in Literary History," *Comparative Literature* 1, no. 2 (Spring 1949): 1-23, 147-72.

9. Abrams, *Natural Supernaturalism*, 13.

10. Arthur O. Lovejoy, "The Meaning of Romanticism for the Historian of Ideas," *Journal of the History of Ideas* 2, no. 3 (June 1941): 257-78.

11. Quoted from Friedrich Schelling, *The Philosophy of Art: An Oration on the Relation between the Plastic Arts and Nature* (London: Chapman, 1845), 3. Scott Masson explores the significance of this shift in sensibility from a sensory to a supersensory hermeneutic, as it ramifies through art, politics, and epistemology in the early nineteenth century, in *Romanticism, Hermeneutics and the Crisis of the Human Sciences* (Burlington, VT: Ashgate, 2004). The lack of a basis for truth-claims was both liberating and problematic for the Romantic project, a situation addressed in the late nineteenth century through psychical research.

12. Michael Carrithers, "An Alternative Social History of the Self," in Michael Carrithers, Steven Collins, and Steven Lukes, eds., *The Category of the Person: Anthropology, Philosophy, History* (Cambridge: Cambridge University Press, 1985), 237.

13. A recent wave of scholarship looks at the correlation between types of textuality and styles of cognition, as well as the historicity of how reading conditions self-representation and the hermeneutics of representing "the other." For a good survey, see Alan Richardson, "Studies in Literature and Cognition: A Field Map," in Alan Richardson and Ellen Spolsky, eds., *The Work of Fiction: Cognition, Culture, and Complexity* (Aldershot, UK: Ashgate, 2004), 1-30; Lisa Zunshine, *Why We Read Fiction: Theory of Mind and the Novel* (Columbus: Ohio State University Press, 2006); William Nelles, "Jane's Brains: Austen and Cognitive Theory," *Interdisciplinary Literary Studies* 16, no. 1 (2014): 6-29.

14. For more on Seigel's construction of these categories, see Jerrold Seigel, "Problematizing the Self," in Victoria E. Bonnell and Lynn Hunt, eds., *Beyond the Cultural Turn* (Berkeley: University of California Press, 1999), 281-314; and Seigel, *The Idea of the Self: Thought and Experience in Western Europe since the Seventeenth Century* (Cambridge: Cambridge University Press, 2005), 3-44,.

15. Seigel, *The Idea of the Self*, 603-50.

16. The "egoistical sublime" was originally Keats's term for Wordsworth's aesthetics, a commentary on the latter's self-absorbed, subjective materialism. The theory of anti-self-consciousness belonged to Thomas Carlyle. What is interesting is how once this self-positing subject emerges in full force, stabilizing one's gaze outward became an issue for those like John Stuart Mill suffering from toxic "self-consciousness" and depression. For more on this fascinating kink in the tale of modern self-awareness, see Geoffrey Hartman, "Romanticism and 'Anti-Self Consciousness,'" *Centennial Review* 6, no. 4 (Fall 1962): 553-65.

17. Excerpted from Wordsworth's poem, "Climbing Mount Snowden," quoted in Forest Pyle, *The Ideology of Imagination: Subject and Society in the Discourse of Romanticism* (Stanford: Stanford University Press, 1995), 72.

18. For more on this biological stratum of literary cognition and human textuality see Joseph Carroll, "Introduction," in *Evolution and Literary Theory* (Columbia: University of Missouri Press,

1995), 1–40; Ellen Spolsky, "Darwin and Derrida: Cognitive Literary Theory as a Species of Post-Structuralism," *Poetics Today* 23, no. 1 (2002): 43–62. For a more brick-and-mortar analysis of consciousness itself, see F. Kessell, P. Cole, and D. Johnson, eds., *Self and Consciousness: Multiple Perspectives* (Hillsdale, NJ: Erlbaum, 1992). A good description of what "literary cognitive theory" is and aspires to be is provided by Mary Thomas Crane and Alan Richardson, "Literary Studies and Cognitive Science: Toward a New Interdisciplinarity." *Mosaic* 32, no. 2 (1999):123–40; Elizabeth Hart, "The Epistemology of Cognitive Literary Studies," *Philosophy and Literature* 25, no. 2 (2001): 314–34; and Patrick Hogan, *Cognitive Science, Literature, and the Arts: A Guide for Humanists* (New York: Routledge, 2003), 1–58.

19. Norbert Elias, *The Civilising Process: Sociogenetic and Psychogenetic Investigations*, trans. Edmund Jephcott (Oxford: Blackwell, 2000), 449–83; Reddy, "Against Constructionism."

20. Gerald Edelman and Giulio Tononi, *A Universe of Consciousness: How Matter Becomes Imagination* (New York: Basic Books, 2000). Edelman's dynamic core hypothesis posits that the unified subjective experience of consciousness arises from the temporary coherence of individual neurons into a "functional cluster." The former lies beyond the realm of empirical inquiry, though it is entirely coterminous with the latter, a physical process knowable to science. Despite the brain's complexity and the ongoing process of differentiation, each brain-state is experienced as a single point of view, or "quale," "be it primarily a sensation, an image, a thought or even a mood." Edelman's model of consciousness is emphatically integrated, thereby guaranteeing the unity of experience, while at the same time profoundly differentiated, linking the mind to the moment. The "self" is not an entity but a kind of a process, always taking on new information and inhabiting more and more complex physical/experiential information states.

21. Edelman and Tononi, *A Universe of Consciousness*, 221.

22. Michelle Z. Rosaldo, "Toward an Anthropology of Self and Feeling," in Richard A. Shweder and Robert A. LeVine, eds., *Culture Theory: Essays on Mind, Self, and Emotion* (Cambridge: Cambridge University Press, 1995), 143.

23. Clifford Geertz, "Religion as a Cultural System," in his *The Interpretation of Cultures* (New York: Basic Books, 1973), 113.

24. Marcel Mauss, "A Category of the Human Mind: The Notion of Person, the Notion of Self," in *The Category of the Person: Anthropology, Philosophy, History* (Cambridge: Cambridge University Press,1985), 1–25.

25. John Beer, "Myers' Secret Message," in *Providence and Love: Studies in Wordsworth, Channing, Myers, George Eliot, and Ruskin* (Oxford: Clarendon Press, 1999), 116–84.

26. Myers, *Fragments of Inner Life*, 7.

27. Myers, *Fragments of Inner Life*, 7.

28. Hartman, "Romanticism and 'Anti-Self Consciousness,'" 290.

29. Myers, *Fragments of Inner Life*, 7.

30. Beer, "Myers' Secret Message," in *Providence and Love*, 117.

31. Myers, *Fragments of Inner Life*, 5.

32. Quoted in Alan Gauld, *Founders of Psychical Research* (New York: Schocken Books, 1968), 41.

33. Quoted in Ernest Jones, *The Life and Work of Sigmund Freud*, vol. 1 (New York: Basic Books, 1957), 5. Both Freud and Myers were in Munich in August 1896, at the time of the Third International Congress of Psychology. Freud did not present a paper there, but mentions attending a lecture by Lipps where he could well have met Myers—too briefly, no doubt, to inspire Freud's theory, but certainly Myers is an interesting type. Raised by a genteel, stay-at-home mother, that mid-Victorian generation of "angels in the home" provided the doting domesticity that might create such "conquerors." See Günter Gödde, "Freud and Nineteenth-Century Philosophical Sources of the Unconscious," in Angus Nicholls and Martin Liebscher, *Thinking the Unconscious: Nineteenth-Century German Thought* (Cambridge: Cambridge University Press, 2010), 274.

34. Myers Papers, Box 10, Wren Room, Trinity College Library, Cambridge University.

35. Quoted in Gauld, *Founders of Psychical Research*, 39.

36. For a deeply personal sense of the grieving experience of Victorians across the century, largely told through personal accounts, see Pat Jalland, *Death in the Victorian Family* (Oxford: Oxford University Press, 2000). Absorbing death in this profoundly shifting cultural framework, with the new interiority and sentimentality of this literate, introspective generation of early and mid-Victorians, in addition to the expectations of "progress" that worked against traditional resignation to loss, created a perfect storm in which grief could root itself. We note that Myers was born into an intensely evangelical setting that, for him, abruptly shifted; John Stuart Mill himself was agonized by "self-consciousness.

37. Gauld, *Founders of Psychical Research*, 40.

38. Myers, *Fragments of Inner Life*, 8.

39. Myers, *Fragments of Inner Life*, 16.

40. Myers, *Fragments of Inner Life*, 17.

41. Frank Turner, *The Greek Heritage in Victorian Britain* (New Haven, CT: Yale University Press, 1981), 373.

42. Turner, *The Greek Heritage in Victorian Britain*, 370.

43. Hartman quoted in Peter Cochrin, *Byron's Religions* (Cambridge: Cambridge Scholars Publishing, 2011), 244.

44. David Ferris, *Silent Urns: Romanticism, Hellenism, Modernity* (Stanford: Stanford University Press, 2000).

45. Frederic Myers, *Wordsworth* (New York: Harper & Brothers, 1885), 127.

46. Myers, *Fragments of Inner Life*, 10.

47. Gauld, *Founders of Psychical Research*, 90.

48. Julia M. Wright, "Growing Pains," in Joel Faflak and Julia M. Wright, eds., *Nervous Reactions* (New York: SUNY Press, 2004), 163–88.

49. Myers, *Fragments of Inner Life*, 9.

50. Frederic Myers to William Whewell, September 26, 1863; quoted in Beer, *Providence and Love*, 125.

51. Georges Poullet, "Timelessness and Romanticism," *Journal of the History of Ideas* 15, no. 1 (January 1954): 10.

52. Poullet, "Timelessness and Romanticism," 8.

53. Frederic Myers to William Whewell, September 26, 1863; quoted in Beer, *Providence and Love*, 125.

54. Beer, *Providence and Love*, 130.

55. Quoted in Beer, *Providence and Love*, 126.

56. Myers, *Fragments of Inner Life*, 28.

57. Myers, *Fragments of Inner Life*, 18.

58. See Gauld, *Founders of Psychical Research*, 90–91; and Beer, *Providence and Love*, 143, for this specific suggestion of homosexuality. Linda Dowling, *Hellenism and Homosexuality in Victorian Oxford* (Ithaca, NY: Cornell University Press, 1994), makes the case (by broadly circumstantial evidence) that, to a certain extent, homosexuality was part of the cultural fabric of the Oxford classics department. While this is an obvious backdrop against which to interpret the relationship between Lang and William Pater, classicists together at Oxford, Arthur Sidgwick and Myers might yet fit the bill at Cambridge. They did to a certain extent "pair off" as intimate friends while at Cambridge, and in part identified that bond with a shared love of the classics.

59. Myers, *Fragments of Inner Life*, 36.

60. Unidentified manuscript, chap. 1, Myers Papers, Wren Room, Trinity College Library, Cambridge University.

61. Unidentified manuscript of Myers biography (J. C. Broad?), Myers Papers, Wren Room, Trinity College Library, Cambridge University.

62. Myers, *Fragments of Inner Life*, 13 (emphasis added).

63. Richard Jebb, journal entry for February 26, 1866, quoted in Gauld, *Founders of Psychical Research*, 95.

64. This letter is reprinted in Eveleen Myers's bowdlerized reissuing of her husband's autobiography, *Fragments of Prose and Poetry*, and dated May 5, 1865. The references to Mrs. Butler have been removed. Alan Gauld reproduces the text in full, including the reference to Mrs. Butler, but he accepts Mrs. Myers's erroneous date, which may be unclear in the original copy. Since Myers did not become intimate with Mrs. Butler until his return from America, which was in the fall of 1865, I suspect this letter dates to May 5, 1866, when Myers would have been fully in the grip of his Butlerian infatuation. Apparently he had casually known Mrs. Butler several years earlier, before her own Christian enthusiasm. In the unpublished manuscript of his "Account of Friendship with Henry Sidgwick. Oct 18, 1873," Myers writes, "In 1866 my acquaintance with Mrs. Butler which had been for some years been interrupted was revived."

65. Frederic Myers to Arthur Sidgwick, May 5, 1866; quoted in Gauld, *Founders of Psychical Research*, chap. 4. See note 64 for the proper dating of this item.

66. Myers, *Fragments of Inner Life*, 14.

67. Myers, *Fragments of Inner Life*, 15.

68. Gauld, *Founders of Psychical Research*, 121.

69. See Adrian Desmond, *The Politics of Evolution: Morphology, Medicine, and Reform in Radical London* (Chicago: University of Chicago Press, 1989).

70. Myers, *Fragments of Inner Life*, 15.

71. Myers composed these private memoirs in 1887 (though he did not print them until 1893 and did not have them distributed to friends until his death in 1901). He writes with some distance, therefore, to the chronological epicenter of the "crisis of faith," and notably, his rhetoric does seem to have neatly reduced in the intervening years to the somewhat standardized form, "essence of crisis." Nonetheless, it is clear from his contemporary accounts of this period (letters and journal entries) that Myers's actual intellectual concerns and social experiences were true to that form.

72. See Turner, *Between Science and Religion*, 8-37.

73. Frederic Myers, *Science and a Future Life* (London: Macmillan, 1901), 1. This essay was originally written in 1890.

74. Masson, *Romanticism, Hermeneutics and the Crisis of the Human Sciences*.

75. "Account of Friendship with Henry Sidgwick. Oct 18, 1873."

76. Myers, *Fragments of Inner Life*, 14.

77. See Myers's essay collection, *Science and a Future Life*, as well as his autobiography, *Fragments of Inner Life*, for more on this anxiety regarding moral decay as it is specifically linked to evolution and the social and cultural philosophies it inspires. Sidgwick wrestles with this directly in his article "The Theory of Evolution in Its Application to Practice," for *Mind* 1, no.1 (January 1876): 52-67, as well as in his *Essays on Ethics and Methods* (1874), which challenged the egoism and utilitarianism associated with evolutionary philosophy, while remaining within a secular framework.

78. There was an early round of this outrage directed at spiritualism in the 1860s by religionists in such works as Rev. James M. Cosh, *The Supernatural in Relation to the Natural* (1862); Rev. John Tulloch, *The Christ of the Gospels and the Christ of Modern Criticism* (1864); Isaac Taylor, *The Restoration of Belief* (1864); and Horace Bushnell, *Nature and the Supernatural as Together Constituting the One System of God* (1861).

79. Oppenheim, *The Other World*, 123-34. The Sidgwick group formed the core constituency of what would later become the SPR, including Eleanor and Henry Sidgwick, Frederic Myers, Edmund Gurney, Walter Leaf, Arthur James Balfour, and Richard Hodgson.

80. For a good history of mediumship from a scholarly insider, see Alan Gauld, *Mediumship and Survival: A Century of Investigations* (London: Heinemann, 1982), especially the first half of the book, which pertains to the period under discussion.

81. Myers, *Fragments of Inner Life*, 15.

82. Myers to Henry Sidgwick, December 10, 1875, Myers Papers, Wren Room, Trinity Library, Cambridge University.

83. Myers to Henry Sidgwick December 10, 1875. Myers Papers, Wren Room, Trinity Library, Cambridge University.

84. Myers, *Science and a Future Life*, 50.

85. Myers, *Fragments of Inner Life*, 101.

86. Gurney was a classics fellow in the 1870s at Cambridge, where he eventually took a medical degree in 1881, but not in order to practice. Myers also devoted himself to the formal study of medicine, but neither had status as a research professional until psychologists began to take notice of their work around 1884.

87. Myers was ultimately not made to testify, but the new dragnet being cast was now going after dupes (presumably Myers) as well as frauds, by entangling the former in the prosecution of the latter. Myers confided in a letter to Sidgwick concerning Slade that "he had been extremely impressed by him." In slate writing, the medium took two "blank" slates and bound them together, after which spirit writing would miraculously appear. Slade had been definitely caught in the act of prefabrication, much to Myers's mortification. Myers to Sidgwick, July 26, 1876, Myers Papers, Wren Room, Trinity Library, Cambridge University.

88. Quoted from James Braid, *Neurypnology; or, The Rationale of Nervous Sleep, Considered in Relation with Animal Magnetism* (Edinburgh: Charles Black, 1843), 4. Braid unambiguously rejected all mysterious physical influences and extraordinary faculties, making his "shock to the nerves" the most viable option going forward in terms of conservation physics. Elliotson and Eisdale never got quite free of their mesmeric stigma. Magnetism and electricity may have been "natural" in the 1840s, but they were not mechanical. This was an unwelcome vital influence avoided only by Braid. Eisdale was also tied to a more esoteric form of mesmerism, behaving much like telepathy and rooted in the mind, while Elliotson was further tainted by phrenology. For more on Elliotson, see Adela Pinch, *Thinking about Other People in Nineteenth-Century British Writing*, Cambridge Studies in Nineteenth-Century Literature and Culture 73 (Cambridge: Cambridge University Press, 2010), 35-45; and Winter, *Mesmerized*, 32-107.

89. This was essentially the point put forward by W. B. Carpenter in unconscious muscular cerebration. Gauld, *A History of Hypnotism*, 287-305. Carpenter's theory was still regarded with suspicion because it allowed for a form of peripheral intention outside of conscious oversight to direct human action. This more complex view of brain intelligence, which allowed the body to subsume reason, put Carpenter more in line with the future of evolutionary psychology, despite his resistance to the "spiritual" trance.

90. *The Proceedings for the Psychological Society of Great Britain, 1875-79* (London: Privately printed, 1880) contains the entire publication of history and membership roles for the society and will bear out this analysis. Myers's silence speaks volumes, as does Barrett's lack of direct involvement. In general, historians have assumed too close a connection between Cox's organization and the SPR.

91. Luckhurst, *The Invention of Telepathy*, 47-51.

92. For this early modern alchemical carryover into Mesmer's work, see S. Schaffer, "The Astrological Roots of Mesmerism," *Studies in History and Philosophy of Biological and Biomedical Sciences* 41, no. 2 (June 2010): 41158-68. doi: 10.1016/j.shpsc.2010.04.011.

93. Richardson has labeled this "Romantic brain science" to connote the convergence of literary and neurophysiological themes and persons in the period between 1810 and 1830, highlighting how brain dissection offered a productive alternative to the traditional introspection of Descartes and Locke in the study of the mind. According to Richardson, Coleridge and others actively pursued correspondence with Erasmus Darwin and Charles Bell to make science a part of their humanistic project, rooting the mind more firmly in the body, according to the more synthetic

agenda of Romanticism, even as they spiritualized it. See Alan Richardson, *British Romanticism and the Science of Mind* (Cambridge: Cambridge University Press, 2005), 1–92.

94. Myers, *Fragments of Inner Life*, 101.

CHAPTER FOUR

1. Désiré Magloire Bourneville and Paul Regnard, *Iconographie photographique de La Salpêtrière, service de M. Charcot*, vol. 3 (Paris: Aux Bureaux du Progrès Médical, Delahaye & Lecrosnier, 1879–80), 265–347.

2. Quoted in Michel Bonduelle, Toby Gelfand, and Christopher G. Goetz, *Charcot: Constructing Neurology* (New York: Oxford University Press, 1995), 186.

3. The role of medical photography in the evolution of psychiatry is analyzed by Georges Didi-Huberman, *Invention of Hysteria: Charcot and the Photographic Iconography of the Salpêtrière* (Cambridge, MA: MIT Press, 2003), 1–81.

4. Quoted from Pierre Briquet's *Traité* (1859) in Didi-Huberman, *Invention of Hysteria*, 25.

5. See Goldstein, *Console and Classify*, for more on the rise of psychiatry in the nineteenth century as an arm of the modern state's secular approach to managing madness, hitherto a religious domain. As a professional trained and seeking career advancement in this system, Charcot was deeply embedded in its secular politics. By the 1880s, the earlier religious threat had become more sublimated in the wake of a bourgeois bureaucratic victory. It was not clerical interference and its revanchist royal politics that was to be feared in the stewardship of psychology, so much as the residual threat of a gauzy metaphysics come to undercut hard science.

6. For more on Charcot's early research program and his role in articulating a distinct disciplinary narrative for brain science, see David R. Kumar et al., "Jean-Martin Charcot: The Father of Neurology," *Clinical Medicine & Research* 9, no. 1 (2011): 46–49; PMC, May 30, 2018; https://www.ncbi.nlm.nih.gov/pmc/articles/PMC3064755/. doi: 10.3121/cmr.2009.883. See Bonduelle, Gelfand, and Goetz, *Charcot: Constructing Neurology*; and Julien Bogousslavsky, ed., *Following Charcot: A Forgotten History of Neurology and Psychiatry* (Basel: Karger, 2011), for a sense of how Charcot shaped the future progression of knowledge through his publications, disciples, and designated heirs. For a broader, more critical cultural grasp of this project, I recommend the histories by Sander Gilman, Helen King, Roy Porter et al., *Hysteria beyond Freud* (Berkeley: University of California Press, 1993), 226–446; and Didi-Huberman, *Invention of Hysteria*.

7. While Charcot considered himself a neurologist and tended to relate to the inmates of the Salpêtrière as research subjects rather than as patients, given the limited opportunity for autopsy, his principal study was of psychiatric symptoms. His synthetic approach to the experiential and the structural elements of consciousness is best described by the modern term "neuropsychiatry"—a discipline directly descended through Charcot's neurology. For more on this, see Bogousslavsky, *Following Charcot*, 1–23.

8. Didi-Huberman, *Invention of Hysteria*, 1–81; and Olivier Walusinski, "Jean-Martin Charcot (1825–1893): A Treatment Approach Gone Astray?" *European Neurology* 78, nos. 5–6 (October 2017): 296–306. Edmond de Goncourt (1822–96) was a follower of Charcot who later rejected him, stating, "What is curious in Charcot's scientific activities is that he combined genius with charlatanism." Quoted from Walusinki, "Jean-Martin Charcot," 297.

9. "Psycho-physics" was initiated by Gustav Fechner in 1860 with his two-volume introductory work, *Elemente der Psychophysik*, documenting his efforts to find a mathematical formula for correlating sensation with stimulus. For more on the history of psychophysics, see Robert H. Wozniak, *Mind and Body: René Descartes to William James* (Washington, DC: American Psychological Association, 1992), 1–15, 31–44. Wozniak explains Fechner's philosophy as a "dual-aspect monism," where the objective and subjective were parallel aspects of the whole of existence, accessing one reality

from two points of view. Myers found encouragement for his psychology in Fechner's thinking because of this reification of subjectivity. Under Wundt in the 1870s, panpsychism lost this duality, making this mental element a quality of matter, not a distinct phase of reality. While Wundt was wary of the term "psycho-physics" because of this dualistic legacy, that label was widely applied to his work (then and now) to distinguish his quantitative methods from the more neuropsychiatric approach associated with the Salpêtrière. While some may take issue with this use, Wundt himself did not, certifying the use of Fechner's term in his introduction to *Physiological Psychology*: "Insofar as physiological psychology investigates relationships between physical and mental events, the term, coined by Fechner, can be applied to it . . . quite without metaphysical assumptions about the relation between body and mind." Quoted in Robert Rieber and David K. Robinson, eds., *Wilhelm Wundt and the Making of a Scientific Psychology* (New York: Plenum Publishers, 2001), 161. William James and Frederic Myers also referred to Wundt's physiological psychology as "psycho-physics." See James quote in Robert Rieber and David K. Robinson, eds., *Wilhelm Wundt and the Making of a Scientific Psychology* (New York: Plenum Publishers, 2001), 161; and Myers, "Introduction," in Gurney, Myers, and Podmore, *Phantasms of the Living*, 1:xli–xlii.

10. According to Henri Ellenberger, Pierre Janet in 1887 was the first to use the term "subconscious" to designate that region of the mind cut off from the awareness of personality, yet capable of constituting its own discrete psychological experience. For more on the development of this concept in the late 1880s, see Ellenberger, *The Discovery of the Unconscious*, 133–417. However, Myers actively used this term as an adjective as early as 1884, describing subconscious processes in his discussion of automatism. See Frederic Myers, "On the Telepathic Explanation of Some Phenomena Normally Classified as Spiritualistic," *PSPR* 2 (November 1884): 219.

11. Charcot had already cooled on this research well before his death in 1893. The Salpêtrière continued on as a research center under the disciples he had trained, although neurology shifted away from hysteria and hypnosis, whose narratives had already begun to slip from Charcot's control after 1885, compromising their utility. For more on this, see Bogousslavsky, ed., *Following Charcot*, 1–161.

12. As James Braid stated in *Neurypnology*, 4: "I have now entirely separated Hypnotism from Animal Magnetism. I consider it to be merely a simple, speedy, and certain mode of throwing the nervous system." Heidenhain further updated hypnotism in the context of physiological psychology with *Der sogenannte thierische Magnetismus. Physiologische Beobachtungen* (1880), which helped further legitimize hypnotism in time for Charcot's public release of the *Iconographie photographique* in 1881.

13. For a contemporary account detailing hypnotic techniques at Charcot's clinic and elsewhere, see George Kingsbury, *The Practice of Hypnotic Suggestion, Being an Elementary Handbook for the Use of the Medical Profession* (London: Simpkin Marshall Hamilton Kent & Co, 1891), 27.

14. Anne Harrington, "Hysteria, Hypnosis, and the Lure of the Invisible: The Rise of Neo-Mesmerism in Fin-de-Siècle French Psychiatry," in William Bynum, Roy Porter and Michael Shepherd, eds., *The Anatomy of Madness: Essays in the History of Psychiatry*, vol. 3 (London: Routledge, 2004), 226–42.

15. According to historians J. Poirier, P. Ricou, and V. Leroux-Hugon, "Charcot began to study hypnosis under [Richet's] influence." See Bogousslavsky, ed., *Following Charcot*, 200. Eugene Taylor writes, "Charles Richet, experimental physiologist in Charcot's inner circle, had started it all by allowing the hypnotist Burq into the Salpêtrière to first introduce hypnosis to the patients." See Eugene Taylor, "Charcot's Axis," in *The Mystery of Personality: A History of Psychodynamic Theories* (New York: Springer-Verlag, 2009), 20. Régine Plas also notes that Richet was the point man for the early Société (while Ribot and Paul Janet were given the honor of the vice presidencies). See Régine Plas, "Psychology and Psychical Research in France around the End of the 19th Century," *History of the Human Sciences* 25 (2012): 91–107.

16. Staff Author, "Charcot and Hypnotism," *British Medical Journal*, 2, no. 1704 (August 1893): 480.

17. Arthur T. Myers (1851-94), who was eight years younger than Frederic, attended Cambridge during his brother's fellowship and was a member of the extended Sidgwick group. Arthur, eight years younger than Frederic, was at Cambridge during his brother's fellowship and was a member of the extended Sidgwick Group. Unlike his brother, he sat both the classics and the natural science tripos (Myers only prepared for the latter), and since 1874 had been involved in various forms of medical training before receiving his medical degree from Cambridge in 1881.

18. Myers, *Fragments of Prose and Poetry*, 101, 65.

19. Myers, *Fragments Prose and Poetry*, 65.

20. Myers, *Fragments of Prose and Poetry*, 65.

21. Myers, *Fragments of Prose and Poetry*, 65.

22. Frederic Myers, "Automatic Writing II," *PSPR* 3 (January 1885): 30.

23. Myers, *Fragments of Prose and Poetry*, 65.

24. For more on this meeting, see Richard Noakes, "'The Bridge Which Is between Physical and Psychical Research': William Fletcher Barrett, Sensitive Flames, and Spiritualism," *History of Science* 42, no. 4 (2004): 419-64.

25. Edmund Dawson Rogers and William Barrett provided the initial impetus for the SPR and are thus often considered its official founders. However, Myers ensured the success of this venture with his zealous recruitment of his Cambridge associates and other elites who brought social and intellectual cachet to the cause.

26. Staff Author, *PSPR* 1 (1882): 3.

27. William K. Clifford quoted by Myers in *Human Personality and Its Survival of Bodily Death* (London: Longman's, Green & Co., 1903), 297.

28. Reverend Moses was invited for personal and strategic reasons more than philosophical ones. Eleanor Sidgwick expressed the general relief at the resolution of this contretemps: "I think the spiritualists had better go. It seems to me if there is truth to spirit, their attitude and state of mind distinctly hinder its being found out . . . Their spirit is theological not scientific, and it is so difficult to run theology and science in harness together." Letter quoted in Oppenheim, *The Other World*, 140n98.

29. The politics of scientific authority is an important context for understanding the SPR's mix of disciplinary deference and elite insurgence. For more on this as a specifically professional boundary dispute, see Lightman, *Victorian Popularizers of Science*. Also, for more on the many contexts and ways in which science was valued as a source of meaning and authority, see Bernard Lightman, ed., *Victorian Science in Context* (Chicago: University of Chicago Press, 1997), especially "Introduction" by Lightman and "Defining Knowledge" by George Levine, 1-23.

30. For more on this confrontational tone and personalization of scientific authority, see Sir William Crookes, "Spiritualism Viewed by the Light of Modern Science," *Quarterly Journal of Science* (July 1870): 316-20.

31. This was effectively the compromise eventually made by parapsychology, which radically pruned the psychical agenda in an effort to remain in conversation with academic psychology (with limited success). For an insider's review of the most important parapsychological studies of recent decades, see Dean Radin, *Supernormal: Science, Yoga, and the Evidence for Extraordinary Psychic Abilities* (New York: Deepak Chopra, 2013), 119-286. While most scientific reviewers tend to be critical of such accounts, particularly what they deem to be a kind of misappropriation and deployment of its legitimacy, that tension illuminates exactly the politics parapsychology was tasked to navigate. The expansive psychical curiosity of the 1880s and 1890s was radically limited to something more akin in spirit to Barrett's approach of physical experimentation: isolate, quantify, and otherwise verify the phenomena with little theoretical or cultural filtration.

32. Psychology was not recognized with its own section (J) until 1913, coming first under Section D for biology (which included zoology, botany, anatomy, and physiology—and later just physiology, which broke away to form its own section in 1893). See *Report of the Meeting of the British Association for the Advancement of Science* 74 (1904).

33. There was a great deal of popular/professional competition in Britain to control the public reception of novelty and wonders within science (among which was telepathy or "mind-reading"), which contributed to the unique success of the psychical project in that setting. For more on this, see Roger Luckhurst and Josephine McDonagh, *Nineteenth Century* (Manchester, UK: Manchester University Press, 2002), especially 96-150. The French context, however, was more problematic, and efforts to establish a designated society for psychical research were stymied through the early twentieth century.

34. Myers discusses the discrediting assumptions of the SPR's critics at several places in his autobiography (see Myers, *Fragments of Prose and Poetry*, 20, 75, 83, 98, 131). Henry Sidgwick also discussed this problem at length in his presidential address at the general meeting of May 28; see Henry Sidgwick in *PSPR* 2 (1884): 152-56.

35. W. F. Barrett, "Note from the Editor," *JSPR* (1884): 1.

36. For more on the cultural negotiation of scientific legitimacy among scientists, educators, publishers, popularizers, and infotainment profiteers in the mid- to late nineteenth century, see Lightman, *Victorian Popularizers of Science*.

37. See Ian Hacking, "Telepathy: Origins of Randomization in Experimental Design," *Isis* 79, no. 3 (September 1988): 427-51.

38. William James, "Frederic Myers' Service to Psychology," *PSPR* 17 (1901-3): 13-14.

39. W. Stainton Moses, in *JSPR* 2 (December 1886): 488.

40. Balfour Stewart reflects on the state of the society in his "President's Address," *PSPR* 3 (1885): 64-68.

41. Eleanor Sidgwick, "Results of a Personal Investigation into the Physical Phenomena of Spiritualism," *JSPR* 2 (1886): 266-68. Her paper was read before the May 1886 meeting, a synopsis of which appears in these pages.

42. Frederic Myers, "Objects of the Society," *PSPR* 1 (1882-83): 3. See also Edward T. Bennett, *Twenty Years of Psychical Research 1882-1901* (London: Brimley & Johnson, 1904), 7-8. Bennett lists the six committees to be constituted as follows: "Thought-Reading, Mesmerism, Reichenbach's Experiments, Apparitions, Haunted Houses, Physical Phenomena, in addition to which a Literary Committee was appointed."

43. Editor, *JSPR* 1 (February 1885): 268.

44. Myers, *Fragments of Prose and Poetry*, 78.

45. *PSPR* 1 (1882-83): 22.

46. William F. Barrett, Edmund Gurney, and Frederic W.H. Myers, "First Report of the Committee on Thought-reading," *PSPR* 1 (1882-83): 13-32.

47. See Lodge's autobiography, *Past Years* (New York: Charles Scribner's Sons, 1932) for more on Lodge's conflicted relationship with science and spiritualism, which made him initially suspicious of, yet clearly susceptible to, the appeal made by Gurney and Myers.

48. Oliver Lodge, "An Account of Some Experiments in Thought Transference," *PSPR* 2 (1884): 189-200, especially 195.

49. Henry Sidgwick, "Proceedings of the General Meeting, May 28, 1884," *PSPR* 2 (1884): 152.

50. Sidgwick, "Proceedings of the General Meeting, May 28, 1884," 153.

51. As Richard Noakes has argued, the early focus on telepathic experiment, designed by Barrett, had put physics in the lead on questions regarding the mind. Noakes, "'The Bridge Which Is between Physical and Psychical Research,'" 419-40.

52. Edmund Gurney gave a summary of the Mesmeric Committee's early research with "The Problems with Hypnotism," *PSPR* 2 (1884): 265-92. The discussion of animal magnetism and electrobiology comes from the first report. Three reports were issued in the first two years, making this the most active branch of research other than the Committee on Thought Transference, which issued four reports during that period. The Mesmeric Committee also included Richard Hodgson, Arthur Myers, Henry Ridley, W. H. Stone, George Wyld, and C. Lockhart.

53. Krister Dylan Knapp, *William James: Psychical Research and the Challenge of Modernity* (Chapel Hill: University of North Carolina Press, 2007), 220-22. Knapp establishes that a caution was issued by James through an analysis of Gurney's side of the correspondence for 1883-84.

54. James's curiosity regarding altered states had brought him to the Salpêtrière from America in 1882; in the wake of that visit he befriended Gurney and Myers, and eventually embraced psychical investigation. For more on these early influences on James's theories of consciousness, see Knapp, *William James*, 211-47; and Eugene Taylor, *Beyond the Margins of Consciousness* (Princeton: Princeton University Press, 2011), 3-81. For more on the French context as it relates to Pierre Janet, whose research was instrumental in helping Myers advance his model of the unconscious, see Christian Kerslake, *Deleuze and the Unconscious* (London: Continuum International Publishing Group, 2007), 5-48; Onno Van der Hart and Rutger Horst, "The Dissociation Theory of Pierre Janet," *Journal of Traumatic Stress* 2, no. 4 (1989): 397-412.

55. For more on the midcentury exploration of mesmerism and spiritualism in France, see Monroe, *Laboratories of Faith*, 15-149, particularly on the use of the planchette and other forms of spirit writing traveling through mesmerism as well. Myers points out in his review of Richet's work that the topic had not been touched for over twenty years.

56. This independent, sentient will was what distinguished planchette phenomena from "unconscious muscular cerebration." For William Carpenter, this "divided" cerebration was still homologous with one's ordinary subjectivity, only it was intentionality that had drifted into the margins of one's field of awareness. Thus, "unconscious muscular cerebration" was in fact conscious activity of which one was unaware or "unconscious." Carpenter took a strictly physiological view of consciousness, equating all mental states to brain states. Carpenter develops this idea in his *Principles of Mental Physiology* (1874).

57. Gurney's interest in automatic writing was sparked by his belief that hypnotic suggestion was the key to understanding the subconscious, preferring the paradigm of Liébeault to that of Charcot. In his study of hypnotic memory, he noted that the planchette could pour forth suggestions made in a trance state even after he woke, so long as his conscious awareness remained alienated from the submerged agency operating his hands beyond the screen. The fascination here, for both Gurney and Myers, was the complete dissociation between these two active tracks of intelligence: the self-aware writer, holding the planchette, and then the hidden, subconscious author, furnishing the actual content, of whom the writer remained utterly unaware. For more on this see Knapp, *William James*, 218-25.

58. Myers, "On the Telepathic Explanation of Some Phenomena Normally Classified as Spiritualistic," 218.

59. The first published medical case of *dédoublement* dated back to 1876 and involved Felida X, a patient of Eugene Azam's, who had originally come to the doctor as a single entity in 1858 but later cleaved in two under the effects of hypnosis and hysteria. As Ian Hacking argues in *Rewriting the Soul*, 142-82, this sensational new diagnosis of multiple personality disorder reflected the ambition of psychiatry to bring issues of human identity into its own, nonreligious sphere of control in alignment with a new secular politics. The self becomes anchored not in discourses of the soul but rather in biologically based, experiential memory and placed under the care of a state-sponsored psychiatry.

60. Myers, "On the Telepathic Explanation of Some Phenomena Normally Classified as Spiritualistic," 219.

61. Myers, "On the Telepathic Explanation of Some Phenomena Normally Classified as Spiritualistic," 228.

62. Myers, "On the Telepathic Explanation of Some Phenomena Normally Classified as Spiritualistic," 231.

63. While Myers later admitted to conducting hundreds of such trials (in *Human Personality*, vol. 2 [1904], 93), here, in a follow-up discussion of Mr. A. (see Myers, "Automatic Writing II"),

Myers described himself as only an "ordinary insensitive person," limiting his experience to an episode of "low-grade autographia" in 1875. No mention of any spiritualistic context was made, though it would have been likely. John Gray suggests that "the friend" may have been Myers himself. The experiment took place in 1883, when he was already involved in his long attempt to contact his dead love, Annie Marshall. Edmund Gurney is another likely candidate (in addition to Myers or his brother Arthur). He was the first to experiment with automatic writing, and the naiveté of Mr. A. could represent that earliest phase of Gurney's research, now presented by Frederic Myers with the more authoritative understanding worked out by the Committee on Mesmerism. See John Gray, *The Immortalization Commission* (New York: Farrar, Straus & Giroux, 2011), 99.

64. Myers, "Automatic Writing II," 219.

65. This was the work of Susan M. Marsh and David C. Taylor, "Hughlings Jackson's Dr. Z: The Paradigm of Temporal Lobe Epilepsy Revealed," *Journal of Neurology, Neurosurgery, and Psychiatry* 43 (1980): 758-67.

66. Though Arthur is a good candidate, Gurney is perhaps better, as the timing is right for his initial exploration of automatic writing, first with a pen, then with the planchette. I doubt Myers would commit himself to the blatant falsehood of calling his brother "a friend."

67. In his diary, Myers would later note after a trip to Nancy that all Liébeault would talk about was "Gurney, Gurney, Gurney"—suggesting that psychical research was more actively thought about than mentioned in those early years.

68. Richet's "La suggestion mentale et le calcul des probabilités" first appeared in the *Revue philosophique* (December 1884) and was given a detailed write-up a few weeks later by Edmund Gurney in "Researches in Thought Transference," *PSPR* 2 (December 30, 1884): 249-64. This was not a translation but a very thorough account. Up to that point, no mention of psychical research had appeared in any French academic journal, based on my own comprehensive search of the Bibliothèque Nationale's digitized Gallica catalogue, using multiple criteria: http://gallica.bnf.fr.

69. Lodge sent Gurney a letter at the last minute challenging Richet's statistical model with suggestions for how his formula could be improved. Gurney went ahead and published it, though under duress, which speaks highly of the psychical commitment to peer review. Richet was a very big fish, and a French one at that, while Lodge, who had only recently been landed, was also someone the society would not wish to offend. The letter was published under "Note," appended to the end of "Researches in Thought Transference," right after the reprinting of Richet's "La suggestion mentale et le calcul des probabilités," so the two would be read together. Gurney would later work with the statistician Francis Edgeworth in modeling the data for *Phantasms of the Living*, ensuring the utmost technical accuracy be reflected in a work published with the official sanction of the council of the SPR. Gurney's efforts to enhance psychical methodology are further discussed in Andreas Sommer, "Professional Heresy: Edmund Gurney (1847-88) and the Study of Hallucinations and Hypnotism," *Medical History* 55, no. 3 (2011): 383-88.

70. Myers, "Automatic Writing II."

71. For more on the centrality of Charcot in shaping the emerging landscape of the fin-de-siècle psychiatry of the subconscious, see Taylor, "Charcot's Axis," in *The Mystery of Personality*, 19-51. Brower, *Unruly Spirits*, 1-44, gives psychical research a rich French context. This includes its early historical entwinements with Allan Kardec's spiritism and mesmerism, as well as its development through academic publications like *La revue métaphysique* and medical elaboration through practitioners like Ribot, Janet, and Richet in the context of French neuropsychiatry. For a sense of the French cultural narratives surrounding occultism in the second half of the nineteenth century, with an emphasis on midcentury spiritualism and mesmerism, see Monroe, *Laboratories of Faith*.

72. Charcot's empire of neurology was partly maintained by the extent of the publications he controlled, allowing him to promote certain discussions, and even adjust the volume of that discourse up or down at will. Charcot's skilled management of medical media is explored in "The Birth and Death of Charcot's Scientific Journals" by J. Poirier, P. Ricou, and V. Leroux-Hugon, in

Bogousslavsky, ed., *Following Charcot*, 187–201. Michel Bonduelle, Toby Gelfand, and Christopher Goetz have identified Charcot as the sole founder of the *Bulletins de la Société de Psychologie Physiologique*, which served "not to disseminate Charcot's views to a wide readership, but rather consolidated an elite group of investigators from eclectic backgrounds." See their *Charcot: Constructing Neurology*, 96.

73. Courtenay Raia, "The Search for a Transcendent Madness: A Journey in Brief across *Les Bulletins de la Société de Psychologie Physiologique* (1885–1887)," paper presented at the Interdisciplinary Nineteenth Century Studies Conference: Serials, Cycles, Suspensions, held in San Francisco, March 2018. It became clear in my reading of the *Bulletins* and related documents that Richet had tapped a pent-up curiosity, largely kept from official view. It was "re-"pressed by Charcot back into the bottle, but to interpret this as a lack of French interest is to accept the sanctioned perspective on the phenomena, not the actual one.

74. M. Gley, Ch. Richet, and M. Rondeau, "Notes sur le Haschich"; M. Guéroult, *Le raisonnement inconscient dans les localisations auditives*; M. Beaunis, "Influence de la durée de l'expectation sur le temps de réaction des sensations visuelles," and "Suggestion à 172 jours d'intervalle," in *Bulletins de la Société de Psychologie Physiologique* 1 (1885): 9–20.

75. For an epidemiological and epistemological account of representations of hysteria in thie period, see J. P. Luauté, "Fin-de-Siècle' Epidemiology of Hysteria," and E. Medeiros De Bustos, S. Galli,· E. Haffen, and T. Moulin, "Clinical Manifestations of Hysteria: An Epistemological Perspective or How Historical Dynamics Illuminate Current Practice," both in J. Bogousslavsky, ed., *Hysteria: The Rise of an Enigma* (Basel: S. Karger, 2014), 20–43.

76. The substance of this hostility is laid out in J. Bogousslavsky and B. Piechowski-Jozwiak, "Hypnosis and the Nancy Quarrel," in Bogousslavsky, ed., *Hysteria*, 56–64.

77. The essays submitted to the *Bulletins* never directly challenged hysterical nosology, but often an almost supplicant language was added toward the end about other paradigms (usually suggestion or alienism) from which to further consider the physiological phenomena under review. See Raia, "The Search for a Transcendent Madness."

78. In 1858, Eugène Azam began treating hysterics in Bordeaux, inspired, like Liébeault, by the works of James Braid, which had yet to be translated. But Azam's attempts to gain institutional support for neurohypnotism failed with both the Société Nationale de Chirgurie and the Académie des Sciences, and this brief surge of official magnetic interest quickly receded. (For more on Azam and this early formal interest, see Brower, *Unruly Spirits*, 20–25.) Liébault, however, quietly continued with his practice, though he was largely ignored, selling only six copies of his textbook, *Du sommeil et des états analogues* (1866). That being said, he must be credited for the later neo-mesmeric revival, having prepared a workable model in advance, ready at hand when hypnotism's cultural moment struck.

79. Myers specifically referenced Théodule Ribot's statement that "a voluntary act is only a reflex act of the whole organism," suggesting that all choices were the result of conditioned neural processes. Such arguments were in vogue at the time of his visit. Ribot is quoted in Frederic Myers, "Human Personality in the Light of Hypnotic Suggestion," *PSPR* 4, no. 10 (1886): 12. This paper was first read at the general meeting of the society on October 29, 1885, to summarize the continental outlook after his trip. It first appeared in print as "Human Personality," *The Fortnightly Review* 38 (July–December 1885): 637–55. Before its publication in the *PSPR*, Myers wrote an addendum to his original paper that included exciting developments in *psychologie physiologique* that began to take off in the wake of Richet's November visit.

80. See Michael Gazzaniga, *Who's in Charge? Free Will and the Science of the Brain* (New York: HarperCollins, 2011), 1–220, for a readable survey of the contemporary neuroscience of decision making and moral will as it relates to these earlier models.

81. During this trip to the Salpêtrière, Myers also met with Charles Féré (already a member of the SPR), who endorsed Liégeois's depressing assessment of personal responsibility under the

effects of hypnosis, though Liégeois was an affiliate of the Nancy school. While hypnotic passivity could be exploited therapeutically, Liégeois argued it could be exploited criminally as well, offering himself as a defense witness in several sensational trials, referenced by Myers in "Human Personality in the Light of Hypnotic Suggestion." See Alfred Binet and Charles Féré, "Hypnotisme et responsabilité," *Revue philosophique de la France et de l'étranger* (March 19, 1885): 265-79.

82. Letter from Frederic Myers to Eveleen Myers, August 30, 1885, item no. 209, Box 7, Myers Papers, Wren Room, Trinity College Library, Cambridge University.

83. Myers, "Human Personality in the Light of Hypnotic Suggestion," 3.

84. Myers, "Human Personality in the Light of Hypnotic Suggestion," 2.

85. Myers, "Human Personality in the Light of Hypnotic Suggestion," 4. The idea of "colonial consciousness" was formally elaborated by Théodule Ribot and served as a shorthand for a generally accepted notion among evolutionary naturalists that consciousness was an emergent property arising from the aggregation of cells into increasingly complex information systems. Charcot himself formally embraced this idea as a cornerstone of his neurology, depicting disease as the unwinding of this accretion.

86. Myers, "Human Personality in the Light of Hypnotic Suggestion," 3.

87. Myers, "Human Personality in the Light of Hypnotic Suggestion," 1.

88. Frederic Myers, "Further Notes on the Unconscious Self, Part I," *JSPR* 2 (December 1885): 122-31.

89. This was worked out in a detailed analysis of *Les Bulletins de la Société de Psychologie Physiologique* (1885-87) for a conference paper (Raia, "The Search for a Transcendent Madness"). When we take all the papers of *Les Bulletins* (printed in no particular order for each year) and rearrange them consecutively according to their actual meeting dates, the outline of Richet's machinations becomes patently clear.

90. Frederic Myers's Second Diary, Box M14, Myers Papers, Wren Room, Trinity College Library, Cambridge University.

91. Frederic Myers, "De certaines formes d'hallucinations," in *Les Bulletins de la Société de Psychologie Physiologique*, vol. 1 (Paris: Félix Alcan, 1885), 51-52. The psychical language of "hallucination" acknowledged the complex composition of the mental imagery, which often synthesized an exterior image of the person along with elements of that person's point of view. Such nuanced discussions allowed the phenomenon itself to enter more into the fashionable debates about how perception arises.

92. See *Les Bulletins de la Société de Psychologie Physiologique*, vol. 1, containing Pierre Janet, "Note sur quelques phénomènes de somnambulisme," 24-32; Charles Richet, "Un fait de somnambulisme à distance," 33-34; M. Héricourt, "Un cas de somnambulisme à distance," 35-38; and M. Beaunis, "Un fait de suggestion mentale," 39-40.

93. Héricourt, "Un cas de somnambulisme à distance," 35 (my translation).

94. Beaunis, "Un fait de suggestion mentale," 40 (my translation).

95. *Présidence* can refer to a presidency, but also to the individual who takes charge or presides over a meeting. In this case, it is also the latter. The same phrase is used to indicate that Vice President Ribot is chairing the meeting: "Présidence de M. Ribot, Vice Président." (See *Bulletins de la Société de Psychologie Physiologique* 2 (1886): 113.)

96. While Hippolyte Bernheim's polemical book, *De la suggestion et de ses applications à la thérapeutique* (1886), had yet to be released, the battle lines were already drawn by 1884 when Bernheim began to publicly oppose Charcot's authority, challenging him at the very site of his fame, the public demonstration. Bernheim concluded that the effects Charcot produced in these living experiments were the result of suggestion, with patients continuously primed at the hospital to meet medical expectations. Given the deference to which Charcot had become accustomed and the subversive nature of this critique, animus was inevitable. It is useful to consider the rising challenge of Nancy as part of the context for the Société de Psychologie Physiologique, and

for Charcot's willingness to concede such a wide experimental latitude to its members so long as any theoretical challenge was kept at bay. For more on the relationship between Nancy and the Salpêtrière, see Bogousslavsky, *Hysteria*, 56–64; and Andreas Mayer, "The Controversy between Paris and Nancy over Hypnotic Suggestion," in Mayer, *Sites of the Unconscious*, translated by Christopher Barber (Chicago: University of Chicago Press, 2013).

97. Bourru and Burot, "Les premières expériences sur l'action des médicaments à distance"; and Charles Richet, "L'action des substances toxiques et médicamenteuses à distance," *Bulletins de la Société de Psychologie Physiologique* 2 (1886): 10–20, 31–33. Bourru and Burot's paper was presented at the meeting for December 28, 1885, but it was included in the *Bulletins* for 1886.

98. "Louis V." was an abbreviation offered to protect the privaty of the patient, Louis Vivet. The case was no doubt already known to Myers in December when he published "Further Notes on the Unconscious Self I," and, by its content, he was clearly inspired by it. He gave a report of the case at the general meeting for March 8, 1886, updating members on its progress, and mentioned it in the appendix of the April printing of "Human Personality in the Light of Hypnotic Suggestion." Multifocality continues to be the focus as well in "Further Notes on the Unconscious Self II." He does not, however, mention Louis Vivet by name anywhere in either part of "Further Notes," refraining from intruding on a highly conspicuous case over which he had no direct knowledge or authority. It was enough that Myers was beginning to ruminate theoretically. He would not be ready to put the theory together with argumentative data until the following year with his breakout essay, "Multiplex Personality," delivered before the SPR in October 1886.

99. Myers, "Human Personality in the Light of Hypnotic Suggestion," 21.

100. Myers, "Further Notes on the Unconscious Self II," 234-47. 243.

101. Myers, "Further Notes on the Unconscious Self II," 243. Myers made a point of distancing himself from any spiritual hypothesis in this article, insisting that in representing even these stranger phenomena, "the influence of other minds is what we do say." He also defends *psychologie physiologique* to his more sentimental critics: "Physiology is an indispensable factor in every one of the problems which we are here discussing." See Myers, "Further Notes on the Unconscious Self II," 237 and 239.

102. Myers, "Further Notes on the Unconscious Self II," 243.

103. Myers, "Further Notes on the Unconscious Self I," 129.

104. Myers, "Human Personality in the Light of Hypnotic Suggestion," 20.

105. Myers, "Human Personality in the Light of Hypnotic Suggestion," 20.

106. There was a meeting on May 18, and possibly one earlier in the month, which is likely when Myers presented his report, given that he concluded his stay in Le Havre before the end of April. Only Janet's formal report, presented at the meeting for May 31, 1886, made it into the *Bulletins*, along with a write-up by Ochorowicz. See Pierre Janet, "Deuxième note sur le sommeil provoqué à distance et la suggestion mentale pendant l'état somnambulique," *Bulletins de la Société de Psychologie Physiologique* 2 (1886): 70–80. Myers translated part of Janet's account and included it alongside his own field notes, published together in Frederic Myers, "On Telepathic Hypnotism," *PSPR* 4, no. 10 (August 1886): 127–82.

107. Myers, "On Telepathic Hypnotism," 127.

108. Charles Féré became a member in 1884, at the same time as Richet, making them the first of the French to join the SPR. In these particular experiments, Féré instructed a deeply hypnotized patient to perform certain behaviors, including handwriting, on the right side of her body, and then, secretly, applied a magnet to drag that same routine to the left side of her body. See Edmund Gurney, "Critical Notices: A. Binet and Ch. Féré, *Le magnétisme animal*," *PSPR* 4, no. 10 (1886): 545, for the reference to this trip. Myers also described Féré's private demonstration on a female patient named Witt: Frederic Myers, "Report of the General Meeting," *JSPR* 2 (November 1886): 444–46.

109. For a summary of their work on physical transfer, see Alfred Binet and Charles Féré, "L'hypnotisme chez les hystériques: Le transfert physique," *Revue philosophique* 19 (1884): 1–25.

In 1885, metallotherapy began to morph into a psychological phenomenon as well. Alfred Binet and Charles Féré, "La théorie physiologique de l'hallucination," *Revue scientifique* 35 (January 10, 1885): 49–53; Alfred Binet and Charles Féré, "L'hypnotisme chez les hystériques," *Revue philosophique de la France et de l'étranger* (January 19, 1885): 1–25; and Alfred Binet and Charles Féré, "La polarisation psychique," *Revue philosophique de la France et de l'étranger* (April 19, 1885): 369–402.

110. Staff Author, "Summary of Foreign Reviews," *Scottish Review* 5 (1885): 395.

111. Myers, "Report of the General Meeting," 445.

112. Myers, "On Telepathic Hypnotism," 169.

113. Alfred Binet and Charles Féré, *Animal Magnetism* (New York: Appleton & Co., 1894), 170–210. The original work was published in 1887, but excerpts began appearing the year before. This included a consideration of the recent work of the Société de Psychologie Physiologique, including the "telepathic" action of medicinal and psychoactive substances by Richet, Bourru, and Burot, mentioned previously.

114. Joseph Babinski, "Transfert de certaines manifestations hystériques," *Bulletins de la Société de Psychologie Physiologique* 2 (1886): 113–16. This paper was presented at the meeting held on October 25, 1885.

115. Myers, "On Telepathic Hypnotism," 147. This assumption regarding the effects of hypnotic lethargy on the perception of the magnet rested purely on the researchers' assumptions. Myers made a point of noting this as a kind of reservation, but supported their conclusions and basic assumptions, nonetheless.

116. Myers, "On Telepathic Hypnotism," 137.

117. Frederic Myers, "On the Consciousness of Hysterical Subjects," *PSPR* 6 (1889): 200.

118. For more on patients' psychiatric narratives as theatrical presentations, see Colette Conroy, *Theatre and the Body* (New York: Palgrave Macmillan, 2009), 62–75.

119. Gurney, "Critical Notices," 540.

120. Myers, "On Telepathic Hypnotism," 179.

121. Myers, "Report of the General Meeting," 443.

122. Myers, "Report of the General Meeting," 450.

123. Frank Podmore assessed the dangers posed to Charcot and others attempting the scientific study of telepathy in the introduction to his *Apparitions and Thought-Transference: An Examination of the Evidence for Telepathy* (London: W. Scott, 1894), 4.

124. However, this was a limited-circulation journal (part of Charcot's strategy of containment), for which he never authorized a second printing. It has remained relatively obscure because of this, though that will change with digitization. For more on the journal, see J. Poirier, P. Ricou, and V. Leroux-Hugon, in Bogousslavsky, ed., *Following Charcot*, 187–201.

125. For more on Victorian diseases of the nerves and the culture of modernity, see George Frederick Drinka, *The Birth of Neurosis: Myth, Malady, and the Victorians* (New York: Simon & Schuster, 1984); Ilza Veith, *Hysteria: The History of a Disease* (Chicago: University of Chicago Press, 1965); Susan Ashley, "Railway Brain: The Body's Revenge against Progress," *Proceedings of the Western Society for French History* 31 (2003): 177–96.

126. Charcot's clinic was notable for its early adoption (1878) of the mechanical vibrator in the treatment of its patients. For more on this and the pathology of anorgasmia in Victorian medicine, see Rachel P. Maines, *The Technology of Orgasm: Hysteria, the Vibrator and Women's Sexual Satisfaction* (Baltimore: Johns Hopkins University Press, 2001), 21–99.

127. Frederic Myers, "Multiplex Personality," *PSPR* 4, no. 11 (1887): 514. First read before the society on Friday, October 29, 1886, in the immediate wake of the October meeting of the Société de Psychologie Physiologique. For mention of the scheduled reading, see the meeting notes as published in *JSPR* 2, no. 34 (1886): 443.

128. Louis V.'s traumatic childhood and overwhelming experiences make him one of psychiatry's earliest trauma victims. The evolution of trauma as a psychiatric diagnosis and in relation

to personality and memory are explored in Mark S. Micale and Paul Frederick Lerner, eds. *Traumatic Pasts: History, Psychiatry, and Trauma in the Modern Age, 1870–1930*, Cambridge Studies in the History of Medicine (Cambridge: Cambridge University Press, 2001), 115–39; and Roger Luckhurst, *The Trauma Question* (London: Routledge, 2008), 20–48.

129. Myers, "Multiplex Personality" (1887), 497–98.

130. Myers, "Multiplex Personality" (1887), 501.

131. Myers, "Human Personality in the Light of Hypnotic Suggestion," 3.

132. Myers, "Human Personality in the Light of Hypnotic Suggestion," 3.

133. Frederic Myers, "Multiplex Personality," *Nineteenth Century* (November 1886): 650. This is the same paper Myers read before the general meeting of the SPR on October 29, 1886, and published in *PSPR* 4 (1887). It also appeared in two parts in the *English Mechanic and World of Science*, 1159–60 (June 10 and 17, 1887): 338–40, 360–62. Myers typically campaigned his important articles and ideas outside of the SPR's official publications, which were limited to subscribers.

134. Myers, "Multiplex Personality" (1886), 649.

135. The repressed evolutionary instinct that Freud began sketching in his "Project for Psychology" has a precursor in the evolutionary unconscious explored by French neuropsychiatry, and, indeed, Freud spent 1886 interning at the Salpêtrière. For more on this psychical and psychiatric entwinement with Freudian psychoanalysis, particularly as it pertains to the Salpêtrière and the 1880s, see Mayer, *Sites of the Unconscious*, 1–92.

136. Myers, "Multiplex Personality" (1887), 504.

137. Myers, "Multiplex Personality" (1887), 497.

138. Myers, "Introduction," in Gurney, Myers, and Podmore, *Phantasms of the Living*, xliii.

139. Myers, "Introduction," in Gurney, Myers, and Podmore, *Phantasms of the Living*, xliii.

140. Myers, "Multiplex Personality" (1887), 504.

141. Myers, "Multiplex Personality" (1887), 508.

142. Myers, "Introduction," in Gurney, Myers, and Podmore, *Phantasms of the Living*, xlii.

143. Myers, "Multiplex Personality" (1887), 505.

144. Myers, "Multiplex Personality" (1887), 507.

145. Myers, *Human Personality*, 1:101.

146. For the impact of Kantian epistemology on the development of psychology, see Benjamin Wolman, "Immanuel Kant and His Impact on Psychology," in Wolman, ed., *The Historical Roots of Psychology* (New York: Harper & Row, 1968), 229–47.

147. Myers, "Human Personality in the Light of Hypnotic Suggestion," 20.

148. For more on the different national contexts for psychical research, see Anna Taves, "A Tale of Two Congresses: The Psychological Study of Psychical, Occult, and Religious Phenomena: 1900–1909," *Journal of the History of the Behavioral Sciences* 50, no. 4 (Fall 2014): 376–99. See also Andreas Sommer, "Normalizing the Supernormal: The Formation of the 'Gesellschaft für Psychologische Forschung," *Journal of the History of the Behaviorial Sciences* 49, no. 1 (January 2013): 18–44; Sophi Lachapelle, *Investigating the Supernatural: From Spiritism and Occultism to Psychical Research and Metapsychics in France, 1853–1931* (Baltimore: Johns Hopkins Press, 2011), 7–142. For the difficulty in establishing a German context for psychical research as part of German laboratory psychology, see Heather Wolffram, *The Stepchildren of Science: Psychical Research and Parapsychology in Germany, c. 1870–1939*, Wellcome Series in the History of Medicine 88 (Amsterdam: Rodopi, 2009); and Corinna Treitel, *A Science for the Soul: Occultism and the Genesis of the German Modern* (Baltimore: Johns Hopkins University Press, 2004), 1–131.

149. As Richet put it in his memoir: "[Charcot] was considered as the great hypnotist and was reputed to be a semi-magician by the masses and a creator by scientists. Charcot was the obvious choice for President." This suggests that Richet aggressively importuned Charcot to assume the role, and Charcot may have felt somewhat in his debt. It would also have been awkward to publicly undermine with an explicit refusal the hypnotic curiosity he himself had incited. There

were also incentives: assuming "the presidency" let Charcot deprive Bernheim and other rivals of some of their dominance, maintaining the equation of his name with a final authority. Quoted from Richet's memoirs in Plas, "Psychology and Psychical Research in France around the End of the 19th Century," 96.

150. The sheer volume of papers submitted on the topic compelled the organizers to establish Le Premier Congrès International d' Hypnotisme. This was also meant as a check on a topic threatening to overrun the more orthodox concerns of *psychologie physiologique*. While a few papers on hypnotism found their way to the main congress, the majority of its program dealt with diseases of the nerves, speech impediments, and analysis of perception, among other physiological and psychophysical lines of inquiry.

151. For more on how panpsychism operated in Romanticism versus mechanism, see David Skrbina, *Panpsychism in the West* (Cambridge, MA: MIT Press, 2017), 121–70.

152. This effort to fully naturalize Palladino's phenomena as an organic process gone awry is explored as part of French professional angst in Lachapelle, *Investigating the Supernatural*, 86–112. Andreas Sommer places it in a more transatlantic context in "Psychical Research and the Origins of American Psychology: Hugo Münsterberg, William James and Eusapia Palladino," *History of the Human Sciences* 25, no. 2 (2012): 23–44.

153. Henry Sidgwick, "Proceedings of the General Meeting on December 30, 1884," *PSPR* 2 (1884): 238. Séance mediums were eventually incorporated into the Paris program of the Congrès International de la Société de Psychologie Physiologique to demonstrate their "aberrational phenomena"; however, they hardly considered themselves candidates for diagnosis. The mix did not go over well, causing considerable umbrage among the more conservative members, who threatened to withdraw. For more on this, see Taves, "A Tale of Two Congresses," 10–30; and Brower, *Unruly Spirits*, 23–44.

154. Frederic Myers, "The Work of Edmund Gurney," *PSPR* 5 (1889): 370.

155. For a more detailed exposition of these chapters, see Emily Williams Kelly, "F. W. H. Myers and the Empirical Study of the Mind-Body Problem," in Edward F. Kelly, Emily Williams Kelly et al., *Irreducible Mind: Toward a Psychology for the 21st Century* (Lanham, MD: Rowman & Littlefield, 2006), 47–114. The book aims to rewrite psychical research and the subliminal self back into the history of twentieth-century psychology where, the authors argue, Myers's theoretical propositions and methodologies have continued on as an unrecognized tributary.

156. Poullet, "Timelessness and Romanticism," 3–22.

157. Frederic Myers, "The Subliminal Consciousness," *PSPR* 7 (1891–92): 301–2.

158. This is the general thrust of Kelly, Kelly et al., *Irreducible Mind*, which documents the overlapping concerns between Myers's psychology and psychology's subsequent history, as well as the role he played in mapping out the space of the subconscious, which was colonized and reconfigured by Freud and the Freudians in the twentieth century without proper attribution.

159. Frederic Myers, "Presidential Address," *PSPR* 15 (1900–1901): 126.

160. Frederic Myers, undated notebook, item 59, Box 5, Myers Papers, Wren Room, Trinity College Library, Cambridge University.

161. Myers, "The Possibility of a Scientific Approach to Problems Generally Classed as Religious," April 29, 1898, published in *Papers Read before the Synthetic Society (1896–1908)* (London: Spottiswoode Press, 1909), 187–97.

162. Myers, *Human Personality and Its Survival of Bodily Death*, 278–79.

163. Myers, "The Possibility of a Scientific Approach to Problems Generally Classed as Religious," 187–97.

164. Myers, "The Possibility of a Scientific Approach to Problems Generally Classed as Religious," 190.

165. Myers, "The Possibility of a Scientific Approach to Problems Generally Classed as Religious," 188.

166. Frederic Myers, *Human Personality and Its Survival of Bodily Death*, 2:282.

167. Myers to Arthur Sidgwick, May 5, 1865; reprinted in Myers, *Fragments of Prose and Poetry*, 26–27.

168. Frederic Myers, *Fragments of Inner Life*, 38–39.

169. Charles Taylor, "The Person," in *The Category of the Person: Anthropology, Philosophy, History* (Cambridge: Cambridge University Press,1985): 279.

170. Taylor, "The Person," in *The Category of the Person*, 277.

171. Frederic Myers, "The Disenchantment of France" (1888), in *Science and a Future Life*, 119.

172. Théodore Flournoy, Review of *Human Personality and Its Survival of Bodily Death*, PSPR 18: 42–50.

173. Oliver Lodge to Sonnenschein, May 7, 1914, Lodge Papers, Archives for the Society for Psychical Research, Department of Manuscripts and University Archives, Cambridge University Library.

174. Myers to William James, December 9, 1900, item 71, Box 11, Myers Papers, Wren Room, Trinity College Library. Cambridge University.

CHAPTER FIVE

1. "Section A. Mathematics and Physics, Opening Address by Prof. Oliver J. Lodge, President of the Section," *Nature* 44 (August 20, 1891): 386–87.

2. For an excellent account of the attempt by British physics to develop a mechanical ether around the blueprints of Maxwell's *Treatise*, see Bruce Hunt, *The Maxwellians* (Ithaca, NY: Cornell University Press, 1991), 1–47, 73–107.

3. George Francis Fitzgerald, *The Scientific Writings of the Late George Francis Fitzgerald*, edited by Joseph Larmor (London: Longmans, Green, & Co. 1902), 229–40.

4. Quoted in Peter Rowlands, *Oliver Lodge and The Liverpool Society* (Liverpool: Liverpool University Press, 1990), 19.

5. For an excellent account of the laboratory pursuit of electromagnetic waves in competition with Hertz, see Rowlands, *Oliver Lodge*, 15–39.

6. Paraphrased from the title of William Thomson's often quoted speech, "Nineteenth Century Clouds over the Dynamical Theory of Heat and Light," given at the weekly evening meeting, Friday, April 27, 1900, and published in *Notices of the Proceedings at the Meetings of the Members of the Royal Institution of Great Britain with Abstracts of the Discourses*, vol. 16 (London: W. Nicol, Printer to the Royal Institution, 1902), 363.

7. Bruce Hunt observes that Hertz would otherwise probably not have come to the notice of British science. Part of Fitzgerald's speech is included in Hunt, *The Maxwellians*, 160.

8. Lodge discusses the failures of the Michelson-Morley experiment as well as his own reengineered design in Oliver Lodge, "Aberration Problems: A Discussion concerning the Motion of the Ether near the Earth and concerning the Connexion between Ether and Gross Matter, with Some New Experiments," *Philosophical Transactions of the Royal Society* 184 (1893): 727–804.

9. Lodge explained his rationale for the experiment in Oliver Lodge, *The Ether of Space* (London: Harper & Brothers, 1909), 70–71.

10. Lodge, *Past Years*, 195–207.

11. For a detailed report of Lodge's experiment, see *Philosophical Transactions of the Royal Society of London*, Series A, 184 (October 1893): 727–804; and Series A, 189 (May 1897): 149–66.

12. Recalled by Oliver Lodge, *My Philosophy* (London: Ernest Benn, 1933), 48–49.

13. This consideration was later described by Lodge, "Inertia," *Scientific Monthly* 10, no. 1 (January-June, 1920): 379.

14. Dr. Chiaia, letter to Cesare Lombroso; quoted in Rupert Homes, "Seeing Things: The Most Famous of Living Mediums," *Pearson's Magazine* 20 (1908): 391.

15. "Rapport de la Commission réunie à Milan pour l'étude des phénomènes psychiques," *Annales des sciences psychiques* 3, no. 1 (January–February 1893). The investigation was held in 1892, and the report reprinted in *PSPR* 9 (1893): 223 (quote).

16. Charles Richet wrote Myers from Rome, where he was a member of an investigative team led by M. de Siemiradski researching Eusapia Palladino's phenomena. Letter from Myers to Lodge, April 16, 1894, no. 1394, Lodge Papers, Archives of the Society for Psychical Research, Department of Manuscripts and University Archives, Cambridge University Library.

17. Lodge to Lang, December 22, 1909, no. 1303, Lodge Papers, Archives of the Society for Psychical Research, Department of Manuscripts and University Archives, Cambridge University Library.

18. For an early biography of Eusapia Palladino from a contemporary source acquainted with his subjects, see Hereward Carrington, *Eusapia Palladino and Her Phenomena* (New York: B. W. Dodge & Co., 1909).

19. References to ectoplasm and pseudopods are made in letters as early as August 3, 1894, no. 1407, and August 7, 1894, no. 1408, in Lodge Papers, Archives of the Society for Psychical Research, Department of Manuscripts and University Archives, Cambridge University Library. The August 3 letter even seems to suggest that Myers might have had a sample of ectoplasm, extruded from a person in their sleep prompted by some other party telepathically. Richet stated this claim, along with an assertion that Schrenck-Notzing's phenomena (called teleplasm) were one and the same, in *Thirty Years of Psychical Research*, 515.

20. For a discussion of the origin of the term "ectoplasm," see Marc Demerest, "Spirits of the Trade: Teleplasm, Ectoplasm, Psychoplasm, Ideoplasm," *PsyPioneer Journal* 3 (March 2013): 88–98. While Richet credited himself with a neologism, according to Demerest, Henry Holt exposed Richet's error in his article "A Review of Richet," *Journal of the American Society for Psychical Research* (1922). While other terms, such as "ideoplasm" and "teleplasm," seem to imply some formal taxonomy, the nomenclature, like the substance, is fluid and hard to pin down. Since both Schrenck-Notzing and Ochorowicz were on the island with Richet in 1894, they were all at the mouth of the same stream. The only quibble with Richet's claim of invention is that the term had already been in existence for several years, invoked to describe the hard substance of a cell wall.

21. Lodge, *Past Years*, 302. Though Lodge does not use the term "pseudopod" directly in his original report of 1894, he does provide the textbook definition to explain Eusapia Palladino's limb prolongation as "when an animal has a swelling or protuberance" and further projects "vital activity" in the form of an "off-shoot or bud that detaches itself." Lodge, "Experience of Unusual Physical Phenomena," 335. Correspondence with Myers, however, confirms both these terms guiding their conversation.

22. For more on spirit photographs in the nineteenth and twentieth centuries, see essays by Cloutier, Fischer, and Chéroux, in *The Perfect Medium: Photography and the Occult*, edited by Pierre Apraxine, Denis Canguilhelm, Andreas Fischer, and Sophi Schmidt (New Haven, CT: Yale University Press, 2005). Richet, Ochorowicz, and Albert von Schrenck-Notzing became more extravagant in their pursuit of physical phenomena after the turn of the century, exploring a new kind of séance teleplasm violently produced out of the mouth, with gossamer faces floating through it. Though of great interest to many spiritualists, this drew condemnation from peers. The phenomena were considered the work not of spirits, but of minds and bodies. Yet these psychical phenomena had ceased to be camouflaged in scientific reports and had become visible artifacts, accessible to the judgments or ridicule of the public, an unacceptable professional liability. And yet, the sensational and absurd could be partnered with professional scrutiny, as when Shrenck-Notzing burned Eva C.'s teleplasm to analyze its chemical composition, in 1913. Albert von Schrenck-Notzing, *Phenomena of Materialisation*, translated by E. E. Fournier d'Albe (New York: Dutton & Co., 1923), 246–47.

23. The spread of Darwinian naturalism across the biological, social, and human sciences in the 1860s and 1870s resulted in a predictable correction across disciplines in the closing decades of

the century as intellectuals sought to address its limitation as both a moral and causal principle. A good contemporary source is George J. Romanes, "The Darwinism of Darwin, and of Post Darwinian Schools," *The Monist* 6, no. 1 (October 1895): 1-27; for a good history, see Bowler, *Reconciling Science and Religion*.

24. Quoted from Oliver Lodge, *Why I Believe in Personal Immortality* (London: Doubleday, Doran, 1929), 80. Lodge's letters to Myers on the subject of ectoplasm, in the immediate wake of the séance, were destroyed, but much can be inferred from Myers's half of the conversation, as well as Lodge's own contemporary report on Eusapia Palladino's island phenomena.

25. See Richard Noakes for more on Cromwell Varley's spiritual technology and technological rhetoric, in "Instruments to Lay Hold of Spirits, 125-64; as well as his "Cromwell Varley FRS, Electrical Discharge and Victorian Spiritualism."

26. Lodge, *Past Years*, 298. Richet gives this assessment in *Thirty Years of Psychical Research*, 43, 70, 209.

27. For more on this, see John Ryan Haule, "Pierre Janet and Dissociation: The First Transference Theory and Its Origins in Hypnosis," *American Journal of Clinical Hypnosis* 29, no. 2 (October 1986): 86-94; and Rutger Hors and Onno van der Hart, "The Dissociation Theory of Pierre Janet," *Journal of Traumatic Stress* 2, no. 4 (October 1989): 397-412.

28. This characterizes the Sidgwick Group, but a note about W. F. Barrett is warranted since he was not just a founding member but one of the instigators. He was a deeply religious nonconformist who brought faith and action into the center of his life, seeing science and spiritualism as a continuation of that nearness to God. To the extent that Lang, Lodge, Myers, Gurney, and Sidgwick were spiritual seekers on the path of rational inquiry, the theological content of religion was disqualifying because "belief" mediated an intellectual's relationship to information—and truth was sacred to them. This was a religion of the heart and did not require the same ideological reconciliation. He does seem sometimes the odd man out, not part of the "club," because whatever his "archetype," it was not characterologically intellectual. For a good monograph on Barrett, see Noakes, "The "Bridge Which Is between Physical and Psychical Research," 419-64.

29. Lodge, "Experience of Unusual Physical Phenomena occurring in the Presence of an Entranced Person (Eusapia Palladino)," 310-12.

30. Lodge, *Past Years*, 301.

31. Lodge, *Past Years*, 301.

32. Lodge discusses conservation laws in relation to Eusapia Palladino's biomass as well as her vital energy in Lodge, "Experience of Unusual Physical Phenomena Occurring in the Presence of an Entranced Person (Eusapia Palladino)," 306-60.

33. Lodge, "Experience of Unusual Physical Phenomena Occurring in the Presence of an Entranced Person (Eusapia Palladino)," 336.

34. Myers to Lodge, June 21, 1894, no. 1400 (copy), Lodge Papers, Archives of the Society for Psychical Research, Department of Manuscripts and University Archives, Cambridge University Library.

35. "I was satisfied by the phenomena I had seen on the island and remained so satisfied." Lodge, *Past Years*, 309.

36. Lodge addresses this issue briefly in a letter read at the general meeting of the society on October 11, 1895, which appeared in the *JSPR* (October 1895): 133-35.

37. Alex Owen offers an interesting discussion of what motivated spiritual testimony across a range of class, gender, and professional considerations in *The Darkened Room*.

38. Lodge, "Experience of Unusual Physical Phenomena occurring in the Presence of an Entranced Person (Eusapia Palladino)," 325; and Oliver Lodge, "Account of Sittings with Mrs. Piper. Formal Report," *PSPR* 6 (1889-90): 443.

39. Myers to Lodge, June 25, 1890, no. 1306, Lodge Papers, Archives of the Society for Psychical Research, Department of Manuscripts and University Archives, Cambridge University Library.

40. Lodge, *Past Years*, 107.

41. Lodge, *Past Years*, 75.

42. Lodge, *Past Years*, 76.

43. Lodge, *Past Years*, 76.

44. John Atkinson Hobson, *The Crisis of Liberalism: New Issues of Democracy* (London: P. S. King & Son, 1909), 2–23.

45. In *Reconciling Science and Religion*, 59–86, Peter Bowler gives an account of "the new idealism" that marked the first few decades of the twentieth century as a counterpoint to the neopositivism of Bertrand Russell and Sigmund Freud. This metaphysical climate was partly enabled by relativity in physics and a resurgent strain of Lamarckianism in evolutionary biology and psychology, still awaiting the clarification to be later provided by DNA.

46. Quoted from Lodge, *Past Years*, in W. P. Jolly, *Oliver Lodge* (Rutherford, NJ: Fairleigh Dickinson University Press, 1975), 217.

47. Lodge, *Past Years*, 168. "For hitherto during my time in London, I had been obviously under the influence of Huxley, Tyndall and W. K. Clifford."

48. John Tyndall, "Address, John Tyndall FRS Professor of Natural Philosophy," in *Report of the Forty-fourth Meeting of the British Association for the Advancement of Science; Held at Belfast in August 1874* (London: John Murray, Albemarle Street Office of the Association, 1875), 65–97.

49. Lodge, *Past Years*, 110.

50. Quoted in Gauld, *Founders of Psychical Research*, 121. For a sense of the cultural politics involved in the public advocacy of institutional science and Huxley's role in the shaping of popular beliefs, see chaps. 1, 7, and 8 in Lightman, *Victorian Popularizers of Science*.

51. Lodge, *Past Years*, 139.

52. *Times*, September 18, 1876, quoted in Palfreman, "Between Skepticism and Credulity," 224.

53. Peter Guthrie Tait, in *Report of the Meeting of the British Association for the Advancement of Science* 41 (1872): 121.

54. See chapter 2, above, for a detailed account of Crookes's professional travails with spiritualism.

55. Lodge to Myers, October 21, 1890, no. 1309, Lodge Papers, Archives of the Society for Psychical Research, Department of Manuscripts and University Archives, Cambridge University Library. Myers wrote a letter to Lodge about discouraging his nomination to the presidency until the passing of Lord Kelvin so "as not to ruin his chances in Glasgow." Myers to Lodge, October 16, 1893, no. 1374, Lodge Papers.

56. The term *odium scientificum* was used by a staff author for the *Spectator* to describe the professional prejudice keeping psychical research at arm's length, in "Professor Oliver Lodge on Time," *The Spectator*, August 29, 1891, 284.

57. Staff Author, "Psychical Research," *Pall Mall Gazette*, October 21, 1882, 2.

58. Oliver Heaviside to Lodge, January 1, 1895, Lodge Papers, University College London, Manuscripts Room, London University.

59. Using Kelvin's linkage between the velocity of light and the rate of rotation of the vortex atom, Lodge explained why this might be the absolute speed at which light could travel. If energy was added to the rotation in excess of the speed of light, it was expressed not in more rapid rotation but as an increase in electrical inertia or "mass" (Einstein's equation, $E = mc^2$). Or, it increased the locomotion of the transverse waves—that is, the distance such waves traveled. Thus, as in relativity, space and time become relative, not the speed of light. Oliver Lodge, *A Century's Progress in Physics, Second of a Series of Centenary Addresses, Delivered on Monday, March 14, 1927* (London: University of London Press, 1927), 28.

60. Lodge, *Past Years*, 138.

61. J. L. Heilbron, *Electricity in the 17th and 18th Centuries: A Study of Early Modern Physics* (Berkeley: University of California Press, 1979), 68.

62. Lodge discussed this point early on in "The Relation between Electricity and Light," *Nature* (January 27, 1881): 303. Faraday's magnetic lines of force existed independently in space and did not demand an ether, but he had opened the door to considerations of an ether field, helping to resolve the mysterious provenance of force energies in a vacuum.

63. Barbara Giusti Doran, "Origins and Consolidation of Field Theory in Nineteenth Century Britain: From the Mechanical to the Electromagnetic View of Nature," in *Historical Studies for the History of Science*, vol. 6, edited by Russell McCormmach (Princeton, NJ: Princeton University Press, 1976), 238-55.

64. From a lecture delivered by Thomson in 1860, quoted in Doran, "Field Theory in Nineteenth Century Britain," 255.

65. Oliver Lodge, "The Ether and Its Functions," *Nature* 27 (January 25 and February 1, 1883): 304-6, 328-30.

66. Oliver Lodge, "Fundamental Notions," in *Modern Views of Electricity* (London: Macmillan, 1889), 1-17.

67. Lodge gave an account of this demonstration in his article "On the Theory of Lightning Conductors," *Philosophical Magazine* 26, no. 5 (August 1888): 217-30. The method differed from Hertz's design as well as from what Lodge had been previously trying with induction wires and Leyden jars.

68. Oliver Lodge, "The Foundations of Dynamics," *Philosophical Magazine* 36 (July 1893): 6-7.

69. Oliver Lodge, "On Action at a Distance and the Conservation of Energy," *Philosophical Magazine* 11 (June 1881), 533.

70. Pierre Duhem, *The Aim and Structure of Physical Theory*, trans. Philip Weiner (Princeton, NJ: Princeton University Press, 1954).

71. Geertz, "Religion as a Cultural System," in *The Interpretation of Cultures*, 87-125.

72. Crosbie Smith and Norton Wise, *Energy and Empire: A Biographical Study of Lord Kelvin* (New York: Cambridge University Press, 1989), 173, 470.

73. Oliver Lodge, "An Attempt at a Systematic Classification of the Various Forms of Energy," *Philosophical Magazine* 3 (October 1879): 277.

74. For a more detailed discussion of this epistemological debate, see D. H. Mellor, "Models and Analogies in Science: Duhem versus Campbell," *Isis* 59, no. 3 (Autumn 1968): 282-90.

75. Lodge, "The Foundations of Dynamics," 6-7.

76. From page 8 of a privately printed letter from Oliver Lodge to Mr. Lund (November 3, 1888), regarding a discussion at his home of *Robert Elsmere*, Mary Ward's bestseller about a young man in religious crisis. The meeting was sponsored by the Philalethean Club, and the letter was meant to be publicly circulated.

77. Lodge, "The Foundations of Dynamics," 5-6.

78. William Thomson, quoted in Smith and Wise, *Empires and Energy*, 429.

79. Letter from Myers to Lodge concerning séance, December 12, 1889, no. 1299, Lodge Papers, Archives of the Society for Psychical Research, Department of Manuscripts and University Archives, Cambridge University Library.

80. Oliver Lodge's formal report describing the November 30 séance was submitted to the SPR in the immediate wake of that first séance, but was only published later, after Lodge had completed several months of private investigation. He included it under the separate heading, "An Account of Sittings with Mrs. Piper, Formal Report," as the first page of "Certain Observations Regarding the Phenomena of Trance," *PSPR* 6 (1889-90):443-559. This monumental account of Lodge's privately held Liverpool séances included detailed field notes from multiple participants, as well as commentary from Lodge throughout. Myers and Walter Leaf also submitted reports, bringing the entirety to over two hundred pages.

81. Lodge, "Certain Observations Regarding the Phenomena of Trance," 464. In explaining his reasons to continue with the Piper sittings, he stated his main object as being to "discriminate

between unconscious thought transference and direct clairvoyance plainly establishing in his summary that the criteria for clairvoyance had been met (464, 525). While Mrs. Piper later fell under suspicion of fraud with subsequent SPR investigators, Lodge's conviction in her phenomena never wavered, as confirmed in his autobiography written forty years later. See Lodge, *Past Years*, 278-79.

82. Lodge, "Account of Sittings with Mrs. Piper," in "Certain Observations Regarding the Phenomena of Trance," 459.

83. Lodge, "Account of Sittings with Mrs. Piper," in "Certain Observations Regarding the Phenomena of Trance," 464.

84. Jolly, *Oliver Lodge*, 94.

85. Jolly, *Oliver Lodge*, 88.

86. Oliver Lodge, "Section A. Mathematics and Physics, Opening Address by Prof. Oliver J. Lodge, President of the Section," *Nature* 44 (August 20, 1891): 384.

87. Lodge, "Opening Address," 385.

88. Lodge, "Opening Address," 389.

89. Lodge, "Opening Address," 387.

90. Lodge, "Opening Address," 385.

91. Oliver Lodge, "Letters to the Editor," *Nature* (March 1, 1888): 416.

92. See chapter 1 of this book for a discussion of midcentury science and scientific values.

93. Lodge addressed this statement to his fellow physicist James Jean who publicly advocated this more phenomenological stance in science. Lodge, *My Philosophy*, 125.

94. By the time Lodge's *My Philosophy* was going to press, logical positivism had already begun to promote verification in science as a linguistic construct that had to be rationally consistent with both itself and observable phenomena. This further abstracted science into a specialized form of discourse asserting a criterion for meaning in reference to science, not reality.

95. Oliver Lodge, "The Interstellar Ether," *Littell's Living Age*, 5th series, 83, no. 2557 (Boston: Littell & Co., July 1893): liii, 856; Lodge, "Note concerning Thought and Reality," *Mind* (April 1905): xiv, 294.

96. *Philosophical Transactions* (February 1889). Lodge brings this up in "The Foundations of Dynamics," 30.

97. Lodge, "Opening Address," 385.

98. Staff Author, "Oliver Lodge on Time," *Spectator* (1891): 285.

99. Lodge, "The Foundations of Dynamics," 12.

100. Sir John Pentland Mahaffy, *Kant's Critical Philosophy for English Readers: The Æthestic and Analytic* (London: Longman Green, 1874), 60.

101. Lodge, "The Nature of Time," in *Modern Problems* (London: Methuen & Co, 1912), 14.

102. Oliver Lodge, "The Foundations of Dynamics, Part I and II," *Philosophical Magazine*, 5th series, 36 (July 1893): 1-36. Lodge is referring to the address by James Gordon MacGregor, which appeared in *Transactions of the Royal Society Canada* 10, no. 3 (1892), which made a splash across the Atlantic. MacGregor specifically addressed Lodge's attack with "On the Hypothesis of Dynamics," *Philosophical Magazine* 36, no. 220 (September 1893).

103. Lodge, "The Foundations of Dynamics," 7-9.

104. Lodge, "The Interstellar Ether," 859.

105. Oliver Lodge, "Electrical Theory of Vision," *Nature* (June 21, 1894): 172.

106. Lodge, "The Foundations of Dynamics," 5-6.

107. Lodge, *A Century's Progress in Physics*, 23.

108. For a discussion of the religious implications of Lodge's monism, see John D. Root, "Science, Religion, and Psychical Research: The Monistic Thought of Sir Oliver Lodge," *Harvard Theological Review* 71 (July-October 1978): 245-63.

109. Oliver Lodge, *Science and Immortality* (New York: Moffat, Yard & Co., 1908), 262.

110. Quoted in Hunt, *The Maxwellians*, 98–99. Fitzgerald's lecture was given in 1896 and made a public issue, but these were ideas he had entertained for some time and with which Lodge, a close friend of his, was already familiar.

111. Hunt, *The Maxwellians*, 98–99. The lecture was given in 1896, but these ideas had been entertained for some time, perhaps since Fitzgerald's college days. They would certainly have come up during the long course of his and Lodge's friendship, marking eighteen years by 1896.

112. Lodge, "The Interstellar Ether," 857.

113. Hunt, *The Maxwellians*, 98–99.

114. Lodge, "Note concerning Thought and Reality," *Mind* 14 (April 1905): 295.

115. Lodge, *Continuity: The Presidential Address to the British Association, Birmingham MCMXIII* (London: J. M. Dent & Sons, 1913), 90.

116. "I was satisfied by the phenomena I had seen on the island and remained so satisfied." Lodge, *Past Years*, 309.

117. Lodge, *Mind and Matter: Address Delivered in the Town Hall, Birmingham, on Wednesday, Oct 12, 1904* (Birmingham: Council of the Birmingham & Midland Institute, 1904), 25.

118. Oliver Lodge, "Mind and Matter," *The Hibbert Journal* (January 1905), 317.

119. Lodge, "Account of Sittings with Mrs. Piper," in "Certain Observations Regarding the Phenomena of Trance," 505.

120. Lodge, *Mind and Matter*, 45.

121. Lodge, "The Interstellar Ether," 862.

122. Oliver Lodge, "Letter Addressed to W. Ward, September, 1897," in *Papers Read before the Synthetic Society*, 148.

123. Lodge recaps Canon Gore's arguments in Oliver Lodge, "On the Possibility of a Logical Proof of Religious Doctrines: An Answer to Canon Gore. To be read January 29, 1897," in *Papers Read before the Synthetic Society*, 59. Also see Lodge, "Argument from Design"; and Lodge, "Is Obscurity or Infinitude the Well Spring of Religion," presented to the Synthetic Society the following year (1898).

124. Oliver Lodge, "On the Nature of Proof, May 28 1897," in *Papers Read before the Synthetic Society*, 129.

125. Oliver Lodge, "Materialism and Christianity, May 9, 1905," in *Papers Read before the Synthetic Society*, 494.

126. Oliver Lodge, "The Outstanding Controversy between Science and Faith," *The Hibbert Journal* 1 (1902): 54–55.

127. Lodge, "The Nature of Proof, May 28, 1897," 58–59.

128. Oliver Lodge, "Faith and Knowledge, April 1902," in *Papers Read before the Synthetic Society*, 375.

129. Lodge was as critical of scientific skepticism as he was of skepticism about science. Both materialism and the denial of realism are frequent themes throughout his Synthetic Society essays. For this earlier period, see Oliver Lodge, "In Reply to Hutton's Reply, June 6, 1896;" "Response to Wilfrid Ward, February 14, 1897"; "On the Nature of Proof, May 28 1897"; and "Letter Addressed to W. Ward, September 1897," all in *Papers Read before the Synthetic Society* 56–59, 77–87, 128–29, 148–54.

130. Oliver Lodge, "Response to R. H. Hutton, Why Is the Universe Intelligible to Us? April 1896," in *Papers Read before the Synthetic Society*, 34.

131. Lodge is quoting Charles Bigg in Lodge, "Is Obscurity or Infinitude the Well Spring of Religion, And If Obscurity Should It Therefore Be Encouraged? (June 10, 1898)," in *Papers Read before the Synthetic Society*, 218. The address referred to here was made to Section A while Lodge was section president. See Lodge, "Section A. Mathematics and Physics, Opening Address," 386–87.

132. Lodge, "Response to R. H. Hutton, Why Is the Universe Intelligible to Us? April 1896," 34.

133. Oliver Lodge, "First Principles of Faith, with Reference to *The Hibbert Journal* for July and October, 1906" (Privately printed, 1906), 5.

134. Lodge's suspect Christianity is discussed in E. A. Sonnenschein, "The New Stoicism," *The Hibbert Journal* 5 (1906-7): 544; and Professor Curtis, "A Modern Mirage of the Gospel," *Current Literature* 43 (July-December 1907): 654.

135. Lodge, "First Principles of Faith, with Reference to *The Hibbert Journal* for July and October," 1.

136. Lodge, "First Principles of Faith, with Reference to *The Hibbert Journal* for July and October," 2.

137. Lodge, "First Principles of Faith, with Reference to *The Hibbert Journal* for July and October," 2.

138. The Reverend R. J. Campbell raises a discussion he had with Lodge regarding the World Soul probably in the 1890s in the context of discussing Myers's work. According to Campbell, Lodge confessed that "the hypothesis of a World-Soul intimately and immediately concerned with ours is the best explanation of things as they are." See "Personal Immortality," *The Homiletic Review* 45 (January-June 1903): 420. Because it was a potentially troubling notion for theists, Lodge did not exploit this term in public, though his description of a God at once fully transcendent and fully immanent would better fit this framework than his other preferred terms like "the Director" or "the Manager" or "the Supreme Being."

139. Oliver Lodge, "Self and the Universe: The Rudiments of Philosophy as Bearing on Christian Dogma: Being an Explanatory Catechism or Glossary for Elder Children Only" (Privately Printed, 1907), 11.

140. Lodge, *Continuity*, 100.

141. Lodge, "Is Obscurity or Infinitude the Well Spring of Religion, And If Obscurity Should It Therefore Be Encouraged? (June 10, 1898)," 220.

142. Of these men (William Thomson, Macquorn Rankine, James Clerk Maxwell, Peter Guthrie Tait, and Fleeming Jenkin) only Balfour Stewart later joined the SPR; the rest disapproved or predeceased its mission. Stewart, who had partnered earlier with Peter Guthrie Tait to write *The Unseen Universe* (1875), had already ventured a scientific defense of religion with this book, incurring the disapproval of Thomson then as now with psychical research.

143. Joseph McCabe, "The Origin of Life, A Reply to Sir Oliver Lodge" (London: Watts & Co., 1906), 7.

144. Staff Author, "Review of Joseph McCabe, 'The Religion of Sir Oliver Lodge,'" *The Homiletic Review* 69 (January-June 1915): 88. Lodge wrote some 170 titles addressing the relationship between science and religion, most of which were published between 1896, when Lodge joined the Synthetic Society, and 1915, when his son Raymond died. See Root, "Science, Religion, and Psychical Research," 245.

145. Lodge, "The Nature of Proof, May 28, 1897," 136.

146. Lodge, "Letter Addressed to W. Ward, September 1897," in *Papers Read before the Synthetic Society*, 154.

147. J. McCabe, "The Origin of Life, A Reply to Sir Oliver Lodge," 7.

148. William Thomson, "Lord Kelvin on Science and Theism," *The Living Age* 238 (July-September 1903): 116.

149. Oliver Lodge, "Lord Kelvin on Creative Purpose," *Popular Science Monthly* 65 (1903): 279-80.

150. Oliver Lodge, "Our Place in the Universe, An Address to the Members of the Nelson Street Adult School, Sunday Morning, August 17th, 1902" (Birmingham: John A. Hodges, Printer, 1902), 6.

151. Lodge, "First Principles of Faith, with Reference to *The Hibbert Journal* for July and October, 1906," 3.

152. His next few books, *The Immortality of the Soul* (1908), *Man and the Universe; A Study of the Influences of the Advances in Scientific Knowledge upon Our Understanding of Christianity* (1908), *Survival of Man; A Study in Unrecognized Human Faculty* (1909), and *Reason and Belief* (1910)], wove religious, evolutionary, psychical, and philosophic concerns together in a consistently messaged

"worldview." There were plenty of critics, particularly those with institutional loyalties, who remained publicly opposed to Lodge's "scientific religion," which generally hinged on one's attitude toward psychical research. As one reviewer put it, "His New Christianity is rejected by Christians as non-Christian" and "scientists will have none of the new science as vouched for by Sir Oliver with its unwarranted or unproved injection of spiritual phenomena." Staff Author, "Review of Joseph McCabe, 'The Religion of Sir Oliver Lodge,'" 88.

153. Lodge, *Self and the Universe*, 11.

154. Lodge, *Self and the Universe*, 24.

155. Lodge, *My Philosophy*, 76. Oliver Lodge, "Lord Kelvin on Creative Purpose," *Popular Science Monthly* 65 (1903): 279-80.

156. Lodge, "Ecce Deus: The Essential Element in Humanity" (Privately Printed, May 1905), 6. This essay appeared the following year in *The Hibbert Journal* 4 (October-July 1905-6): 652-59, as a subsection of "Christianity and Science: The Divine Element in Christianity."

157. Lodge, "Ecce Deus: The Essential Element in Humanity," 3.

158. Lodge, "Ecce Deus: The Essential Element in Humanity," 5.

159. Lodge, "Ecce Deus: The Essential Element in Humanity," 4.

160. Lodge, "Our Place in the Universe," 5.

161. Oliver Lodge, "Sin," *The Hibbert Journal* 3 (October 1903-July 1904):15.

162. Lodge, *Modern Problems*, 39-40.

163. Lodge, *Modern Problems*, 39-40.

164. Lodge, "Ecce Deus: The Essential Element in Humanity," 5.

165. Oliver Lodge, *The Immortality of the Soul* (Boston: Ball Publishing Company, 1908), 44-50. Lodge credits Professor Hoffding of Copenhagen with originating this theory, though Lodge makes his own uses of it.

CHAPTER SIX

1. Andrew Lang, "Realism and Romance," *Contemporary Review* 52 (1887): 683.

2. For more on the rise of a distinctly "popular" fiction in the nineteenth century, see Christine Berberich, ed., *The Bloomsbury Introduction to Popular Fiction* (London: Bloomsbury, 2015), 1-27; and David Franklin Mitch, *The Rise of Popular Literacy in Victorian England: The Influence of Private Choice and Public Policy* (Philadelphia: University of Pennsylvania Press, 1992), 1-42, especially 17 for the expression "penny packets of poison," coined in 1874 by critic James Greenwood.

3. Rider Haggard, "About Fiction," *Contemporary Review* 51 (1887): 172-73.

4. This number is cited by Eleanor de Selms Langstaff, *Andrew Lang* (Boston: Twayne Publishers, 1978), 119.

5. Henry James, "The Art of Fiction," *Longman's Magazine* 4 (1884): 502-21, 507 (quote).

6. Lang, "Realism and Romance," 693.

7. Lang, "Realism and Romance," 684-85.

8. Ed Block, Jr., "James Sully, Evolutionist Psychology, and Late Victorian Gothic Fiction," *Victorian Studies* 25, no. 4 (Summer 1982). For a more comprehensive look at the pervasiveness of evolutionary themes across popular and public culture, see Bernard V. Lightman and Bennett Zon, eds., *Evolution and Victorian Culture*, Cambridge Studies in Nineteenth-Century Literature and Culture 92 (Cambridge: Cambridge University Press, 2014).

9. Lang, "Realism and Romance," 688.

10. Andrew Lang, "A Dip in Criticism," *Contemporary Review* 54 (1888): 497.

11. Andrew Lang, "Mr. Lang on the Art of Mark Twain," *The Critic* 19, no. 395 (July 25, 1891): 45.

12. From a letter to Robert Louis Stevenson, July 31, 1888, excerpted in William R. Veeder and Susan M. Griffin, eds., *The Art of Criticism: Henry James on the Theory and the Practice of Fiction* (Chicago: University of Chicago Press, 1986), 193.

13. Roger Lancelyn Green, *Andrew Lang: A Critical Biography with a Short-Title Bibliography of the Works of Andrew Lang* (London: Bodley Head, 1962), 157. As for this charge of puerility, it can at least be said that, unlike Gosse or James, Lang did have a sense of humor.

14. Stevenson and Haggard were active members of the SPR, while Kipling moved in its social milieu and incorporated its themes in his literature. All three were close with Lang.

15. Martin Green, *Deeds of Adventure, Dreams of Empire* (New York: Basic Books, 1979), 1.

16. Lang, "Realism and Romance," 384.

17. Robert Michalski, "Towards a Popular Culture: Andrew Lang's Anthropological and Literary Criticism," *Journal of American Culture* 18, no. 3 (Fall 1995): 13.

18. Andrew Lang, "Emile Zola," *Fortnightly Review* 37 (April 1882): 443.

19. Lang, "Emile Zola," 445.

20. Lang, "Emile Zola," 445.

21. Walter Besant, "The Art of Fiction," lecture presented at the weekly evening meeting, Friday, April 25, 1884, printed in *Notices of the Proceedings at the Meetings of the Members of the Royal Institution* 11 (1887): 75. James concurred with this suggestion in his own essay of that title, while Zola was known to recommend a similar equipage.

22. Lang, "Emile Zola," 445.

23. John Tosh, "Middle Class Masculinities in the Era of Women's Suffrage Movement, 1860–1914," in his *Manliness and Masculinities* (London: Pearson Longman, 2005).

24. For more on this literary and cultural motif of degeneration and pollution, particularly as it pertains to a feminist threat, see Elaine Showalter, *Sexual Anarchy: Gender and Culture at the Fin de Siècle* (New York: Penguin Press, 1990), 4–67.

25. Lang, "Realism and Romance," 690.

26. Thomas Watson, "The Fall of Fiction," *Fortnightly Review* 50 (1888): 336.

27. Lang, "Realism and Romance," 691.

28. Andrew Lang, "She," *Book News* 5, no. 55 (March 1887): 235.

29. Before writing imperial fiction, Haggard had already taken it upon himself to report the facts. While stationed at Natal in his early twenties, Haggard contributed an article describing a Zulu war dance to the *Gentleman's Magazine* and another to Macmillan's about a delegation visit to Chief Secocceni. These were followed by the monograph *Cetewayo and his White Neighbors* (1882).

30. Watson, "The Fall of Fiction," 330.

31. Robert Louis Stevenson, "A Humble Remonstrance," *Longman's Magazine* 5, no. 26 (December 1884): 142.

32. Brantlinger defines the imperial gothic as a literary subgenre linking imperialism, atavism, and the occult; Patrick Brantlinger, *Rule of Darkness: British Literature and Imperialism, 1830-1914* (Ithaca, NY: Cornell University Press, 1988), 227–99. Feminist scholars have emphasized the primacy of misogyny in this literary trope, a valid but somewhat overstated thesis. See Kelly Hurley, *The Gothic Body: Sexuality, Materialism, and Degeneration at the Fin de Siècle* (Cambridge: Cambridge University Press, 1997), 55–64; Rebecca Stott, "The Dark Continent: Africa as Female Body in Haggard's Adventure Fiction," *Feminist Review* 32 (Summer 1988): 69–89; and Showalter, *Sexual Anarchy*. The orientalist trope proposing black/female/irrational as interchangeable designations of "inferiority" does not adequately express the imperial subjectivity of a Lang or Haggard who rejected the corresponding metonymic chain of white/male/rational to signify their preferred valence of masculine power.

33. This idea of woman-as-landscape was first argued by Stott in "The Dark Continent."

34. Lang, "Realism and Romance," 690.

35. Watson, "Fall of Fiction," 330.

36. Lang, "Realism and Romance," 689.

37. For an exploration of this mimetic desire in imperial fiction, see Gail Ching-Liang Low, *White Skins/Black Masks: Representation and Colonialism* (London: Routledge, 1996).

38. Homi K. Bhabha, "The Other Question: The Stereotype and Colonial Discourse," in K. M. Newton, ed., *Twentieth Century Literary Theory* (London: Macmillan, 1997), 293–301.

39. Ali Behdad, *Belated Travelers: Orientalism in the Age of Colonial Dissolution* (Durham, NC: Duke University Press, 1994), 21.

40. Homi Bhabha, *The Location of Culture* (London: Routledge, 1994), 85–92. As language shifts between sign, a more specific reference, to the less stable, more open-ended form of symbol and back again, the constant process of transcription creates a heterogeneity of potential meanings within the colonial discourse, spawning multiple orientalism across the competing intentions of participants in the discourse. It is potentially productive because it opens up alternative self-images within the dominant discourse, one potentially more congenial to Lang and his values.

41. Henry Rider Haggard, *Days of My Life* (London: Longman's, Green & Co., 1926), chap. 6.

42. Green, *Andrew Lang: A Critical Biography*, 43.

43. John Buchan, "Andrew Lang and the Border, Andrew Lang Lecture Delivered before the University of St. Andrews, October 17th, 1932," in *Concerning Andrew Lang: Lang Lectures 1927–1937* (Oxford: Clarendon Press, 1949), 4–22.

44. Lang, "She," 235.

45. Alan Gauld, "Andrew Lang as Psychical Researcher," *JSPR* 52 (1983–84): 161–207.

46. Green, *Andrew Lang: A Critical Biography*, 43.

47. Joseph Weintraub, "Andrew Lang, Critic of Romance," *English Literature in Transition, 1880–1920* 18, no. 1 (1975): 5–15.

48. Stewart Kelly, "Andrew Lang, the Life and Times of a Prolific Talent," *The Scotsman*, January 30, 2012, http://www.scotsman.com/heritage/people-places/andrew-lang-the-life-and-times-of-a-prolific-talent-1-2085486 (accessed May 14, 2016).

49. Robert Crawford, "Pater's Renaissance, Andrew Lang, and Anthropological Romanticism," *ELH* 53, 4 (Winter, 1986): 849–79.

50. Walter Pater, *Studies in the History of the Renaissance* (1873).

51. Andrew Lang, *Adventures among Books* (1905; Freeport, NY: Books for Libraries Press, 1970), 37–38.

52. For more on the metrical construction of racial inferiority in science, see Stephen Jay Gould, *The Mismeasure of Man* (New York: W. W. Norton, 1996), 120–40. The objectification of indigenous peoples in peep shows, museum displays, and exhibitions of empire is extensively documented in Sadiah Qureshi, *Peoples on Parade: Exhibitions, Empire, and Anthropology in Nineteenth Century Britain* (Chicago: University of Chicago Press, 2011).

53. Johannes Fabian, *Out of Our Minds: Reason and Madness in the Exploration of Central Africa* (Los Angeles: University of California Press, 2000), 180–208.

54. George Stocking, *Victorian Anthropology* (New York: Free Press, 1987), remains the most comprehensive account of this field, tracing anthropology's multiple strains from the Enlightenment discourses of the eighteenth century. See his chap. 6, "Victorian Cultural Anthropology and the Image of Savagery," for more on the rise of "culture" as an object of anthropological study.

55. Lang, *Myth, Ritual, and Religion*, 1:35.

56. Lang, "Realism and Romance," 693.

57. Lang, "Realism and Romance," 693.

58. Lang, "Realism and Romance," 693.

59. Crawford, "Pater's Renaissance, Andrew Lang, and Anthropological Romanticism," 854–62.

60. Andrew Lang, "Charles Kingsley," in *Essays in Little* (London: Henry & Co., 1891), 153–59.

61. For more on the ways in which the Oxford aesthetic movement intersected with Victorian values and the cultural politics of homosexuality, see Dowling, *Hellenism and Homosexuality in Victorian Oxford*; Turner, *The Greek Heritage in Victorian Britain*; and Richard Jenkyns, *The Victorians and Ancient Greece* (Oxford: Basil Blackwell, 1980). Lang is a peripheral figure here, partly because he covered his tracks upon leaving and subsequently distanced himself from this homo-aesthetic sphere in his persona as a social scientist and champion of manly romance.

62. Green, *Andrew Lang: A Critical Biography*, 201: "The missing chapter in the development of his personality is that concerned with his emotional relations with women at the time when his character was still forming."

63. Among the interpretations placed upon the male adventure quest was the notion that it was a flight from the oppressive state of Victorian matrimony. See Tosh, *Manliness and Masculinities*.

64. Tosh, *Manliness and Masculinities*, 207.

65. Peter Bowler explores how the developmental model informed debates about human identity and the character of modernity. Lang's more cyclical understanding of progress had, according to Bowler, a more conservative valence as opposed to the positivistic approach of Tylor. See Bowler, *The Invention of Progress*.

66. Lang, *Adventures among Books*, 37-38.

67. Lang, *Adventures among Books*, 37.

68. Richard Mercer Dorson, *The British Folklorists: A History* (Chicago: University of Chicago Press, 1968), 206-24.

69. Lang, *Adventures among Books*, 37.

70. Alan Menhennet, *The Romantic Movement* (New York: Routledge, 2016), 40-45, 65-80.

71. Andrew Lang, *Custom and Myth* (London: Longmans, Green & Co., 1884), 242.

72. Lang, *Custom and Myth*, 213, 217.

73. Lang, *Myth, Ritual, and Religion*, 1:327-28.

74. Lang, *Myth, Ritual, and Religion*, 36.

75. Lang, *Myth, Ritual, and Religion*, 1:103-4.

76. Andrew Lang, *The Making of Religion* (London: Longman's, Green & Co., 1898), 50.

77. Andrew Lang, *Cock Lane and Common-Sense* (London: Watts, 1894), 6-8.

78. Lang, *Cock Lane and Common-Sense*, 77.

79. Lang, *Cock Lane and Common-Sense*, 173.

80. Lang, *Cock Lane and Common-Sense*, 99, 357.

81. Antonius Petrus Leonardus de Cocq, *Andrew Lang, a Nineteenth Century Anthropologist* (Tilburg: Zwijsen, 1968), 53.

82. Lang, *The Making of Religion*, 50.

83. Andrew Lang, "Mr. F. Podmore's Studies in Psychical Research," *PSPR* 13 (1897-98): 606.

84. February 10, year unknown, no. 1038, Lodge Papers, Archives for the Society for Psychical Research, Department of Manuscripts and University Archives, Cambridge University Library.

85. Lang, *The Making of Religion*, 216-17.

86. Andrew Lang, "Science and Superstition," in *Magic and Religion* (London: Longmans, Green & Co., 1901), 4.

87. Lang, "The Supernatural in Fiction," in *Adventures among Books*, 279.

88. Quoted in de Cocq, *Andrew Lang*, 24.

89. Quoted in Gauld, "Andrew Lang as Psychical Researcher," 173.

90. Lang, "Savage Supreme Beings," in *The Making of Religion*, 229.

91. Lang, *Myth, Ritual, and Religion* (2nd ed., 1898), xiii. Lang was citing Darwin's *Descent of Man* (1871), 68.

92. Lodge, *Continuity*, 100.

93. Lang, *The Making of Religion*, 199.

94. Lang, *The Making of Religion*, 292.

95. Lang, *The Making of Religion*, 334.

96. Lang, *The Making of Religion*, 335.

97. Lang, *The Making of Religion*, 186.

98. Lang, *Cock Lane and Common-Sense*, 337-38.

99. Andrew Lang, "Mythology," in *A Dictionary of Arts, Sciences and General Literature*, vol. 17 (Chicago: R. S. Peale & Co., 1890), 136.

100. Lang, *The Making of Religion*, 50.

101. Lang, *The Making of Religion*, 335.

102. While Lang does not factor in his study, Peter Bowler broadens the framework in which Lang's psycho-folklore and the Synthetic Society can be understood, showing how questions of a spiritual nature were being posed within science in subtle but pervasive ways in the early twentieth century, signaling a "new idealism" in science and a turn toward a new natural theology. See Bowler, "Evolution and the New Natural Theology," in *Reconciling Science and Religion*, 122-90.

103. Friedrich Nietzsche, *On the Genealogy of Morals: A Polemic*, translated by Douglas Smith (New York: Oxford University Press, 1998), 11.

104. Lang, *The Making of Religion*, 98.

105. Haggard, "About Fiction," 175.

106. Haggard, "About Fiction," 175.

107. Lang, "Science and Superstition," in *Magic and Religion*, 4.

108. Lang, *Cock Lane and Common-Sense*, 63.

109. Green, *Andrew Lang: A Critical Biography*, 109-10.

110. Lang, *Cock Lane and Common-Sense*, 6.

CHAPTER SEVEN

1. Sidgwick and Sidgwick, *Henry Sidgwick: A Memoir*, 3:259.

2. Retrieved from the Parapsychology Association website, http://www.parapsych.org/section /34/university_education_in.aspx (accessed January 10. 2019).

3. Paul Devereux, "Arthur Koestler: A Look Back at One of the Researchers into Coincidence and Parapsychology," October 2005, *Fortean Times Magazine*, http://www.forteantimes.com/fea tures/profiles/118/arthur_koestler.html (accessed August 15, 2015).

4. Koestler joined resistance fighters in Spain to fight against Franco and personally traveled to Russia to investigate the Ukrainian famine and Moscow show trials.

5. Kathleen Nott, "The Trojan Horses: Koestler and the Behaviorists," in Harold Harris, ed., *Astride Two Cultures: Arthur Koestler at Seventy* (London: Hutchinson, 1975), 162-74.

6. David Dickson, "Edinburgh Sets up Parapsychology Chair" *Science*, n.s. 223, no. 4642 (March 23, 1984): 1274.

7. Arthur Koestler, interview by Duncan Fallowell, "Arthur Koestler, The Art of Fiction No. 80," *Paris Review*, August 1984, http://www.theparisreview.org/interviews/2976/the-art-of-fiction-no -80-arthur-koestler (accessed August 17, 2012).

8. Arthur Koestler, *Darkness at Noon* (New York: Bantam Books, 1968).

9. Timothy Snyder, "Hitler vs. Stalin: Who Killed More?" *New York Times Review of Books*, March 10, 2011. http://www.nybooks.com/articles/archives/2011/mar/10/hitler-vs-stalin (accessed August 17, 2012).

10. "Everything for the state . . ." was a popular slogan in Mussolini's Fascist state. The other quote begins F. Scott FitzGerald's short story, "The Rich Boy" (1926) and is most commonly referenced to express a kind of envy and exaltation of the rich, though this was not its original context. See http://www.nytimes.com/1988/11/13/books/l-the-rich-are-different-907188.html.

11. Arthur Koestler, *The Act of Creation* (New York: Macmillan, 1964), 320-58.

12. Myers, "Charles Darwin and Agnosticism," 75.

13. Myers, "Charles Darwin and Agnosticism," 74-75.

14. Myers, *Human Personality and Its Survival of Bodily Death*, 2:279.

15. Myers, "In Memory of Henry Sidgwick," *PSPR* 15 (1901): 456.

16. August Comte, *The Positive Philosophy of August Comte*, translated by Harriet Martineau (London: Trubner & Co., 1875), 1:2.

17. Myers, "Tennyson as Prophet (1889)," in *Science and a Future Life*, 143. Myers notes a cultural shift in tone here as the best-laid plans of reconciliation came to naught.

18. Myers, "Tennyson as Prophet" (1889), in *Science and a Future Life*, 143-44.

19. Ernst Mach, *The Analysis of Sensations, and the Relation of the Physical to the Psychical*, translated by C. M. Williams, rev. Sydney Waterlow (Chicago: Open Court Publishing, 1914), 365.

20. Mach, *The Analysis of Sensations*, 18.

21. Mach, *The Analysis of Sensations*, 18.

22. Judith Ryan, *The Vanishing Subject: Early Psychology and Literary Modernism* (Chicago: University of Chicago Press, 1991), 21.

23. For more on modernism in relation to the history of science, particularly psychology, see Martin Halliwell, *Romantic Science and the Experience of Self: Transatlantic Crosscurrents from William James to Oliver Sacks* (Brookfield, VT: Ashgate, 1999), 1-70; Mark S. Micale, ed., *The Mind of Modernism: Medicine, Psychology, and the Cultural Arts in Europe and America, 1880-1940* (Stanford: Stanford University Press, 2004); Paul Peppis, *Sciences of Modernism: Ethnography, Sexology, and Psychology* (New York: Cambridge University Press, 2014), 197-280; Dorothy Ross, ed., *Modernist Impulses in the Human Sciences, 1870-1930* (Baltimore: Johns Hopkins University Press, 1994), 1-53, 128-51, 190-254.

24. This is different from the first-generation Romantic science of Friedrich Schelling's *Naturphilosophie* (1797) and *System of Transcendental Idealism* (1800) where, in the most supreme aesthetic acts, the artist merged with nature's creative will, and this subjective, embodied force was ultimately framed by an absolute mind, giving rise to nature's laws.

25. Andrew Lang, "The Comedies of Shakespeare," *Harper's New Monthly Magazine* 86, no. 513 (February 1893): 327.

26. Oliver Lodge, *Beyond Physics or the Idealisation of Mechanism* (London: George Allen & Unwin, 1930), 69.

27. Lodge, *Beyond Physics*, 171.

28. Ross, *Modernist Impulses in the Human Sciences, 1870-1930*, 2.

29. Margaret Washburn, "Review: Contributions to the Analysis of the Sensations by Dr Ernst Mach, Translated by CM Williams," in *Philosophical Review* 6 (1897): 565.

30. Myers, *Fragments of Inner Life*, 69. "About 1875 a great revival of hypnotism began in France with Charcot and Richet in Paris and spread from another focus the persistent labours of Dr Liébeault to the Professors at Nancy." Also see Ellenberger, *The Discovery of the Unconscious*, 53-182. Ellenberger puts the start of Charcot's hypnotic investigations at 1878. Myers had already joined the Psychological Society in 1875, inspired by Crookes's psychic force. Interest in French neomesmerism also began around that time, gaining on the fringes of medicine by 1875.

31. Myers, "Science and a Future Life," in *Science and a Future Life*, 19.

32. Mark S. Micale, "The Modernist Mind: A Map" and "The Modernist Mind: A Timeline" (with Jesse Wegman), in *The Mind of Modernism*, 1-70.

33. Myers, "Science and a Future Life," 37.

34. Freud's "Quantitative Project for Psychology, 1895" lays out a psychophysical analysis of human impulsive behavior, mapping how stimulus (sensory and endogenous) charges the brain's cells with unmanageable quantities of neural energy requiring discharge (that is, thermodynamics becomes psychodynamics). For further analysis, including excerpts from Freud's script, see Zvi Lothane, "Freud's 1895 Project: From Mind to Brain and Back Again," *Annals of the New York Academy of Sciences* 843 (May 1996): 43-65. doi:10.1111/j.1749-6632.1998.tb08204.

35. Myers, *Science and a Future*, 95.

36. Myers, *Science and a Future*, 94.

37. Myers, "Charles Darwin and Agnosticism," in *Science and a Future Life*, 75.

38. David Hume, *A Treatise of Human Nature* (1739), Section 6, Part IV, Book I; reprinted in John Perry, ed., *Personal Identity* (Los Angeles: University of California Press, 1975), 69.

39. Mach, *The Analysis of Sensation*, 23.

40. Henry Sidgwick, "The Philosophy of Common Sense, An Address Delivered to the Glasgow Philosophical Society on Jan. 10, 1895," *Mind* 4, no. 14 (April 1895): 145–58.

41. Oliver Lodge, *Life and Matter* (London: Williams & Norgate, 1905), 7.

42. Gurney, Myers, and Podmore, *Phanstasms of the Living*, 1:xv.

43. Oliver Lodge, *Reason and Belief* (New York: Moffat & Co., 1910), 17.

44. Oliver Lodge, *Science and Immortality* (New York: Moffat, Yard & Co., 1908), 189; Lang, *The Making of*, 334; Myers, *Fragments of Inner Life*, 46.

45. Frank Angell, "Introduction" to John Edgar Coover, *Experiments in Psychical Research at Leland Stanford Junior University* (Stanford: Stanford University Press, 1917), xxi–xxii.

46. Angell, "Introduction," xx.

47. John Beloff, *Parapsychology: A Concise History* (London: Palgrave Macmillan, 1997), 125–52. Walter J. Levy, MD, and the mathematician Samuel Soal were two of the most celebrated parapsychologists of the early 1970s, generating a growing consensus about the reality of "psi" around their groundbreaking research. When they were both exposed as frauds (notably, by parapsychological colleagues), their public disgrace had a contaminating effect on the field.

48. Dean Radin, *Entangled Minds: Extra Sensory Experiences in a Quantum Reality* (New York: Simon & Schuster, 2006), 131–46.

49. Daryl Bem, "Feeling the Future: Experimental Evidence for Anomalous Retroactive Influences on Cognition and Affect," *Journal of Personality and Social Psychology* 100, no. 3 (2011): 407–25.

50. Fueling the anxiety, YZ and A were able to verify an absurd proposition by exploiting the kinds of hacks to which even mainstream psychology was prone.

51. Daryl Bem, Patrizio Tressoldi, Thomas Rabeyron, and Michael Duggan, "Feeling the Future: A Meta-analysis of 90 Experiments on the Anomalous Anticipation of Random Future Events," *F1000Res* (October 30, 2015; revised January 29, 2016). doi: 10.12688/f1000research.7177.2.

52. Robert Krulwich, "Could It Be? Spooky Experiments That 'See' The Future," *Krulwich Wonders*, NPR, January 2011, http://www.npr.org/blogs/krulwich/2011/01/04/132622672/could-it-be-spooky-experiments-that-see-the-future.

53. Louie Savva, "Why I Quit Parapsychology," from his blog "Everything is Pointless," November 2006, everythingispointless.com/2006/11/why-i-quit-studying-parapsychology.html (accessed on March 17, 2017).

54. Myers, *Fragments of Inner Life*, 45.

55. Susan Blackmore, "Do We Need a New Psychical Research?" *JSPR* 55 (1987–88): 49–59.

56. Myers, *Fragments of Inner Life*, 100.

57. M. Mitchell Waldrop, "Religion: Faith in Science," *Nature* 470 (February 2011): 323–25.

58. Jerry Coyne, evolutionary biologist at the University of Chicago, quoted in Waldrop, "Religion: Faith in Science," 323.

59. Dorothy Nelkin, "God Talk: Confusion between Science and Religion," *Science, Technology, & Human Values* 29, no. 2 (Spring 2004): 139–52.

60. John Templeton, *Wisdom from World Religions: Pathways toward Heaven on Earth* (West Conshohocken, PA: Templeton Press, 2002), xxii.

61. Jerry Coyne, "Faraday and Templeton Brainwash British kids," *Why Evolution Is True*, January 2012, https://whyevolutionistrue.wordpress.com/2012/01/04/faraday-and-templeton-brainwash-british-kids/.

62. Sam Parnia, "AWARE Study Abstract," Horizon Research Foundation, http://www.horizonresearch.org/research-zone/aware-study/aware-study-abstract/.

63. John Martin Fischer quoted by Betty Miller, *UCR Today*, University of California, Riverside, July 2012, http://ucrtoday.ucr.edu/7496.

64. Between 2002 and 2008, the award was titled "The Templeton Prize for Progress toward Research or Discoveries about Spiritual Realities," but it was toned down after his death, assuming a more secular framework of asking "big questions" of a moral and cosmological nature.

65. John Martin Fischer and Benjamin Mitchell-Yellin, *Near-Death Experiences: Understanding Our Visions of the Afterlife* (Oxford: Oxford University Press, 2016), 171.

66. Fischer and Mitchell-Yellin, *Near-Death Experiences*, viii, 15, 157, 173 for quotes on p. 129.

67. Fischer and Mitchell-Yellin, *Near-Death Experiences*, 156, 179 for quotes on p. 130.

68. Charles Q. Choi, "Peace of Mind: Near-Death Experiences Now Found to Have Scientific Explanations," September 12, 2011, https://www.scientificamerican.com/article/peace-of-mind -near-death/.

69. Fischer and Mitchell-Yellin, *Near-Death Experiences*, 164.

70. Myers, *Science and a Future Life*, 1.

71. Wundt expressed doubts regarding that "purely chemical elements could give rise to a mental process," as quoted in the *Catholic Encyclopedia*, vol. 11 (New York: Universal Knowledge Foundation, 1911), 446.

72. Edmund Husserl published his declaration of a new scientific era based on phenomeno-logical psychology in 1913, under the title *Ideen zu einer reinen Phänomenologie und phänomenolo-gischen Philosophie*.

73. Lawrence M. Krauss, "All Scientists Should Be Militant Atheists," *New Yorker*, September 8, 2015.

74. Empirical cognition was the anchor fixed at the center of this whirl of unpredictable sub-jectivities, creating a formal basis for an intersubjective truth and the publication of knowledge. At the core of its program was a defense of the real existence of the self, a clearing made in the mind for a secular soul with an open-ended theology.

Bibliography

PRIMARY SOURCES

Manuscripts and Collections

British Library Reading Room

Lang, Andrew. Letter to G. K. Chesterton. n.d. Add. 73238, f. 50.
———. Letter to T. A. Guthrie. 1901. Add. 54264, f. 39.
———. Letter to W. E. Gladstone. 1882. Add. 44477, f. 104.
———. Letters to Sir E. W. Gosse. 1907. Ashley. 5739, ff. 276-82.
———. Letters to C. C. Bell. 1901. Add. 42711, ff. 208-12.
———. Letters to E. B. Tylor. 1883. Add. 50254, ff. 130, 211-19b.
Lodge, Oliver. Correspondence with Arthur James Balfour, Repository British Library. Add. 49798, ff. 88-157.
———. Correspondence with G. K. Chesterton. Add. 73238, ff. 77-79v.
———. Correspondence with Society of Authors. Add. 56739, ff. 1-47v.
———. Correspondence with A. R. Wallace. 1891-1913. Add. 46437, ff. 30, 138, 263, 272; Add. 46438, ff. 271, 281b; Add. 46439, f. 240.
———. Letters to Bishop W. Boyd Carpenter. 1903-18. Add. 46724, ff. 52b, 58, 93, 99b, 103b, 151, 155, 160.
———. Letter to ____ Stokes. 1909. Add. 42581, f. 61.

Cambridge University

Lodge, Oliver, Papers. Archives for the Society for Psychical Research, Department of Manuscripts and University Archives, Cambridge University Library. Boxes 1-11.
Myers, Frederic. "Account of Friendship with Henry Sidgwick. Oct 18, 1873." Myers Collection, Special Collections, The Wren Library, Cambridge University. Box 13.
Myers, Frederic, Papers. Wren Room, Trinity College Library. Boxes 1-34.

Harvard University

James, William, Papers. Houghton Library, Harvard College Library. Am 1092.1.
Lang, Andrew, Papers. Houghton Library, Harvard College Library. Eng 1292.1-1292.9.

London University

Lodge, Oliver, Papers. University College London, Manuscripts Room. Boxes 1–7.

Newspapers, Magazines, and Journals

Academy
Athenaeum
Blackwood's Edinburgh Magazine
British Journal of Medicine
Bulletin de la Société de Psychologie Physiologique
The Christian Commonwealth
Contemporary Review
Cosmopolitan Journal
The Eclectic Magazine of Foreign Literature
Harmsworth History of the World
Herapath's Journal and Railway Magazine
The Hibbert Journal
The Homiletic Review
Illustrated London News
Journal of the Anthropological Institution of Great Britain and Ireland
Journal of the Society for Psychical Research (JSPR)
The Ladies Repository
London Times
Longman's Magazine
Manchester Guardian
Mind
Nature
New-Hampshire Journal of Medicine
The Nineteenth Century
North American Review
Notices of the Proceedings at the Meetings of the Members of the Royal Institution of Great Britain with
    Abstracts of the Discourse
Pearson's Magazine
Pharmaceutical Journal: A Weekly Record of Pharmacy and Allied Sciences
Philosophical Magazine
Philosophical Transactions of the Royal Society
Popular Science Monthly
Proceedings of the Society for Psychical Research (PSPR)
"Rapport de la Commission réunie à Milan pour l'étude des Phénomènes Psychiques," Annales des Sciences
    Psychiques
The Spiritual Magazine

Published Works

Angell, Frank. "Introduction" to John Edgar Coover, Experiments in Psychical Research at Leland
    Stanford Junior University. Stanford: Stanford University Press, 1917. xxi–xxii.
Babinski, Joseph. "Transfert de certaines manifestations hystériques." Bulletin 2 (1886): 113–16.

Balfour, Arthur. *A Defense of Philosophic Doubt: Being an Essay on the Foundation of Belief.* London: Macmillan, 1879.

Barrett, William Fletcher. "In Memory of Sir William Crookes, O.M., Etc." *PSPR* 31 (1920-21): 26-27.

———. "Note on the Existence of a Magnetic Sense." *PSPR* 2(January 1884): 56-60.

———. *Psychical Research.* London: Williams & Norgate, 1911.

———, Edmund Gurney, and Frederic W.H. Myers. "First Report of the Committee on Thought-reading." *PSPR* 1 (1882-83): 13-32.

Bennett, Edward T. *Twenty Years of Psychical Research 1882-1901.* London: Brimley & Johnson, 1904, 7-8.

Besant, Walter. "The Art of Fiction." *Notices of the Proceedings at the Meetings of the Members of the Royal Institution* 11 (1887): 75.

Bergson, Henri. *Creative Evolution.* Translated by Arthur Mitchell. London: Macmillan, 1911.

Binet, Alfred, and Charles Féré. *Animal Magnetism.* New York: Appleton & Co., 1894, 170-210.

———, and Charles Féré. "L'hypnotisme chez les hystériques." *Revue philosophique de la France et de l'étranger* (January 19, 1885): 1-25.

———, and Charles Féré. "Hypnotisme et responsabilité." *Revue philosophique de la France et de l'étranger* (March 19, 1885): 265-79.

———, and Charles Féré. "La polarisation psychique," *Revue philosophique de la France et de l'étranger* (April 19, 1885): 369-402.

———, and Charles Féré. "La théorie physiologique de l'hallucination," *Revue scientifique* 35 (January 10, 1885): 49-53.

Board, George M. "The Psychology of Spiritualism." *North American Review* 129 (July 1879): 70-87.

Braid, James. *Neurypnology; or, The Rationale of Nervous Sleep, Considered in Relation with Animal Magnetism.* Edinburgh: Charles Black, 1843.

Bray, Charles. *Manual of Anthropology.* Oxford: Oxford University, 1871.

Britten, Emma Hardinge. *Report on Spiritualism: Of the Committee of the London Dialectical Society.* London: J. Burns, 1873.

Bushnell, Horace. *Nature and the Supernatural: As Together Constituting the One System of God.* London: Alexander Strahan, 1864.

Callaway, Henry. "On Divination and Analogous Phenomena among the Natives of Natal." *Journal of the Anthropological Institution of Great Britain and Ireland* 1 (1872): 163-85.

Campbell, R. J. "Personal Immortality." *Homiletic Review* 45 (January-June 1903): 417-22.

Carpenter, W. B. *Mesmerism, Spiritualism, Historically and Scientifically Considered.* New York: D. Appleton, 1877.

Carrington, Hereward. *Eusapia Palladino and Her Phenomena.* New York: B. W. Dodge & Co., 1909.

———. *The Problems of Psychical Research: Experiments and Theories in the Realm of the Supernormal.* New York: W. Rickey & Co., 1914.

Clifford, William K. *Lectures and Essays.* Edited by Leslie Stephen and Frederick Pollock. London: Macmillan, 1886.

Comte, Auguste. *The Positive Philosophy of Auguste Comte.* Translated by Harriet Martineau. London: Trubner & Co., 1875.

Cosh, Rev. James M. *The Supernatural in Relation to the Natural.* Cambridge: Macmillan, 1862.

Cowan, Charles. *Thoughts on Satanic Influence; or, Modern Spiritualism Considered.* London: Seelys, 1854.

Cox, Edward William. *Spiritualism Answered by Science: With Proofs of a Psychic Force.* London: Longman & Co., 1872.

Crookes, William. "Part of the Presidential Address Delivered to the British Association at Bristol, Sept., 1898." *PSPR* 14 (1898-99): 2-3.

———. *Radiant Matter, A lecture Delivered to the British Association for the Advancement of Science at Sheffield Friday, August 22, 1879*. London: Davey, 1879.

———. *Researches in the Phenomena of Spiritualism*. London: J. Burns, 1874.

———. "Spiritualism Viewed by the Light of Modern Science." *Quarterly Journal of Science* (July 1870): 316-20.

Crowe, Catherine. *Spiritualism and the Age We Live In*. London: T. C. Newby, Publisher, 1859.

Dessoir, Max. "Hypnotism in France." *Science* 9, no. 226 (June 3, 1887): 541-45.

Doyle, Arthur Conan. *The History of Spiritualism*. London: Cassel, 1926.

Duhem, Pierre. *The Aim and Structure of Physical Theory*. Translated by Phillip Weiner. Princeton, NJ: Princeton University Press, 1954.

Evrett, Isaac, and Joel Tiffany. *Modern Spiritualism Compared with Christianity in a Debate between Joel Tiffany Esq of Painesville and Rev. Isaac Evrett of Warren*. London: George Adams, 1855.

Faraday, Michael. "The Table-Turning Delusion." *New-Hampshire Journal of Medicine* 4-5 (January-December 1854): 46-48.

Ferris, W. H. "The Theology of Modern Spiritualism: Its Infidelity." *Ladies Repository* 16 (January 1856): 364-70.

Fitzgerald, George Francis. *The Scientific Writings of the Late George Francis Fitzgerald*. Edited by Joseph Larmor. London: Longmans, Green & Co., 1902.

Freud, Sigmund. "Project for a Scientific Psychology." In *The Standard Edition of the Complete Psychological Works of Sigmund Freud, Volume I (1886-1899): Pre-psychoanalytical Papers and Unpublished Manuscripts* (New York: Vintage, 2001), 298-357.

Gasparin, Count Agenor de. *Science versus Spiritualism: A Treatise on Turning Tables, the Supernatural in General, and Spirits*. New York: Kiggins & Kellogg, 1856.

Gurney, Edmund. "Critical Notices: A. Binet and Ch. Féré, *Le magnétisme animal*." *PSPR* 4, no. 10 (1886): 540-54.

———. "Researches in Thought Transference." *PSPR* 2 (December 30, 1884): 249-64.

———. *Tertium Quid*. London: K. Paul, Trench & Co., 1887.

———, Frederic Myers, and Frank Podmore. *Phantasms of the Living*. 2 vols. London: Trubner, 1886.

Haeckel, Ernst. *Confessions of Faith of a Man of Science*. Translated by J. Gilchrist. London: A. & C. Black, 1903.

Haggard, Rider. "About Fiction." *Contemporary Review* 51 (1887): 172-80.

———. *Days of My Life*. London: Longmans, Green & Co., 1926.

Hammond, W. A. *Spiritualism and Allied Causes and Conditions of Nervous Derangement*. New York: G. P. Putnam, 1876.

Harris, Thomas Lake. *Modern Spiritualism, Its Truths and Its Errors: A Sermon*. New York: Church Publishing Association, 1860.

Hobson, John Atkinson. *The Crisis of Liberalism: New Issues of Democracy*. London: P. S. King & Son, 1909.

Hodgson, Richard. "Report of the Committee Appointed to Investigate Phenomena Connected with the Theosophical Society." *PSPR* 2 (1884): 201-369.

Homes, Rupert. "Seeing Things: The Most Famous of Living Mediums." *Pearson's Magazine* 20 (1908): 389-96.

Howitt, Mary Botham. *Margaret Howitt: An Autobiography*, vol. 2. Cambridge: Cambridge University Press, 2010.

Hume, David. *A Treatise of Human Nature*. Section 6, Part IV, Book I (1739). Reprinted in John Perry, ed., *Personal Identity* (Los Angeles: University of California Press, 1975).

Huxley, T. H. "Possibilities and Impossibilities (1891)." In *Science and Christian Tradition: Essays* (New York: D. Appleton, 1896), 192-208.

Hyslop, James. *Borderland of Psychical Research*. Boston: H. B. Turner, 1906.

James, Henry. "The Art of Fiction." *Longman's Magazine* 4 (1884): 502-21.

James, William. "Frederic Myers Service to Psychology." *PSPR* 17 (1901-3): 13-23.

Kingsbury, George. *The Practice of Hypnotic Suggestion, Being an Elementary Handbook for the Use of the Medical Profession*. London: Simpkin Marshall Hamilton Kent & Co., 1891.

Lang, Andrew. *Adventures in Books*. 1905; Freeport, NY: Books for Libraries Press, 1970.

———. "The Art of Mark Twain." *Illustrated London News*, February 14, 1891, 222.

———. "Charles Kingsley." In *Essays in Little* (London: Henry & Co., 1891), 153-59.

———. *Cock Lane and Common-Sense*. London: Watts, 1894.

———. "A Comparative Study of Ghost Stories." *Nineteenth Century* 17 (1885): 623-32.

———. *Custom and Myth*. London: Longmans, Green & Co., 1884.

———. "A Dip in Criticism." *Contemporary Review* 54 (1888): 495-503.

———. "Émile Zola." *Fortnightly Review* 37 (April 1882):439-52.

———. *The Making of Religion*. London: Longmans, Green & Co., 1898.

———. *Myth, Ritual, and Religion*. London: Longmans, Green & Co., 1887.

———. "Realism and Romance." *Contemporary Review* 52 (1887).

———. "Science and Superstition." In *Magic and Religion* (1901; New York: Greenwood Press, 1969), 1-14.

———. "She." *Book News* 5, no. 55 (March 1887): 235-36.

———. "The Supernatural in Fiction." In *Adventures among Books* (New York: Scribners, 1905), 271-80.

———. "Tendencies in Fiction." *North American Review* 465 (August 1895): 153-160.

Lodge, Oliver. "Aberration Problems: A Discussion concerning the Motion of the Ether near the Earth and concerning the Connexion between Ether and Gross Matter, with Some New Experiments." *Philosophical Transactions of the Royal Society* 184 (1893): 727-804.

———. "An Account of Some Experiments in Thought Transference." *PSPR* 2 (1884): 189-200.

———. "Address to Wilifred Ward, September, 1897." In *Papers Read before the Synthetic Society* (London: Spottiswoode Press, 1909), 148-54.

———. "An Attempt at a Systematic Classification of the Various Forms of Energy." *Philosophical Magazine* 3 (October 1879): 277.

———. *Beyond Physics, or the Idealization of Mechanism*. London: George Allen & Unwin, 1930.

———. *A Century's Progress in Physics, Second of a Series of Centenary Addresses*. Delivered on Monday, March 14, 1927. London: University of London Press, 1927.

———. "Certain Observations regarding the Phenomena of Trance." *PSPR* 6 (1889-90): 443-557.

———. *Continuity: The Presidential Address to the British Association, Birmingham MCMXIII*. London: J. M. Dent & Sons, 1913.

———. "Earth and Heaven, A Catechism for Children." London: Privately Published, 1907.

———. "Ecce Deus, The Essential Element in Christianity." London: Privately Published, 1907.

———. *Ecce Homo, Birmingham MCMXIII*. London: J. M. Dent & Sons, 1913.

———. "Electrical Theory of Vision." *Nature* 50 (June 21, 1894): 172.

———. *Elementary Mechanics including Hydrostatics and Pneumatics*. London: Chambers, 1881.

———. "The Ether and Its Functions II." *Nature* 27 (February 1, 1883): 308-10.

———. *The Ether of Space*. London: Harper & Brothers, 1909.

———. "Experience of Unusual Physical Phenomena Occurring in the Presence of an Entranced Person (Eusapia Palladino)." *JSPR* 6 (November 1894): 306-60.

———. "Faith and Knowledge, April 1902." In *Papers Read before the Synthetic Society* (London: Spottiswoode Press, 1909), 381-83.

———. "First Principles of Faith." *Christian Commonwealth* 26 (December 20, 1906).

———. "The Foundations of Dynamics, Part I and II." *The Philosophical Magazine* 5th series 36 (July 1893):1-36.

———. "Free Will and Determinism." *Cosmopolitan Journal* (December 1903).

———. "Inertia," *Scientific Monthly* 10, no. 1 (January–June, 1920): 379.

———. *Interaction between the Mental and the Material Aspect of Things, Read before the Synthetic Society in London on February 20, 1903*. London: Spottiswoode & Co., 1904.

———. "The Interstellar Ether." *Fortnightly Review* (June 1893): liii, 862.

———. "Is Obscurity or Infinitude the Well Spring of Religion, and If Obscurity Should It Therefore Be Encouraged? (June 10, 1898)." In *Papers Read before the Synthetic Society* (London: Spottiswoode, 1909), 218–21.

———. "Is There Possibility of Proof for Religious Doctrines." In *Papers Read before the Synthetic Society* (London: Spottiswoode, 1909), 78.

———. "The Life of Crookes." *PSPR* 34 (1924): 313.

———. *Life and Matter: A Criticism of Professor Haeckel's "Riddle of the Universe."* London: Williams & Norgate, 1905.

———. "Lord Kelvin on Creative Purpose." *Popular Science Monthly* 65 (1903): 279–80.

———. "Mass and Inertia." *Nature* 39 (January 17, 1889): 270–71.

———. "Matter and Spirit." London: Privately Printed, 1904.

———. "Mind and Matter." *Hibbert Journal* 3 (October 1904–July 1905): 315–31.

———. *Modern Problems*. London: Methuen & Co., 1912.

———. *Modern Scientific Ideas, Especially the Idea of Discontinuity*. London: Ernest Benn, 1927.

———. *Modern Views of Electricity*. London: Macmillan, 1889.

———. "Modern Views of Matter." *Popular Science Monthly* (August 1903): lxiii, 289–303.

———. *Modern Views of Matter*. London: Methuen & Co., 1912.

———. *My Philosophy*. London: E. Benn, 1933.

———. "The Nature of Proof, May 28, 1897." In *Papers Read before the Synthetic Society* (London: Spottiswoode Press, 1909), 128–36.

———. "The Nature of Time." In *Modern Problems* (London: Methuen & Co., 1912), 5–23.

———. "Note concerning Thought and Reality." *Mind* 14 (April 1905): 294.

———. "On Action at a Distance and the Conservation of Energy." *Philosophical Magazine* 11 (June 1881): 529–54.

———. "On the Possibility of a Logical Proof for Religious Doctrines, January 29, 1897." In *Papers Read before the Synthetic Society* (London: Spottiswoode, 1909), 70–76.

———. "On the Theory of Lightning Conductors." *Philosophical Magazine* 26, no. 5 (August 1888): 217–30.

———. "Our Place in the Universe." An address to the members of the Nelson Street adult school, Sunday morning August 17th, 1902. Birmingham: John A. Hodges, Printer.

———. "The Outstanding Controversy between Science and Faith." *Hibbert Journal* 1 (1902): 46–61.

———. *Past Years*. New York: Charles Scribner's Sons, 1932.

———. *Phantom Walls*. New York: G. P. Putnam, 1927.

———. "Propositions to Be Maintained, March 14, 1907." *Papers Read before the Synthetic Society* (London: Spottiswoode Press, 1909), 540–41.

———. *Reason and Belief*. New York: Moffat, Yard & Co., 1910.

———. "Reconciliation between Science and Faith." *Hibbert Journal* 1, no. 2 (1903): 209–27.

———. "The Relation between Electricity and Light." *Nature* (January 27, 1881): 302–4.

———. "Religion, Science and Miracle ." *Contemporary Review* 86 (July–December 1904): 798–807.

———. "Reply to Henry Sidgwick, Argument from Design, February 25, 1898." In *Papers Read before the Synthetic Society* (London: Spottiswoode Press, 1909), 180–81.

———. "Response to Wilfrid Ward, February 14, 1897." In *Papers Read before the Synthetic Society* (London: Spottiswoode Press, 1909), 86–88.

———. *Science and Immortality*. New York: Moffat, Yard & Co., 1908.

———. "Section A. Mathematics and Physics, Opening Address by Prof. Oliver J. Lodge, President of the Section." *Nature* 44 (August 20, 1891): 383-87.

———. "Self and the Universe, The Rudiments of Philosophy as Bearing on Christian Dogma, Being an Explanatory Catechism or Glossary for Adult Children Only." (Privately Printed, 1907).

———. "Sin." *Hibbert Journal* 3 (October 1904-July 1905): 1-25.

———. "Special Experiment on Ethereal Viscosity." In *The Ether of Space* (London: Harper & Brothers, 1909) 70-87.

———. "Supplements to the Discussion of Mr. Balfour's Paper, June 1900." In *Papers Read before the Synthetic Society* (London: Spottiswoode Press, 1909), 334-40.

———. "The Wave Theory of Light." *Engineer* (May 22, 1885): lix, 405.

———. *Why I Believe in Personal Immortality*. London: Doubleday, Doran, 1929.

London Dialectical Society. *Report on Spiritualism of the Committee of the London Dialectical Society Together with the Evidence, Oral and Written and a Selection from the Correspondence*. London: Longmans, Green, Reader & Dyer, 1871.

———. *Report on Spiritualism of the Committee of the London Dialectical Society*. London: J. Burns, 1873.

Mach, Ernst. *The Analysis of Sensations, and the Relation of the Physical to the Psychical*. Translated by C. M. Williams, revised by Sydney Waterlow. Chicago: Open Court Publishing, 1914.

MacGregor, James Gordon. "On the Hypothesis of Dynamics." *Philosophical Magazine* 36, no. 220 (September 1893).

Mahaffy, Sir John Pentland. *Kant's Critical Philosophy for English Readers: The Æsthetic and Analytic*. London: Longmans Green, 1874.

McCabe, Joseph. *Spiritualism: A Popular History from 1847*. London: T. F. Unwin, 1920.

Mee, J. A., Arthur Hammerton, and A. D. Innes, eds. *Harmsworth History of the World*, vol. 7. London: Carmelite House, 1907-9.

Muirhead, J. H. "Survival of the Soul." *Contemporary Review* 84 (1903): 112-21.

Myers, Eveleen Tennant, ed. *Fragments of Prose and Poetry by Frederic W. H. Myers*. London: Longman's Green & Co., 1904.

Myers, Frederic. "Automatic Writing II." *PSPR* 3 (January 1885): 30.

———. "Binet on the Consciousness of Hysterical Subjects." *PSPR* 6 (1889): 200-206.

———. "Chapter 10: Epilogue." In *Human Personality and Its Survival of Bodily Death* (London: Longmans, Green & Co., 1918), 340-55.

———. "Charles Darwin and Agnosticism." In *Science and a Future Life* (London: Macmillan, 1901), 51-75.

———. "A Defence of Phantasms of the Dead." *PSPR* 6 (1889-90): 314-57.

———. "Ernst Renan." In *Essays Classical and Modern* (London: Macmillan, 1921), 389-460.

———. "Fourth Report of the Committee on Thought Transference." *PSPR* 2 (November 1883): 1-11.

———. "Fourth Report of the Literary Committee." *PSPR* 2 (July 1884): 157-86.

———. *Fragments of Inner Life: An Autobiographical Sketch*. London: Society for Psychical Research, 1961.

———. "Further Notes on the Unconscious Self, Part I," *JSPR* 2 (December 1885): 122-31.

———. "Further Notes on the Unconscious Self II." *JSPR* 2 (April 1886): 234-47.

———. *Human Personality and Its Survival of Bodily Death, Volume 1*. London: Longmans, Green & Co., 1903.

———. *Human Personality and Its Survival of Bodily Death, Volume 2*. London: Longmans, Green & Co., 1904.

———. *Human Personality and Its Survival of Bodily Death*. London: Longmans, Green & Co., 1918.

———. "Human Personality in the Light of Hypnotic Suggestion," *PSPR* 4, no. 10 (1886): 1-25.

———. "Introduction." In Edmund Gurney, Frederic Myers, and Frank Podmore, eds., *Phantasms of the Living* (London: Trubner & Co., 1886).

———. "Multiplex Personality." *Nineteenth Century* (November 1886).

———. "Multiplex Personality." *PSPR* 4, no. 11 (1887): 496-514.

———. "Obituary: Robert Louis Stevenson." *JSPR* 7 (January 1895).

———. . "Objects of the Society." *PSPR* 1 (1882-83): 3.

———. "On Indications of Continued Terrene Knowledge on the Part of Phantasms of the Dead." *PSPR* 8 (March 1892): 170-252.

———. "On Telepathic Hypnotism." *PSPR* 4, no. 10 (August 1886): 127-82.

———. "On the Consciousness of Hysterical Subjects." *PSPR* 6 (1889): 200.

———. "On the Telepathic Explanation of Some Phenomena Normally Classified as Spiritualistic." *PSPR* 2 (November 1884): 217-37.

———. "The Possibility of a Scientific Approach to Problems Generally Classed as Religious." April 29, 1898. In *Papers Read before the Synthetic Society* (London: Spottiswoode Press, 1909), 187-97.

———. "Preliminary Report of the Reichenbach Committee." *PSPR* 1 (1882-83): 99-100.

———. "Presidential Address." *PSPR* 15 (1900-1901): 110-27.

———. "Report of the General Meeting." *JSPR* 2 (November 1886): 444-46.

———. *Science and a Future Life*. London: Macmillan, 1901.

———. "Second Report of the Committee on Thought-Transference." *PSPR* 1 (1882-83): 70-98.

———. "The Subliminal Consciousness, Chapter I: General Characteristics of Subliminal Messages." *PSPR* 7 (1891-92): 298-327.

———. "The Subliminal Self Chapter II: The Mechanism of Suggestion." *PSPR* 7 (1891-92): 327-53.

———. "The Subliminal Self Chapter III: The Mechanism of Genius." *PSPR* 8 (1892): 333-61.

———. "The Subliminal Self Chapter IV: Hypermnesic Dreams." *PSPR* 8 (1892): 362-404.

———. "The Subliminal Self Chapter V: Sensory Automatism and Induced Hallucinations." *PSPR* 8 (1892): 436-535.

———. "The Subliminal Self Chapter VI: The Mechanism of Hysteria." *PSPR* 9 (1893-94): 3-25.

———. "The Subliminal Self Chapter VII: Motor Automatism." *PSPR* 9 (1893-94): 26-128.

———. "The Subliminal Self Chapter VIII: The Relation of the Supernormal to Time: Retrocognition." *PSPR* 11 (1896-97): 334-407.

———. "The Subliminal Self Chapter IX: The Relation of the Supernormal to Time: Precognition." *PSPR* 11 (1896-97): 408-593.

———. "Telepathic Hypnotism and Its Relation to Other Forms of Hypnotic Suggestion." *PSPR* 4 (November 1886): 127-88.

———. *Wordsworth*. New York: Harper & Brothers, 1885.

———. "The Work of Edmund Gurney in Experimental Psychology." *PSPR* 5 (1888-89): 359-73.

———, William F. Barrett, and Edmund Gurney. "Third Report of the Committee on Mesmerism." *PSPR* 2 (November 1883):12-19.

———, and Edmund Gurney. "Second Report of the Committee on Mesmerism." *PSPR* 1 (April 1883): 251-62.

———, and Henry Sidgwick. "The Second International Congress of Experimental Psychology." *PSPR* 8 (1892): 601-11.

Podmore, Frank. *Apparitions and Thought-Transference: An Examination of the Evidence for Telepathy*. London: W. Scott, 1894.

———. *Modern Spiritualism, A History and a Criticism*. London: Methuen, 1902.

Ramsay, William. *Spiritualism a Satanic Delusion and a Sign of the Times*. Rochester, NY: H. L. Hastings, 1857.

Richet, Charles Robert. *Thirty Years of Psychical Research: Being a Treatise on Metapsychics*. New York: Macmillan, 1923.

Romanes, George J. "The Darwinism of Darwin, and of Post-Darwinian Schools." *Monist* 6, no. 1 (October 1895): 1-27.

Schelling, Friedrich. *The Philosophy of Art: An Oration on the Relation between the Plastic Arts and Nature*. London: Chapman, 1845.

Sidgwick, Arthur, and Eleanor Mildred Sidgwick. *Henry Sidgwick, A Memoir*. London: Macmillan, 1906.

Sidgwick, Henry. *Essays on Ethics and Method*. Edited by Marcus G. Singer. Oxford: Clarendon Press, 2000.

———. "The International Congress of Experimental Psychology." *JSPR* 5 (1892) 283-93.

———. *Practical Ethics: A Collection of Addresses and Essays*. Edited by Sissela Bok. Oxford: Oxford University Press, 1998.

Spencer, Herbert. *First Principles*. 4th ed. London: Williams & Norgate, 1880.

Staff Author. "Charcot and Hypnotism." *British Medical Journal* 2, no. 1704 (August 1893): 480.

Staff Author. "Science and Spiritualism." *Pharmaceutical Journal: A Weekly Record of Pharmacy and Allied Sciences*, January 11, 1873, 545.

Stephen, Leslie. "Some Early Impressions. *National Review* 42, no. 247 (September 1903–February 1904): 130-34.

Stevenson, Robert Louis. "A Humble Remonstrance." *Longman's Magazine* 5, no. 26 (December 1884): 139-47.

Stewart, Balfour, and Peter Guthrie Tait. *The Unseen Universe, Or Physical Speculations on a Future State*. New York: Macmillan, 1875.

Tamburini, Augusto. "Spiritismo e telepatia." *Rivista sperimentale di freniatria e di medicina legale* 18, no. 2 (August 1892): 411-34.

Taylor, Isaac. *The Restoration of Belief*. London: Macmillan, 1864.

Thomson, Allen. "Opening Address, Section D (Edinburgh, 1871)." *Report of the Meeting of the British Association for the Advancement of Science* 41 (1872): 121.

Tiffany, Joel. *Lectures on Spiritualism: Being a Series of Lectures on the Phenomena and Philosophy of Development, Individualism, Spirit, Immortality, Mesmerism, Clairvoyance, Spiritual Manifestations, Christianity, and Progress, Delivered at Prospect Street Church, in the City of Cleveland, during the Winter and Spring of 1851*. Cleveland: J. Tiffany, 1851.

Tulloch, Rev. John. *The Christ of the Gospels and the Christ of Modern Criticism*. London: Macmillan, 1864.

Tuttle, Hudson. *Arcana of Spiritualism: A Manual of Spiritual Science and Philosophy*. Boston: Adams & Co., 1871.

Von Schrenck-Notzing, Albert. *Phenomena of Materialisation*. Translated by E. E. Fournier d'Albe. New York: Dutton & Co., 1923.

Wallace, Alfred Russel. *In Defense of Modern Spiritualism*. Boston: Colby & Rich, 1874.

———. *On Miracles and Modern Spiritualism: Three Essays*. London: James Burns, 1875.

———. *The World of Life: A Manifestation of Creative Power, Directive Mind and Ultimate Purpose*. New York: Moffat, Yard & Co., 1911.

Watson, Thomas. "The Fall of Fiction." *Fortnightly Review* 50 (1888): 336.

Wordsworth, William. *Home at Grasmere: Part First, Book First, of "The Recluse."* Ithaca, NY: Cornell University Press, 1977.

SECONDARY SOURCES

Abrams, M. H. *Natural Supernaturalism*. New York: W. W. Norton, 1971.

Allen, David. "Amateurs and Professionals." In *The Cambridge History of Science*, vol. 6: *Modern Life and Earth Science* (Cambridge: Cambridge University Press, 2009), 13-33.

Apraxine, Pierre, Denis Canguilhelm, Andreas Fischer, and Sophi Schmidt, eds. *The Perfect Medium: Photography and the Occult*. New Haven, CT: Yale University Press, 2005.

Arata, Steven. *Fictions of Loss in the Victorian Fin de Siècle*. Cambridge: Cambridge University Press, 1996.

Arendt, Hannah. *The Origins of Totalitarianism*. New York: Meridian Books, 1958.

Ashley, Susan. "Railway Brain: The Body's Revenge against Progress." *Proceedings of the Western Society for French History* 31 (2003): 177–96.

Ashton, Rosemary. *The German Idea: Four English Writers and the Reception of German Thought, 1800–1860*. Cambridge: Cambridge University Press, 1980.

Ayer, Shelly. "Virile Magic." http://proxy.arts.uci.edu/~nideffer/_SPEED_/1.2/ayers.html.

Bannister, Robert. *Social Darwinism: Science and Myth in Anglo-American Social Thought*. Philadelphia: Temple University Press, 2010.

Barnard, William G. *Exploring Unseen Worlds: William James and the Philosophy of Mysticism*. New York: State University of New York Press, 1997.

Barrow, Yogi. *Independent Spirits: Spiritualism and English Plebeians, 1850–1910*. London: Routledge & Kegan Paul, 1986.

Barzun, Jacques. *Classic, Romantic, and Modern*. Chicago: University of Chicago Press, 1975.

Beer, John. *Providence and Love: Studies in Wordsworth, Channing, Myers, George Eliot, and Ruskin*. Oxford: Clarendon Press, 1999.

Behdad, Ali Belated. *Travelers: Orientalism in the Age of Colonial Dissolution*. Durham, NC: Duke University Press, 1994.

Beloff, John. *Parapsychology: A Concise History*. London: Palgrave Macmillan, 1997.

Bem, Daryl. "Feeling the Future: Experimental Evidence for Anomalous Retroactive Influences on Cognition and Affect." *Journal of Personality and Social Psychology* 100, no. 3 (March 2011): 407–25.

Bem, Daryl, Patrizio Tressoldi, Thomas Rabeyron, and Michael Duggan. "Feeling the Future: A Meta-analysis of 90 Experiments on the Anomalous Anticipation of Random Future Events." *F1000Research* 4 (2015):1188; revised, January 29, 2016. doi: 10.12688/f1000research.7177.2.

Berberich, Christine, ed. *The Bloomsbury Introduction to Popular Fiction*. London: Bloomsbury Publishing, 2015.

Berman, Art. "The Dilemmas of Empiricism." In *Preface to Modernism* (Urbana: University of Illinois Press, 1994), 120–98.

Bhabha, Homi. *The Location of Culture*. London: Routledge, 1994.

———. "The Other Question: The Stereotype and Colonial Discourse." In K. M. Newton, ed., *Twentieth Century Literary Theory*. London: Macmillan, 1997, 293–301.

Blackmore, Susan. "Do We Need a New Psychical Research?" *Journal of the Society for Psychical Research* 55 (1987–88): 49–59.

Block, Ed, Jr. "James Sully, Evolutionist Psychology, and Late Victorian Gothic Fiction." *Victorian Studies* 25, no. 4 (Summer 1982).

Bloom, Harold. *Romanticism and Consciousness: Essays in Criticism*. New York: W. W. Norton, 1970.

Bogousslavsky, Julien, ed. *Following Charcot: A Forgotten History of Neurology and Psychiatry*. Basel: S. Karger, 2011.

———. *Hysteria: The Rise of an Enigma*. Basel: S. Karger, 2014.

———, Olivier Walusinski, and Denis Veyrunes. "Crime, Hysteria and Belle Époque Hypnotism: The Path Traced by Jean-Martin Charcot and Georges Gilles de la Tourette." *European Neurology* 62 (2009): 195. doi: 10.1159/000228252. Accessed February 2, 2016.

Bonduelle, Michel, Toby Gelfand, and Christopher G. Goetz. *Charcot: Constructing Neurology*. New York: Oxford University Press, 1995.

Bourneville, Désiré Magloire, and Paul Regnard. *Iconographie photographique de La Salpêtrière, service de M. Charcot*, vol. 3. Paris: Aux Bureaux du Progrès Médical, Delahaye & Lecrosnier, 1879–80.

Bowler, Peter J. *Evolution: the History of an Idea*. 25th ed. Berkeley: University of California Press, 2009.

———. *The Invention of Progress: The Victorians and the Past*. New York: B. Blackwell, 1989.

———. *The Non-Darwinian Revolution: Reinterpreting a Historical Myth*. Baltimore: Johns Hopkins University Press, 1988.

———. *Reconciling Science and Religion: The Debate in Early-Twentieth-Century Britain*. Chicago: University of Chicago Press, 2001.

Bown, Nicola, and Carolyn Burdett, eds. *The Victorian Supernatural*. Cambridge: Cambridge University Press, 2009.

Brake, Laurel. and Marysa Demoor. *Dictionary of Nineteenth-Century Journalism: In Great Britain and Ireland*. Ghent: Academia Press, 2009.

Brantlinger, Patrick. *Rule of Darkness: British Literature and Imperialism, 1830-1914*. Ithaca, NY: Cornell University Press, 1988.

Braude, Ann. *Radical Spirits: Spiritualism and Women's Rights in Nineteenth-Century America*. 2nd ed. Bloomington: Indiana University Press, 2001.

Brock, William H. *William Crookes and the Commercialization of Science*. Farnham, UK: Ashgate Publishing, 2008.

Brook, John, and Geoffrey Cantor. *Reconstructing Nature: The Engagement of Science and Religion*. New York: Oxford University Press, 2000.

Brower, M. Brady. *Unruly Spirits: The Science of Psychic Phenomena in Modern France*. Urbana: University of Illinois Press, 2010.

Buchan, John. "Andrew Lang and the Border, Andrew Lang Lecture Delivered before the University of St. Andrews, October 17th, 1932." In *Concerning Andrew Lang: Lang Lectures 1927-1937* (Oxford: Clarendon Press, 1949): 4-22.

Budd, Susan. *Varieties of Unbelief: Atheists and Agnostics in English Society, 1850-1960*. London: Heinemann Educational Books, 1977.

Burd, Van Akin. *Ruskin, Lady Mount-Temple and the Spiritualists: An Episode in Broadlands History*. London: Brentham Press for Guild of St. George, 1982.

Butterfield, Herbert. *The Origins of Modern Science, 1300-1800*. London: G. Bell & Sons, 1949.

Byrne, Georgina. *Modern Spiritualism and the Church of England, 1850-1939*. Woodbridge, UK: Boydell Press, 2010.

Calinescu, Matei. *Five Faces of Modernity: Modernism, Avant-garde, Decadence, Kitsch, Postmodernism*. Durham, NC: Duke University Press, 1987.

———, and Sally Shuttleworth. *Science Serialized: Representation of the Sciences in Nineteenth-Century Periodicals*. Cambridge, MA: MIT Press, 2004.

Carrithers, Michael. "An Alternative Social History of the Self." In Michael Carrithers, Steven Collins, and Steven Lukes, eds., *The Category of the Person: Anthropology, Philosophy, History* (Cambridge: Cambridge University Press, 1985), 234-56.

Carroll, Joseph. *Evolution and Literary Theory*. Columbia: University of Missouri Press, 1995.

Cerullo, John. *The Secularization of the Soul*. Philadelphia: Institute for the Study of Human Issues, 1982.

Chadwick, Owen. "The Established Church under Attack." In Anthony Symondson, ed., *The Victorian Crisis of Faith* (London: Society for Promoting Christian Knowledge, 1970), 91-105.

Ching-Liang Low, Gail. *White Skins/Black Masks: Representation and Colonialism*. New York: Routledge, 1996.

Clark, J. F. M. "John Lubbock, Science and the Liberal Intellectual." *Notes and Records: The Royal Society Journal of the History of Science* 68 (2014): 65-87.

Cochrin, Peter. *Byron's Religions*. Cambridge: Cambridge Scholars Publishing, 2011.

Cocq, Antonius Petrus Leonardus de. *Andrew Lang, a Nineteenth Century Anthropologist*. Tilburg: Zwijsen, 1968.

Cole, P., F. Kessell, and D. Johnson, eds. *Self and Consciousness: Multiple Perspectives*. Hillsdale, NJ: Erlbaum, 1992.

Conroy, Colette. *Theatre and the Body*. New York: Palgrave Macmillan, 2009.

Coombes, Annie. *Reinventing Africa*. New Haven, CT: Yale University Press, 1994.

Cooter, Roger. *The Cultural Meaning of Popular Science: Phrenology and the Organization of Consent in Nineteenth-Century Britain*. New York: Cambridge University Press, 2005.

Crabtree, Adam. *From Mesmer to Freud*. New Haven, CT: Yale University Press, 1993.

Crane, Mary Thomas, and Alan Richardson. "Literary Studies and Cognitive Science: Toward a New Interdisciplinarity." *Mosaic* 32, no. 2 (1999):123–40.

Crawford, Robert. "Pater's Renaissance, Andrew Lang, and Anthropological Romanticism." *ELH* 53, no. 4 (Winter 1986): 849–79.

Dale, Peter Allan. *In Pursuit of Scientific Culture: Science, Art and Society in the Victorian Age*. Madison: University of Wisconsin Press, 1989.

Daston, Lorraine, and Katharine Park, eds. *The Cambridge History of Science*, vol. 3: *Early Modern Science*. Cambridge: Cambridge University Press, 2006.

Dawson, Gowan, and Bernard V. Lightman, eds. *Victorian Scientific Naturalism: Community, Identity, Continuity*. Chicago: University of Chicago Press, 2014.

Dawson, Graham. *Soldier Heroes: British Adventure, Empire, and the Imaginings of Masculinities*. London: Routledge, 1994.

DeKosky, Robert K. "William Crookes and the Fourth State of Matter." *Isis* 67, no. 1 (March 1976): 36–60.

Demerest, Marc. "Spirits of the Trade: Teleplasm, Ectoplasm, Psychoplasm, Ideoplasm." *PsyPioneer Journal* 3 (March 2013): 88–98.

Desmond, Adrian. *The Politics of Evolution: Morphology, Medicine, and Reform in Radical London*. Chicago: University of Chicago Press, 1989.

Devereux, Paul. "Arthur Koestler: A Look Back at One of the Pioneering Researchers into Coincidence and Parapsychology." *Fortean Times Magazine*, October 2005. http://www.forteantimes .com/features/profiles/118/arthur_koestler.html. Accessed August 15, 2015.

Dickson, David. "Edinburgh Sets Up Parapsychology Chair." *Science*, n.s. 223, no. 4642 (March 23, 1984).

Didi-Huberman, Georges. *Invention of Hysteria: Charcot and the Photographic Iconography of the Salpêtrière*. Cambridge, MA: MIT Press, 2003.

Dijkstra, Bram. *Evil Sisters: The Threat of Female Sexuality and the Cult of Manhood*. New York: Alfred A. Knopf, 1996.

———. *Idols of Perversity: Fantasies of Feminine Evil in Fin-de-Siècle Culture*. New York: Oxford University Press, 1986.

Doran, Barbara Gusti. "Origins and Consolidation of Field Theory in Nineteenth Century Britain: From the Mechanical to the Electromagnetic View of Nature." In *Historical Studies for the History of Science*, vol. 6, edited by Russell McCormmach (Princeton, NJ: Princeton University Press, 1975), 238–55.

Dorson, Richard Mercer. *The British Folklorists; A History*. Chicago: University of Chicago Press, 1968.

Dowling, Linda. *Hellenism and Homosexuality in Victorian Oxford*. Ithaca, NY: Cornell University Press, 1994.

Drinka, George Frederick. *The Birth of Neurosis: Myth, Malady, and the Victorians*. New York: Simon & Schuster, 1984.

Edelman, Gerald, and Giulio Tononi. *A Universe of Consciousness: How Matter Becomes Imagination*. New York: Basic Books, 2000.

Eichner, Hans. "Modern Science and the Genesis of Romanticism." *PMLA* 97, no. 3 (May 1982): 409.

Elias, Norbert. *The Civilising Process: Sociogenetic and Psychogenetic Investigations*. Translated by Edmund Jephcott. 1939; Oxford: Blackwell, 2000.

Ellenberger, Henri F. *The Discovery of the Unconscious: The History and Evolution of Dynamic Psychiatry*. New York: Basic Books, 1970.

Ellis, Peter Berresford. *H. Rider Haggard: A Voice from the Infinite*. London: Routledge & Kegan Paul, 1978.

Engel, S. Morris, and Andrew Pessin. *The Study of Philosophy: A Text with Readings*. Baltimore: Rowman & Littlefield, 2015.

Etherington, Norman. *Rider Haggard*. Boston: G. K. Hall, 1984.

Evans, Chris. "The Long Dream Ended." *New Scientist* (March 20, 1969): 638–40.

Fabian, Johannes. *Out of Our Minds: Reason and Madness in the Exploration of Central Africa*. Los Angeles: University of California Press, 2000.

Faflak, Joel, and Julia M. Wright, eds. *Nervous Reactions*. New York: SUNY Press, 2004.

Fallowell, Duncan. Interview with Arthur Koestler. "Arthur Koestler, The Art of Fiction No. 80." *Paris Review*, August 1984. http://www.theparisreview.org/interviews/2976/the-art-of-fiction-no-80-arthur-koestler. Accessed August 17, 2012.

Felski, Rita. *The Gender of Modernity*. Cambridge, MA: Harvard University Press, 1995.

Ferngren, Gary B., ed. *Science and Religion: An Historical Introduction*. Baltimore: Johns Hopkins University Press. 2002.

Ferris, David. *Silent Urns: Romanticism, Hellenism, and Modernity*. Stanford: Stanford University Press, 2000.

Fischer, John Martin, and Benjamin Mitchell-Yellin. *Near-Death Experiences: Understanding Visions of the Afterlife*. Oxford: Oxford University Press, 2016.

Foucault, Michel. "What Is an Author." In Josué V. Harari, ed., *Textual Strategies: Perspectives in Post-Structuralist Criticism* (Ithaca, NY: Cornell University Press, 1979), 141–60.

Fournier d'Albe, Edmund Edward. *The Life of Sir William Crookes*. London: T. Fisher Unwin, 1924.

Fyfe, Aileen, and Bernard Lightman, eds. *Science in the Market Place*. Chicago: University of Chicago Press, 2007.

Garland, Martha McMackin. "Victorian Unbelief and Bereavement." In Ralph Anthony Houlbrooke, ed., *Death, Ritual and Bereavement* (London: Routledge & Kegan Paul, 1989), 151–67.

Gauld, Alan. "Andrew Lang as Psychical Researcher." *JSPR* 52 (1983–84): 161–76.

———. *The Founders of Psychical Research*. New York: Schocken Books, 1968.

———. *A History of Hypnotism*. Cambridge: Cambridge University Press, 1992.

———. *Mediumship and Survival: A Century of Investigations*. London: Heinemann, 1982.

Gay, Hannah. "Invisible Resource: William Crookes and His Circle of Support, 1871–81." *British Journal for the History of Science* 29, no. 3 (September 1996): 311–36.

Gazzaniga, Michael. *Who's in Charge? Free Will and the Science of the Brain*. New York: HarperCollins, 2011.

Geertz, Clifford. *The Interpretation of Cultures*. New York: Basic Books, 1973.

Gilbert, Sandra M., and Susan Gubar. *No Man's Land: The Place of the Woman Writer in the Twentieth Century*, vol. 2. New Haven, CT: Yale University Press, 1988.

Gillespie, C. C. *Genesis and Geology: A Study in the Relations of Scientific Thought, Natural Theology, and Social Opinion in Great Britain, 1790–1850*. New York: Harper, 1959.

Gilman, Sander L. *Difference and Pathology: Stereotypes of Sexuality, Race, and Madness*. Ithaca, NY: Cornell University Press, 1985.

Gilman, Sander, Helen King, Roy Porter, et al. *Hysteria beyond Freud*. Berkeley: University of California Press, 1993.

Glucklich, Ariel. *The End of Magic*. New York: Oxford University Press, 1997.

Gold, Barri J. *Thermopoetics: Energy in Victorian Literature and Science*. Cambridge, MA: MIT Press, 2010.

Goldstein, Jan. *Console and Classify: The French Psychiatric Profession in the Nineteenth Century*. Chicago: University of Chicago Press, 2001.

Goldstein, Robert. "Inclined toward the Marvelous: Romantic Uses of Clinical Phenomena in the Work of Frederic W. H. Myers." *Psychoanalytic Review* 79, no. 4 (Winter 1992): 577–89.

Gomel, Elana. "Spirits in the Material World: Spiritualism and Identity in the Fin de Siècle." *Victorian Literature and Culture* 35, no. 1 (2007): 189–213.

Gorer, Geoffrey. *Death, Grief, and Mourning.* Garden City, NY: Doubleday, 1965.

Gould, Steven Jay. *The Mismeasure of Man.* New York: W. W. Norton, 1996.

———. "Non-overlapping Magisteria." *Natural History* 106 (March 1997): 16–22.

Gray, John. *The Immortalization Commission.* New York: Farrar, Straus & Giroux, 2011.

Green, Martin. *Deeds of Adventure, Dreams of Empire.* New York: Basic Books, 1979.

Green, Roger Lancelyn. *Andrew Lang: A Critical Biography with a Short-Title Bibliography of the Works of Andrew Lang.* London: Bodley Head, 1962.

Gregory, Frederick. "Intersections of Physical Science and Western Religion in the Nineteenth and Twentieth Centuries." In *The Cambridge History of Science*, vol. 5: *The Modern Physical and Mathematical Sciences* (Cambridge: Cambridge University Press, 2002), 36–39.

———. *Nature Lost? Natural Science and the German Theological Traditions of the Nineteenth Century.* Cambridge, MA: Harvard University Press, 1992.

Griffin, Susan M., and William R. Veeder, eds. *The Art of Criticism: Henry James on the Theory and the Practice of Fiction.* Chicago: University of Chicago Press, 1986.

Grimes, Hilary. *The Late Victorian.* Aldershot, UK: Ashgate, 2011.

Hacking, Ian. *Rewriting the Soul: Multiple Personality and the Sciences of Memory.* Princeton, NJ: Princeton University Press, 1995.

———. "Telepathy: Origins of Randomization in Experimental Design." *Isis* 79, no. 298 (September 1988): 427–51.

Hall, A. Rupert. *The Scientific Revolution, 1500–1800: The Formation of the Modern Scientific Attitude.* Boston: Beacon Press, 1962.

Hall, Trevor. *The Medium and the Scientist: The Story of Florence Cook and William Crookes.* Buffalo: Prometheus Books, 1985.

———. *The Spiritualists.* London: Duckworth, 1962.

Halliwell, Martin. *Romantic Science and the Experience of Self: Transatlantic Crosscurrents from William James to Oliver Sacks.* Brookfield, VT: Ashgate, 1999.

Hanke, Amala. *Spatiotemporal Consciousness in English and German Romanticism: A Comparative Study of Novalis, Blake, Wordsworth, and Eichendorff.* Bern: Peter Lang, 1981.

Harrington, Anne. "Hysteria, Hypnosis, and the Lure of the Invisible: The Rise of Neo-Mesmerism in Fin-de-Siècle French Psychiatry." In William Bynum, Roy Porter, and Michael Shepherd, eds., *The Anatomy of Madness: Essays in the History of Psychiatry*, vol. 3 (London: Routledge, 2004), 226–42.

———. "Metals and Magnets in Medicine: Hysteria, Hypnosis and Medical Culture in Fin-de-Siècle Paris," *Psychological Medicine* 18 (1988): 21–38.

Hart, Elizabeth. "The Epistemology of Cognitive Literary Studies." *Philosophy and Literature* 25, no. 2 (2001): 314–34.

Hartman, Geoffrey. "Romanticism and 'Anti-Self Consciousness.'" *Centennial Review* 6, no. 4 (Fall 1962): 553–65.

Haule, John Ryan. "Pierre Janet And Dissociation: The First Transference Theory and Its Origins in Hypnosis." *American Journal of Clinical Hypnosis* 29, no. 2 (October 1986): 86–94.

Haynes, Renee. *The Society for Psychical Research.* London: Macdonald, 1982.

Hazard, Paul. *The European Mind, 1680–1715.* Translated by Lewis May. New York: Meridian Books, 1963.

Heilbron, J. L. *Electricity in the 17th and 18th Centuries: A Study of Early Modern Physics.* Berkeley: University of California Press, 1979.

Helmstadter, Richard J., and Bernard Lightman, eds. *Victorian Faith in Crisis: Essays on Continuity and Change in Nineteenth-Century Religious Belief.* Houndmills, UK: Macmillan, 1990.

Hempton, David. *Evangelical Disenchantment: Nine Portraits of Faith and Doubt.* New Haven, CT: Yale University Press, 2008.

Hill, Georgina O'Brien. "'Above the Breath of Suspicion': Florence Marryat and the Shadow of the Fraudulent Trance Medium." In Tatiana Kontou, ed., *Women and the Victorian Occult* (New York: Routledge, 2010), 59–73.

Hogan, Patrick. *Cognitive Science, Literature, and the Arts: A Guide for Humanists.* New York: Routledge, 2003.

Holf, Robert R. "The History of the Vitalist-Mechanist Controversy." In Benjamin Wolman, ed., *Historical Roots of Contemporary Psychology* (New York: Harper & Row, 1968).

Hors, Rutger, and Onno van der Hart. "The Dissociation Theory of Pierre Janet." *Journal of Traumatic Stress* 2, no. 4 (1989): 397–412.

Houghton, Walter. *The Victorian Frame of Mind, 1830–1870.* New Haven, CT: Yale University Press, 1957.

Houlbrooke, Ralph Anthony. *Death, Ritual and Bereavement.* London: Routledge & Kegan Paul, 1989.

Hunt, Bruce. *The Maxwellians.* Ithaca, NY: Cornell University Press, 1991.

Hurley, Kelly. *The Gothic Body: Sexuality, Materialism, and Degeneration at the Fin de Siècle.* Cambridge: Cambridge University Press, 1997.

Hustvedt, Asti. *Medical Muses: Hysteria in Nineteenth-Century Paris.* New York: W. W. Norton, 2011.

Huxley, Aldous. *The Doors of Perception.* London: Chatto & Windus, 1953.

Inglis, Alex Brian. *Natural and Supernatural: A History of the Paranormal from Earliest Times to 1914.* London: Hodder & Stoughton, 1977.

Jacob, Margaret. *The Newtonians and the English Revolution 1689–1720.* Brighton, Sussex, UK: Harvester Press, 1976.

Jalland, Pat. *Death in the Victorian Family.* Oxford: Oxford University Press, 2000.

James, Frank A. J. L. "Of 'Medals and Muddles': The Context of the Discovery of Thallium: William Crookes' Early Spectro-Chemical Work." *Notes and Records of the Royal Society of London* 39, no. 1 (September 1984): 65–90.

Jenkyns, Richard. *The Victorians and Ancient Greece.* Oxford: Basil Blackwell, 1980.

Jolly, W.P. *Oliver Lodge.* Rutherford, NJ: Fairleigh Dickinson University Press, 1975.

Jones, Ernest. *The Life and Work of Sigmund Freud,* vol. 1. New York: Basic Books, 1957.

Jung, Karl, and Wolfgang Pauli. *The Interpretation of Nature and the Psyche.* London: Routledge & Kegan Paul, 1955.

Katz, Wendy. *Rider Haggard and the Fiction of Empire: A Critical Study of British Imperial Fiction.* New York: Cambridge University Press, 1987.

Kelly, Edward F., Emily Williams Kelly, Adam Crabtree, Alan Gauld, Michael Grosso, and Bruce Greyson. *Irreducible Mind: Toward a Psychology for the 21st Century.* Lanham, MD: Rowman & Littlefield, 2006.

Kelly, Stewart. "Andrew Lang, the Life and Times of a Prolific Talent." *Scotsman,* January 30, 2012. http://www.scotsman.com/heritage/people-places/andrew-lang-the-life-and-times-of-a-prolific-talent-1-2085486. Accessed May 14, 2016.

Kerslake, Christian. *Deleuze and the Unconscious.* London: Continuum International Publishing Group, 2007.

Kessell, F., P. Cole, and D. Johnson, eds. *Self and Consciousness: Multiple Perspectives.* Hillsdale, NJ: Erlbaum, 1992.

Keynes, Randal. *Darwin, His Daughter, and Human Evolution.* New York: Riverhead, 2002.

Knapp, Krister Dylan. *William James: Psychical Research and the Challenge of Modernity.* Chapel Hill: University of North Carolina Press, 2007.

Koestler, Arthur. *The Act of Creation.* New York: Macmillan, 1964.

———. *Darkness at Noon.* New York: Bantam Books, 1968.

Kontou, Tatiana, ed. *Women and the Victorian Occult*. New York: Routledge, 2010.

Kontou, Tatiana, and Sarah Willburn, eds. *The Ashgate Research Companion to Nineteenth-Century Spiritualism and the Occult Reader*. New York: Ashgate, 2012.

Kumar, David R., et al. "Jean-Martin Charcot: The Father of Neurology." *Clinical Medicine & Research* 1 (2011): 46–49.

Lachapelle, Sophi. *Investigating the Supernatural: From Spiritism and Occultism to Psychical Research and Metapsychics in France, 1853–1931*. Baltimore: Johns Hopkins University Press, 2011.

Laity, Cassandra. *H. D. and the Victorian Fin de Siècle*. Cambridge: Cambridge University Press, 1996.

Lamont, Peter. "Spiritualism and a Mid-Victorian Crisis of Evidence." *Historical Journal* 47, no. 4 (December 2004): 897–920.

Langstaf, Eleanor de Selms. *Andrew Lang*. Boston: Twayne Publishers, 1978.

Larsen, Timothy. *Crisis of Doubt: Honest Faith in Nineteenth-Century England*. Oxford: Oxford University Press, 2007.

Lawson, John H. "The Wordpower of Frederic Myers?" *Psi Researcher* 21 (1996), 24.

Leary, Timothy, Ralph Metzner, and Ram Dass. *The Psychedelic Experience: A Manual Based on the Tibetan Book of the Dead*. New York: Citadel Press, 1964.

Ledger, Sally. *The New Woman: Fiction and Feminism at the Fin de Siècle*. Manchester: Manchester University Press, 1997.

Lerner, Paul Frederick, and Mark S. Micale, eds. *Traumatic Pasts: History, Psychiatry, and Trauma in the Modern Age, 1870–1930*. Cambridge Studies in the History of Medicine. Cambridge: Cambridge University Press, 2001.

Levine, George. "By Knowledge Possessed: Darwin, Nature, and Victorian Narrative." *New Literary History* 24, no. 2 (Spring 1993): 363–91.

Lewis, Linda M. *Elizabeth Barrett Browning's Spiritual Progress: Face to Face with God*. Columbia: University of Missouri Press, 1998.

Lightman, Bernard V. *The Origins of Agnosticism: Victorian Unbelief and the Limits of Knowledge*. Baltimore: Johns Hopkins University Press, 1987.

———. *Victorian Popularizers of Science: Designing Nature for New Audiences*. Chicago: University of Chicago Press, 2007.

———, ed. *Victorian Science in Context*. Chicago: University of Chicago Press, 1997.

———. "Victorian Sciences and Religions: Discordant Harmonies." *Osiris*, 2nd ser., 16 (2001): 343–66.

Lightman, Bernard V., and Bennett Zon, eds. *Evolution and Victorian Culture*. Cambridge Studies in Nineteenth-Century Literature and Culture 92. Cambridge: Cambridge University Press, 2014.

Lindberg, David, and Ronald L. Numbers, eds. *When Science and Christianity Meet*. Chicago: University of Chicago Press, 2003.

Livingston, James C. *Religious Thought in the Victorian Age: Challenges and Reconceptions*. New York: Continuum, 2007.

Lothane, Zvi. "Freud's 1895 Project: From Mind to Brain and Back Again." *Annals of the New York Academy of Sciences* 843 (May 1996): 43–65. doi:10.1111/j.1749-6632.1998.tb08204.

Lovejoy, Arthur O. "The Meaning of Romanticism for the Historian of Ideas." *Journal of the History of Ideas* 2, no. 3 (June 1941): 257–78.

Lubenow, William C. *"Only Connect": Learned Societies in Nineteenth-Century Britain*. Suffolk, UK: Boydell & Brewer, 2015.

Luckhurst, Roger. *The Invention of Telepathy*. Oxford: Oxford University Press, 2002.

———. "Knowledge, Belief and the Supernatural at the Imperial Margin." In *The Victorian Supernatural*. Cambridge: Cambridge University Press, 2004.

———. *Transactions and Encounters: Science and Culture in the Nineteenth Century*. Manchester, UK: Manchester University Press, 2002.

———. *The Trauma Question*. London: Routledge, 2008.

———, and Josephine McDonagh. *Transactions and Encounters: Science and Culture in the Nineteenth Century*. Manchester, UK: Manchester University Press, 2002.

Mackenzie, John M., ed. *Imperialism and Popular Culture*. Manchester, UK: Manchester University Press, 1986.

Maines, Rachel P. *The Technology of Orgasm: Hysteria, the Vibrator and Women's Sexual Satisfaction*. Baltimore: Johns Hopkins University Press, 2001.

Marsh, Susan M., and David C. Taylor. "Hughlings Jackson's Dr. Z: The Paradigm of Temporal Lobe Epilepsy Revealed." *Journal of Neurology, Neurosurgery, and Psychiatry* 43 (1980): 758–67.

Masson, Scott. *Romanticism, Hermeneutics and the Crisis of the Human Sciences*. Burlington, VT: Ashgate, 2004.

Materer, Timothy. *Modernist Alchemy: Poetry and the Occult*. Ithaca, NY: Cornell University Press, 1995.

Mauss, Marcel. "A Category of the Human Mind: The Notion of Person, the Notion of Self." Translated by W. D. Halls. In *The Category of the Person: Anthropology, Philosophy, History* (Cambridge: Cambridge University Press, 1985), 1–25.

Mayer, Andreas. *Sites of the Unconscious: Hypnosis and the Emergence of the Psychoanalytic Setting*. Chicago: University of Chicago Press, 2013.

Medhurst, Richard and Kathleen Goldney. "The Anderson Testimony." *JSPR* 41 (June 1963): 93–97.

———, and Kathleen Goldney. "Sir William Crookes and the Physical Phenomena of Mediumship." *PSPR* 54 (1964): 25–157.

Menhennet, Alan. *The Romantic Movement*. New York: Routledge, 2016.

Mellor, D. H. "Models and Analogies in Science: Duhem versus Campbell." *Isis* 59, no. 3 (Autumn 1968): 282–90.

Melnyk, Julie. *Victorian Religion: Faith and Life in Britain*. Ann Arbor: University of Michigan Press, 2008.

Meyer, Birgit, and Peter Pels, eds. *Magic and Modernity: Interfaces of Revelation and Concealment*. Stanford: Stanford University Press, 2003.

Micale, Mark S., ed.. *The Mind of Modernism: Medicine, Psychology, and the Cultural Arts in Europe and America, 1880–1940*. Stanford: Stanford University Press, 2004.

———. "The Salpêtrière in the Age of Charcot: An Institutional Perspective on Medical History in the Late Nineteenth Century." *Journal of Contemporary History* 20, no. 4 (October 1985): 703–31.

Michalski, Robert. "Towards a Popular Culture: Andrew Lang's Anthropological and Literary Criticism." *Journal of American Culture* 18, no. 3 (Fall 1995): 13–17.

Mitch, David Franklin. *The Rise of Popular Literacy in Victorian England: The Influence of Private Choice and Public Policy*. Philadelphia: University of Pennsylvania Press, 1992.

Monroe, John Warne. *Laboratories of Faith: Mesmerism, Spiritism, and Occultism in Modern France*. Ithaca, NY: Cornell University Press, 2007.

Moore, James R. *The Post-Darwinian Controversies: A Study of the Protestant Struggle to Come*. Cambridge: Cambridge University Press, 1979.

———. "Theodicy and Society: The Crisis of the Intelligentsia." In Richard Helmstadter and Bernard Lightman, eds., *Victorian Faith in Crisis: Essays on Continuity and Change in Nineteenth-Century Religious Belief* (Houndmills, UK: Macmillan, 1990).

Moore-Gilbert, Bart J. *Postcolonial Theory: Contexts, Practices*. London: Verso, 1997.

Morrell, Jack, and Arnold Thackray. *Gentlemen of Science: Early Years of the British Association for the Advancement of Science*. Oxford: Clarendon Press, 1982.

Nelkin, Dorothy. "God Talk: Confusion between Science and Religion." *Science, Technology & Human Values* 29, no. 2 (Spring 2004): 139–52.

Nelles, William. "Jane's Brains: Austen and Cognitive Theory." *Interdisciplinary Literary Studies* 16, no. 1 (2014): 6–29.

Nicholls, Angus, and Martin Liebscher. *Thinking the Unconscious: Nineteenth-Century German Thought*. Cambridge: Cambridge University Press, 2010.

Noakes, Richard. "'The Bridge Which Is between Physical and Psychical Research': William Fletcher Barrett, Sensitive Flames, and Spiritualism." *History of Science* 42, no. 4 (2004): 419–64.

———. "Cromwell Varley FRS, Electrical Discharge and Victorian Spiritualism." *Notes and Records of the Royal Society of London* 61, no. 1 (January 22, 2007): 5–21.

———. "Instruments to Lay Hold of Spirits." In Iwan Rhys Morus, ed., *Bodies/Machines* (New York: BERG, 2002), 125–64.

———. "Telegraphy Is an Occult Art: Cromwell Fleetwood Varley and the Diffusion of Electricity to the Other World." *British Journal for the History of Science* 32, no. 115 (December 1999): 5–21.

Nott, Kathleen. "The Trojan Horses: Koestler and the Behaviorists." In Harold Harris, ed., *Astride Two Cultures: Arthur Koestler at Seventy* (London: Hutchinson, 1975), 162–74.

Oppenheim, Janet. *The Other World: Spiritualism and Psychical Research in England, 1850–1914*. New York: Cambridge University Press, 1985.

Owen, Alex. *The Darkened Room*. Chicago: University of Chicago Press, 1989.

———. *The Places of Enchantment: British Occultism and the Culture of the Modern*. Chicago: University of Chicago Press, 2004.

Palfreman, Jon. "Between Skepticism and Credulity: A Study of Victorian Scientific Attitudes to Modern Spiritualism." In *The Margins of Science* (Keele, UK: University of Keele, 1979), 201–36.

Paradis, James, and Thomas Postlewait, eds. *Victorian Science and Victorian Values: Literary Perspectives*. New Brunswick, NJ: Rutgers University Press, 1985.

Parnia, Sam, with Josh Young. *Erasing Death: The Science That Is Rewriting the Boundaries between Life and Death*. San Francisco: HarperOne, 2013.

Patteson, Richard F. "King Solomon's Mines: Imperialism and Narrative Structure." *Journal of Narrative Technique* 8, no. 2 (Spring 1978): 112–23.

Peppis, Paul. *Sciences of Modernism: Ethnography, Sexology, and Psychology*. New York: Cambridge University Press, 2014.

Phillips, Richard. *Mapping Men and Empire: A Geography of Adventure*. New York: Routledge, 1997.

Pinch, Adela. *Thinking about Other People in Nineteenth-Century British Writing*. Cambridge Studies in Nineteenth-Century Literature and Culture 73. Cambridge: Cambridge University Press, 2010.

Plas, Régine. "The Origins of French Experimental Psychology: Experiment and Experimentalism." *History of the Human Sciences* 9, no. 1 (1996): 73–84.

———. "Psychology and Psychical Research in France around the End of the 19th Century." *History of the Human Sciences* 25 (2012): 91–107.

Porter, Theodore. "The Death of the Object: Fin de Siècle Philosophy of Physics." In Dorothy Ross, ed., *Modernist Impulses in the Human Sciences, 1870–1930* (Baltimore: Johns Hopkins University Press, 1994).

———. "Reason, Faith, and Alienation in the Victorian Fin-de-Siècle." In Hans-Erich Bödeker, Peter Reill, and Jürgen Schlumbohm, eds., *Wissenschaftals kulturelle Praxis, 1750–1900* (Göttingen: Vandenhoek & Ruprecht, 1999), 401–13.

Poullet, Georges. "Timelessness and Romanticism." *Journal of the History of Ideas* 1 (January 1954): 3–22.

Preyer, Robert O. "The Romantic Tide Reaches Trinity." In James Paradis and Thomas Postlewait, eds., *Victorian Science and Victorian Values: Literary Perspectives* (New Brunswick, NJ: Rutgers University Press, 1985).

Proffitt, Edward. "Forum: Modern Science and the Genesis of Romanticism." *PMLA* 97, no. 3 (May 1982): 408–12.

Pyle, Forest. *The Ideology of Imagination: Subject and Society in the Discourse of Romanticism*. Stanford: Stanford University Press, 1995.

Qureshi, Sadiah. *Peoples on Parade: Exhibitions, Empire, and Anthropology in Nineteenth Century Britain*. Chicago: University of Chicago Press, 2011.

Radin, Dean. *Entangled Minds: Extra Sensory Experiences in a Quantum Reality*. New York: Simon & Schuster, 2006.

———. *Supernormal: Science, Yoga, and the Evidence for Extraordinary Psychic Abilities*. New York: Deepak Chopra, 2013.

Raia, Courtenay. "The Search for a Transcendent Madness: A Journey in Brief across *Les Bulletins de la Société de Psychologie Physiologique* (1885–1887)." Paper presented at the Interdisciplinary Nineteenth Century Studies Conference: Serials, Cycles, Suspensions, San Francisco, March 2018.

Reddy, William M. "Against Constructionism: The Historical Ethnography of Emotions." *Current Anthropology* 38, no. 3 (June 1997): 327–51.

———. "Emotional Liberty: Politics and History in the Anthropology of Emotions." *Cultural Anthropology* 14 (1999): 256–88.

———. *The Navigation of Feeling: A Framework for the History of Emotions*. Cambridge: Cambridge University Press, 2001.

Reed, Edward. *From Soul to Mind: The Emergence of Psychology from Erasmus Darwin to William James*. New Haven, CT: Yale University Press, 1997.

Richardson, Alan. *British Romanticism and the Science of Mind*. Cambridge: Cambridge University Press, 2005.

———. "Literary Studies and Cognitive Science: Toward a New Interdisciplinarity." *Mosaic* 32, no. 2 (1999): 123–40.

———. "Studies in Literature and Cognition: A Field Map." In Alan Richardson and Ellen Spolsky, eds., *The Work of Fiction: Cognition, Culture, and Complexity* (Aldershot, UK: Ashgate, 2004), 1–30.

Richardson, Ruth. "Why Was Death So Big in Victorian Britain." In Ralph Anthony Houlbrooke, ed., *Death, Ritual and Bereavement* (London: Routledge & Kegan Paul, 1989), 105–23.

Rieber, Robert, and David K. Robinson, ed. *Wilhelm Wundt and the Making of a Scientific Psychology*. New York: Plenum Publishers, 2001.

Roach, J. P. C., ed. "The University of Cambridge: The Age of Reforms (1800–1882)." In *A History of the County of Cambridge and the Isle of Ely*, vol. 3: *The City and University of Cambridge* (London: Victoria County History, 1959), 235–65.

Robson, John. "The Bridgewater Treatises." In Richard Helmstadter and Bernard Lightman, eds., *Victorian Faith in Crisis: Essays on Continuity and Change in Nineteenth-Century Religious Belief* (Houndmills, UK: Macmillan, 1990).

Root, John D. "Science, Religion, and Psychical Research: The Monistic Thought of Sir Oliver Lodge." *Harvard Theological Review* (July–October 1978): 245–63.

Rosaldo, Michelle Z. "Toward an Anthropology of Self and Feeling." In Richard A. Shweder and Robert A. LeVine, eds., *Culture Theory: Essays on Mind, Self, and Emotion* (Cambridge: Cambridge University Press, 1995), 158–99.

Ross, Dorothy. "Modernism Reconsidered." In Dorothy Ross, ed., *Modernist Impulses in the Human Sciences, 1870–1930* (Baltimore: Johns Hopkins University Press, 1994), 1–25.

Royle, Edward. *Victorian Infidels: The Origins of the British Secularist Movement, 1791–1866*. Manchester, UK: University of Manchester Press, 1974.

Rowlands, Peter. *Oliver Lodge and The Liverpool Society*. Liverpool: Liverpool University Press, 1990.

Russell, Colin A. "The Conflict Metaphor and Its Social Origins." *Science and Christian Belief* 1 (1989): 3–26.

Ryan, Judith. *The Vanishing Subject: Early Psychology and Literary Modernism*. Chicago: University of Chicago Press, 1991.

Rzepka, Charles J. *The Self as Mind: Vision and Identity in Wordsworth, Coleridge, and Keats*. Cambridge, MA: Harvard University Press, 1986.

Sandison, Alan. *The Wheel of Empire: A Study of the Imperial Idea in Some Late Nineteenth and Early Twentieth-Century Fiction*. New York: St. Martin's Press, 1967.

Schmidt, Richard. "The Counter-Enlightenment: Historical Notes on a Concept Historians Should Avoid." *Eighteenth-Century Studies* 49, no. 1 (2015): 83–86.

———. *Persistent Enlightenment* (blog). https://persistentenlightenment.wordpress.com.

Secord, James. *Victorian Sensation*. Chicago: University of Chicago Press, 2000.

Seigel, Jerrold. *The Idea of the Self: Thought and Experience in Western Europe since the Seventeenth Century*. Cambridge: Cambridge University Press, 2005.

———. "Problematizing the Self." In Victoria E. Bonnell and Lynn Hunt, eds., *Beyond the Cultural Turn* (Berkeley: University of California Press, 1999), 281–314.

Showalter, Elaine. *Sexual Anarchy: Gender and Culture at the Fin de Siècle*. New York: Penguin Press, 1990.

Skrbina, David. *Panpsychism in the West*. Cambridge, MA: MIT Press, 2017.

Smith, Crosbie. *The Science of Energy: A Cultural History of Energy Physics in Victorian Britain*. London: Athlone, 1998.

———, and Norton Wise. *Energy and Empire: A Biographical Study of Lord Kelvin*. New York: Cambridge University Press, 1989.

Smith, Roger. *Inhibition: History and Meaning in the Sciences of Mind and Brain*. Berkeley: University of California Press, 1992.

———. "The Physiology of the Will: Mind, Body and Psychology in the Periodical Literature from 1855 to 1875." In G. N. Cantor and Sally Shuttleworth, eds., *Science Serialized: Representation of the Sciences in Nineteenth-Century Periodicals* (Cambridge, MA: MIT Press, 2004), 81–110.

Snyder, Timothy. "Hitler vs. Stalin: Who Killed More?" *New York Times Review of Books*, March 10, 2011.

Sommer Andreas. "Normalizing the Supernormal: The Formation of the 'Gesellschaft für Psychologische Forschung.'" *Journal of the History of the Behaviorial Sciences* 49, no. 1 (January 2013): 18–44.

———. "Professional Heresy: Edmund Gurney (1847-88) and the Study of Hallucinations and Hypnotism." *Medical History* 55, no.3 (2011): 383–88.

———. "Psychical Research and the Origins of American Psychology: Hugo Münsterberg, William James and Eusapia Palladino." *History of the Human Sciences* 25, no. 2 (2012): 23–44.

Spolsky, Ellen. "Darwin and Derrida: Cognitive Literary Theory as a Species of Post-Structuralism." *Poetics Today* 23, no. 1 (2002).

Stanley, Matthew. *Huxley's Church and Maxwell's Demon: From Theistic Science to Naturalistic Science*. Chicago: University of Chicago Press, 2015.

Stocking, George W., Jr. "Animism in Theory and Practice: E. B. Tylor's Unpublished 'Notes on "Spiritualism."'" *Man*, n.s., 6, no. 1 (March 1971): 87–104.

———. *Victorian Anthropology*. New York: Free Press, 1987.

Stott, Rebecca. "The Dark Continent: Africa as Female Body in Haggard's Adventure Fiction." *Feminist Review* 32 (Summer 1988): 69–89.

Street, Brian V. *The Savage in Literature: Representations of "Primitive" Society in English Fiction, 1858-1920*. Boston: Routledge & Kegan Paul, 1975.

Sutin, Lawrence. *Do What Thou Wilt: A Life of Aleister Crowley*. New York: St. Martin's Press, 2002.

Symondson, Anthony. *The Victorian Crisis of Faith*. London: SPCK, 1970.

Taves, Anna. "A Tale of Two Congresses: The Psychological Study of Psychical, Occult, and Religious Phenomena: 1900-1909." *Journal of the History of the Behavioral Sciences* 50, no. 4 (Fall 2014): 376–99.

Taylor, Beverly, and Robert Bain, eds. *The Cast of Consciousness: Concepts of the Mind in British and American Romanticism*. New York: Greenwood Press, 1987.

Taylor, Charles. "The Person." In *The Category of the Person: Anthropology, Philosophy, History* (Cambridge: Cambridge University Press, 1985), 257–81.

———. *Sources of the Self: The Making of Modern Identity*. Cambridge: Cambridge University Press, 1989.

Taylor, Eugene. *Beyond the Margins of Consciousness*. Princeton, NJ: Princeton University Press, 2011.

———. "Charcot's Axis." In *The Mystery of Personality: A History of Psychodynamic Theories* (New York: Springer-Verlag, 2009), 19–51.

———. *William James on Consciousness beyond the Margin*. Princeton, NJ: Princeton University Press, 1996.

Templeton, John. *Wisdom from World Religions: Pathways toward Heaven on Earth*. West Conshohocken, PA: Templeton Press, 2002.

Thorslev, Peter. "Romanticism and Literary Consciousness." *Journal of the History of Ideas* 36, no. 3 (July–September 1975): 563–72.

Thurschwell, Pamela. *Literature, Technology and Magical Thinking, 1880–1920*. Cambridge: Cambridge University Press, 2001.

Tosh, John. *Manliness and Masculinities*. London: Pearson Longman, 2005.

Treitel, Corinna. *A Science for the Soul: Occultism and the Genesis of the German Modern*. Baltimore: Johns Hopkins University Press, 2004.

Tresch, John. *The Romantic Machine: Utopian Science and Technology after Napoleon*. Chicago: University of Chicago Press, 2012.

Tromp, Marlene. *Altered States: Sex, Nation, Drugs, and Self-Transformation in Victorian Spiritualism*. Albany: SUNY Press, 2006.

Tucker, Jennifer. *Nature Exposed: Photography as Eyewitness in Victorian Science*. Baltimore: Johns Hopkins University Press, 2006.

Turner, Frank. *Between Science and Religion*. New Haven, CT: Yale University Press, 1974.

———. *Contesting Cultural Authority*. Cambridge: Cambridge University Press, 1993.

———. *The Greek Heritage in Victorian Britain*. New Haven , CT: Yale University Press, 1981.

———. "The Victorian Conflict between Science and Religion: A Professional Dimension." In Gerald Parsons, ed., *Religion in Victorian Britain*, vol. 4: *Interpretations* (Manchester, UK: Manchester University Press, 1992), 170–97.

———. "The Victorian Crisis of Faith and the Faith That Was Lost." In Richard Helmstadter and Bernard Lightman, eds., *Victorian Faith in Crisis: Essays on Continuity and Change in Nineteenth-Century Religious Belief* (Houndmills, UK: Macmillan, 1990), 9–38.

Urry, James. *Before Social Anthropology: Essays on the History of British Anthropology*. Philadelphia: Harwood Academic Publishers, 1993.

Van der Hart, Onno. "The Dissociation Theory of Pierre Janet." *Journal of Traumatic Stress* 2, no. 4 (October 1989): 397–412.

Veith, Ilza. *Hysteria: The History of a Disease*. Chicago: University of Chicago Press, 1965.

Waldrop, M. Mitchell. "Religion: Faith in Science." *Nature* 470 (February 2011): 323–25.

Wallis, Roy. *On the Margins of Science: The Social Construction of Rejected Knowledge*. Keele, UK: University of Keele, 1979.

Walusinski, Olivier. "Jean-Martin Charcot (1825–1893): A Treatment Approach Gone Astray?" *European Neurology* 78, nos. 5–6 (October 2017): 296–306.

Washburn, Margaret. "Review: Contributions to the Analysis of the Sensations by Dr Ernst Mach, Translated by CM Williams." *Philosophical Review* 6 (1897): 565.

Waters, Karen Volland. *Perfect Gentlemen: Masculine Control in Victorian Men's Fiction, 1870–1901*. New York: P. Lang, 1997.

Webb, James, ed. *The Mediums and the Conjurors*. New York: Arno Press, 1976.

Weintraub, Joseph. "Andrew Lang, Critic of Romance." *English Literature in Transition, 1880–1920* 18, no. 1 (1975): 5–15.

Wellek, René. "The Concept of 'Romanticism' in Literary History." *Comparative Literature* 1, no. 2 (Spring 1949): 1–23, 147–72.

Williams, Emily Kelly. "F. W. H. Myers and the Empirical Study of the Mind-Body Problem." In Alan Gauld, Edward F. Kelly, and Emily Williams Kelly, eds., *Irreducible Mind: Toward a Psychology for the 21st Century* (Lanham, MD: Rowman & Littlefield, 2010), 47–114.

Wilson, Leigh. *Modernism and Magic: Experiments with Spiritualism, Theosophy and the Occult*. Edinburgh: Edinburgh University Press, 2013.

Winter, Allison. *Mesmerized: Powers of Mind in Victorian Britain*. Chicago: University of Chicago Press, 1998.

Wolffram, Heather. *The Stepchildren of Science: Psychical Research and Parapsychology in Germany, c. 1870–1939*. Wellcome Series in the History of Medicine 88. Amsterdam: Rodopi, 2009.

Wolman, Benjamin. "Immanuel Kant and His Impact on Psychology." In Benjamin Wolman, ed., *The Historical Roots of Psychology* (New York: Harper & Row, 1968): 229–47.

Wozniak, Robert H. *Mind and Body: René Descartes to William James*. Washington, DC: American Psychological Association, 1992.

Wright, T. R. *The Religion of Humanity: The Impact of Comtean Positivism on Victorian Britain*. Cambridge: Cambridge University Press, 2008.

Wu, Chau H. "Electric Fish and the Discovery of Animal Electricity: The Mystery of the Electric Fish Motivated Research into Electricity and Was Instrumental in the Emergence of Electrophysiology." *American Scientist* 72, no. 6 (November–December 1984): 598–607.

Yeo, Richard. *Science in the Public Sphere*. Aldershot, UK: Ashgate/Variorum, 2001.

Young, Robert M. *Darwin's Metaphor: Nature's Place in Victorian Culture*. Cambridge: Cambridge University Press, 1985.

———. *The Victorian Crisis of Faith*. London: SPCK, 1970.

Zunshine, Lisa. *Why We Read Fiction: Theory of Mind and the Novel*. Columbus: Ohio State University Press, 2006.

Index